# Korean Crisis
Unraveling of the Miracle in the IMF Era

Also by Donald Kirk

LOOTED: the Philippines After the Bases
(also published by St. Martin's Press)

THE BUSINESS GUIDE to the PHILIPPINES

TELL IT to the DEAD: Stories of a War

KOREAN DYNASTY: Hyundai and Chung Ju Yung

TELL IT to the DEAD: Memories of a War

WIDER WAR: The Struggle for Cambodia, Thailand, and Laos

# Korean Crisis
## Unraveling of the Miracle in the IMF Era

Donald Kirk

St. Martin's Press
New York

ISBN 0-312-22442-7

Library of Congress Cataloging-in-Publication Data

Kirk, Donald, 1938-
    Korean crisis : unraveling of the miracle in the IMF era / Donald Kirk.
        p.  cm.
    Includes bibliographical references and index.
    ISBN 0-312-22442-7 (cloth)
    1. Financial crises—Korea (South)  2. Conglomerate corporations—
        Korea (South)  3. Korea (South)—Economic conditions—1960-
I. Title.
HB3817.5.K57  2000
332'.095195—dc21                                99-41220
                                                            CIP

Design by Binghamton Valley Composition

First edition: March, 2000
10  9  8  7  6  5  4  3  2  1

*For*
*Chang Ki Tak,*
*Lee Nam Bok,*
*and*
*Sung Hee, Sung Eun, Yong Jin, and Han Jin*

# Contents

# Preface

W hen I first visited Korea in 1972 as a correspondent based in Tokyo for the *Chicago Tribune,* one of the people I interviewed was Kim Dae Jung, then under house arrest at his home in the Mapo district of Seoul. There was no problem finding him—I said "Kim Dae Jung" to a taxi driver, and he took me straight to his residence. Kim told me all about the "accident" that had nearly killed him during his presidential campaign the previous year. When I visited Korea again the next year, Kim had another story to tell—a detailed account of the kidnapping from Japan that also almost killed him.

These two interviews illustrated a basic point about Korea. Whenever all is calm, perhaps a little boring, the news gets interesting. On October 26, 1979, this time accredited with London's *Observer* and several American papers, I got the news in Tokyo that President Park Chung Hee had been assassinated. Flights to Korea were canceled for the next two or three days, but I got to Seoul in time to file what the British call a "splash," a huge front-page story, for *The Observer,* a Sunday paper. A few days later, a very dull, ordinary, immensely boring general by the name of Chun Doo Hwan gave a lengthy press conference to which foreign journalists were invited. I was almost as surprised by his ascent to de facto power on December 12, 1979, as I was by Park's assassination.

Soon after my next trip to Seoul, in April 1980, hundreds of thousands of students, joined by many others, took to the streets in downtown Seoul within 200 yards of my hotel. Next, students were taking over the entire city of Kwangju, in the heart of the southwestern Cholla region. I made it into Kwangju in time to interview the leaders of the revolt, who were granting interviews in the city hall. One of them filled out and signed a form for me to carry as my "press card." The next time I was in Kwangju new coffins made of fresh pine were lined up on the ground behind the same city hall. Parents and relatives came around to identify the victims. There was the smell of fertilizer, spread over the bodies and the coffins to mask the stench of death. For years I carried in my wallet the "press card" of the short-lived "revolution."

The story of Korea's economic rise has been just as full of shocks and surprises. Almost none of the predictions has come true. After the Korean War, American aid officials wrote off the country as a "basket case." They said Korea would never be free of the need for vast quantities of foreign economic assistance. Great men, however, were planting the seeds of future industrial success. The names of Chung Ju Yung, Lee Byung Chul, Kim Woo Choong, Koo In Hwoi, and others proved far more important in modern Korean history than those of most of the politicians who have staggered across the national stage. Yet, just as we were all thinking that Korea had reached unassailable economic greatness, the system on which the edifice rested displayed signs of weakness. Suddenly, by mid-1997, foreign bankers and investment analysts began saying publicly that Korea was in trouble.

For Korea to have to appeal to the International Monetary Fund in the fall of 1997 was unthinkable. Could it really be that the Korean miracle was coming to an end? One thing was certain: the shock of the IMF crisis could not have happened at a better time for the former dissident Kim Dae Jung. No one could blame him or his party for what had happened. Since his election two weeks after the signing of the IMF agreement in December 1997, he has cast himself in the role of economic reformer. After two years of economic crisis, however, as the country approaches the fiftieth anniversary of the outbreak of the Korean War in June 1950, how much has really changed?

Several years ago, researching and writing a book about the Hyundai group, I emerged with ambivalent views. One had to admire the unremitting drive of the founder, Chung Ju Yung, some of his family members, and many of their subordinates for creating so vast an empire in such a wide range of fields. One also had to question if not criticize the family's narrow pursuit of wealth, to the exclusion of competition, of creativity, of fairness and equality in dealing with rivals and employees at all levels. Hyundai today epitomizes the strengths and weaknesses of the economy. All pledges to the contrary, it is difficult to believe it will reform.

Such skepticism extends to the entire economy. The Kim Dae Jung administration completed its first two years in a blaze of publicity touting improving statistics, but how much credit was due the government, much less those gargantuan conglomerates known as chaebol, for the changes? For all the reasonable words of many top leaders and bureaucrats, they still lacked the means and motivation to bring about a permanent shift that would truly force the chaebol to mend their ways. The virtual collapse of the Daewoo group in mid-1999 exposed the fragility of reform—and the dangers that still confronted the entire chaebol system.

The leader of every Korean administration within memory has vowed to clean up the garbage left behind by the previous government. Chun Doo Hwan campaigned mightily against the chaebol and against corruption. Kim Young Sam, as the first civilian to become president of Korea in an open election, trapped thousands of bureaucrats in an anticorruption campaign that resulted in jail terms for Chun and his ally and successor, Roh Tae Woo. Many Koreans viewed the trials of these ex-leaders, however, as acts of vengeance. The corruption of Kim Young Sam's son, caught up in another scandal, was one of the first overt signs of impending economic crisis.

What will be the place in history for Kim Dae Jung and those around him? We have to see first if the economy suffers a recurrence of the problems that dragged it down in the first place. Even as it keeps getting better, the danger persists of another severe crisis if the chaebol, the banking system, and the bureaucrats do not show the same urgency for long-lasting reform that they displayed for quick fixes in the first year of DJ's rule. The impression remains that the government may be more interested in showing off for the sake of exports and investment than in actually getting tough. The changes might appear sweeping but turn out to have been perilously superficial.

Like his predecessors, however, Kim Dae Jung faces the question of how far to go without also undermining the finicial system that had vaulted Korea to economic prominence. Did he want to emulate his old nemesis, Park Chung Hee, by forcing business leaders to bow to his will, dictating what sectors of the economy could fall to which group? Could he afford to offend any segment of an elecotrate ranging from old-line conservatives to militant radicals in the run-up to National Assembly elections in the year 2000 that would reveal what Koreans thought of him and his policies—and how much more reform the Assembly would approve? How Kim dealt with such contradictions would be the test of his success. Kim would be the exception if he were to retire gracefully, acclaimed for his achievements rather than reviled for failures.

How the Korean miracle fell apart, then partially came together again, forms one of the great dramas of modern Asian history. My purpose in writing this book has been to trace the evolution of the crisis, including its origins and causes and the struggle of government, business, and labor to overcome it. In gathering material on "the IMF era," I am grateful to innumerable people, many of whom are cited in this book, as well as to the editors of the *International Herald Tribune,* especially Michael Getler, executive editor, and Jonathan Gage, business and finance editor during the Asian economic crisis, for having permitted me to report for them in this critical period. Others who have helped greatly include Mervin Block in New York, who scoured the press for any mention of Korea; Bob Ives, Helen Ives, and Mark Hemhauser of the American University library staff in Washington; and Han Sang Yeon and Hwang Hye Joung of the Seoul Foreign Correspondents' Club. Thanks also to Chang Sung Hee, John Barry Kotch, Hank Morris, and Shim Jae Hoon in Seoul for reading and commenting on the manuscript. Finally, I owe a special debt of gratitude to the entire Chang family, Chang Ki Tak and Lee Nam Bok, Sung Hee, Sung Eun, Yong Jin, and Han Jin.

Donald Kirk
Seoul, 1999

Map of North and South Korea

# Chapter 1

# Crossroads

As they entered the millennium, South Korea's chaebol or conglomerates were at a crossroads. The question was whether these "engines of the economic miracle" had run out of steam or could restructure and modernize for the next stage of Korea's industrial revolution. Factions within the government, within the ruling coalition and the opposition, were at odds over how far to push the chaebol into reform. For that matter, factions among the chaebol and within each individual chaebol also were at odds over how much to reform, and when and how.

Two generations after the Korean War, the chaebol had come to represent the most conservative forces in Korean life. In the days of Korea's worst agony, during and after the war, the men who founded the great chaebol symbolized the can-do spirit of a society battling to overcome the suffering not only of war but also of half a century of Japanese domination. In their reluctance to change, however, they appeared to be the inheritors of centuries of dynastic rule that discouraged the creativity and innovation that had swept the west and finally, in the late nineteenth century, Meiji-era Japan. Considering that historic legacy, one had to venerate the entrepreneurs responsible for Korea's own industrial revolution. Figures such as Chung Ju Yung of Hyundai, Lee Byung Chul of Samsung, Kim Woo Choong of Daewoo, and Koo In Hwoi of Lucky-Goldstar had earned not only billions for themselves and their families but also billions more for hundreds of thousands of employees and for Korean society. They and the founders of the other chaebol had built their empires in an environment of armed protection by the United States and dictatorship by a military leader, Park Chung Hee, desperate for his people and his country to make amends for the past.

The creative drive of the chaebol chieftains, however, had begun to fail them as their groups entered the second generation and power shifted to a new generation of heirs. The numbers caught up with the chaebol in October and November 1997 when they could no longer count on their friends at the banks to come up with the loans needed to pay off their debts. What happened was an inevitable consequence of the habit of the chaebol of overextending in rabid pursuit of new frontiers for an ever-widening range of products. That policy in the 1960s and on into the mid-1990s had seemed innovative and daring. These hard-driving tycoons were willing to risk all as they pursued their dreams of new markets in the most competitive of all marketplaces, the United States, and in the most remote of markets in Latin America, the Middle East, Africa, and Siberia and the Russian Far East. They had gambled, and in most cases they appeared to be winners. The trouble was, some of the basic numbers were getting worse even as Korea's gross national product was growing remarkably when translating the figures from Korean won to dollars before the won began its rapid descent in value in that dark November of 1997.

The devaluation of the won skewed all the numbers so much that it often seemed futile to put them into dollar terms. The exchange rate on November 5, 1997, when I changed money on arrival at Kimpo Airport, was about 900 to one dollar, and, on December 24, when I took off for the holidays, it was more than 1,900 to one. A year later, the won had strengthened to between 1,200 and 1,300 to one dollar, but that was not a good thing for exporters. Other numbers, moreover, demonstrated that disaster was imminent before it struck. The most revealing were for the ratios of debt to equity. Figures at the end of 1996 showed the Hyundai group with total assets of 52.8 trillion won but with equity of shareholders totaling only 9.8 trillion won. That meant that Hyundai, the country's largest chaebol in terms of assets, owed four times more than it had in the bank. The figures for Samsung, the second-largest chaebol in terms of assets, were almost as disturbing. Samsung's total assets, minus its financial firms, came to 50.7 trillion won while shareholders' equity was 13.8 trillion. The ratio for debt to equity for Samsung was nearly 4 to 1—about average, according to Korea's Fair Trade Commission.[1]

The same pattern of debt to equity prevailed in all the top-ranking chaebol, including the third-ranking Daewoo group, fourth-ranking LG, and fifth-ranking SK. Those were the numbers for what were still perceived in 1997 as the relatively healthy chaebol—"too big to die," it was said, as the Daewoo group approached collapse two years later. The numbers got worse at the level of the Kia and Hanwha groups, ranked eighth and ninth, and were totally skewed in the cases of the twelfth-ranked Halla group and nineteenth-ranked Jinro group. The plight of these four chaebol was instructive because it showed what could happen to some of the others, notably Daewoo, if the pattern persisted.

Kia at the end of 1996 had total assets of 14.1 trillion won supported by shareholders' equity of 2.3 trillion won for a ratio of debt to equity of nearly six to one. Some of the individual companies within the Kia group, such as the one that produced specialty steel, were in much worse condition. Well before the economic crisis had forced the country to go to the International Monetary Fund for a bailout in November 1997, Kia had had to apply for court receivership. Under the circumstances, the real miracle about Kia was that it still produced cars for export long after it was put up for auction. When I visited the Kia plant at Asan in December 1997, it was in the second week of production of a new model called the Shuma.

As domestic markets shrank and exports slumped, the most pathetic numbers were those for Halla and Jinro. The Halla group, founded by Chung Ju Yung's next younger brother, Chung In Yung, had assets at the end of 1996 of 6.6 trillion won and shareholders' equity of just 306 billion won. In other words, the ratio of debt to equity at Halla was more than 20 to 1. Chung In Yung's sons, now in charge of the group, had followed a pattern set long ago by their father, bound to a wheelchair but still active in operation of the group. Driven by a younger brother's mad desire to emulate the success of the older brother, and at the same time in conflict with the older brother while shielded and protected by him, Chung In Yung for years had run Halla into ever mounting debt. In the crunch, he counted on his most successful company, Mando Machinery, to make money for the entire group by selling its products, namely components to motor vehicles such as the little motors that run car windows and windshield wipers, to Hyundai Motor Company. Other companies in the group might lose money, but Mando would pay the bills.

So it was that Halla in the early 1990s decided to build the most modern shipbuilding plant in Korea—in emulation of Hyundai Heavy Industries' own enormously successful shipyard, the world's largest producer of merchant ships. Before the shipyard near Mokpo on the southwestern coast had manufactured its first vessel, however, the whole group was in the position of not being able to pay its bills and had to plea for mediation, a step on the way to receivership.

This time around, however, big brother Chung Ju Yung was not quite as ready to come to Chung In Yung's rescue. For one thing, Chung Ju Yung, now 82, had long since assumed the title of honorary chairman of Hyundai and "retired" by handing over the chairmanship to his eldest surviving son. (His fifth son also became chairman in the fall of 1997, making Hyundai the only group with two chairmen.) For another thing, Hyundai was having debt problems of its own. It had had to cancel its plan for one of the founder's most grandiose visions, that of a steel plant rivaling the successful government-invested Pohang Iron and Steel, on which the whole country counted for its steel supply. Hyundai also had had to abandon plans to build an electronics plant in Scotland at a cost of several hundred million dollars and had given up construction of a motor vehicle assembly plant in Indonesia. No way, said Hyundai executives, could they bail out younger brother Chung In Yung and his sons.

As for Jinro, its figures were the worst of any chaebol's. With assets totaling 3.8 trillion won, it had shareholders' equity of 99 billion won for a ratio of debt to equity of an astonishing 40 to 1.[2] The leaders of Jinro, Korea's largest soju maker, had apparently got drunk on their own concoction. Chang Jin Ho, son of the founder, had made the same mistake as the other chaebol heirs. Under his management, the number of Jinro companies had tripled. First they went into competition in department stores with some of Korea's most established names, including Hyundai, Lotte, Midopa, and Shinsegae, among others. Next, they expanded into beer head-to-head with Doosan, Korea's largest beer manufacturer, and Chosun, a small but viable independent company that produced a brew that now outsold Doosan's famed OB (Oriental Brewery) brand. Jinro, like Halla, was in mediation—another term for down but not out.

The problem of the overly ambitious heirs may have been the most acute of all. Take the two other second-tier chaebol, Hanbo and Hanwha. Hanbo's dilemma got the most publicity. The chairman and his executives had passed bribes to Kim Hyun Chul, son of Kim Young Sam, well after the 1992 presidential campaign, in which YS had promised an anticorruption drive that would ensnare the two ex-generals who had preceded him in the Blue House as well as numerous chaebol figures. The Hanbo founder, Chung Tae Soo, had built his empire on bribery, starting, as had many other chaebol, including Hyundai, in construction and winding up most conspicuously with a vastly leveraged steel plant. One of his sons, Chung Bo Keun, was not only chairman of the group but also a friend, at least of expediency, of Kim Hyun Chul. The chairman's son had great plans for turning the steel company into a moneymaker and elevating the group from its number 14 ranking to one of the top ten at least. What better way to guarantee success than to pay off the son of the president? The problem of Hanbo Steel was sometimes blamed by emotional critics for igniting the whole chaebol crisis, though that interpretation was probably too narrow.

Another group that faced a major heir problem was Hanwha, owner of Hanwha Energy, one of Korea's five major oil refineries. Heir Kim Seung Yoon, the son of the founder, fit into the same pattern as the others. He dreamed of expanding his group wildly, building up his oil refinery, going from there into retailing, financing, and machinery, all financed by the original core business of chemicals. Now the group, headquartered in a soaring new office tower in downtown Seoul across the street from the Westin Chosun Hotel, was putting out unlikely stories that Shell, Texaco, and other major companies were interested in its heavily leveraged oil refinery and other portions of the group. (Dissemination of such stories in the local media was a standard negotiating technique to whip up interest in investment.)

Then there were the cases of Dainong and Sammi. Dainong, originally a textile group, made the mistake of expanding into national retailing from a prestigious base at the well-established Midopa Department Store in downtown Seoul. The founder's son, Kim Young Il, was still struggling to revive the group, now in receivership. The heirs of the Sammi group, which had a specialty steel plant in Korea, spent so much on money-losing enterprises in Canada and the United States that Sammi also went into receivership. The older son acquired a chain of chicken restaurants in Canada while the younger struggled to keep the group alive. A Sammi executive, Suh Sang Rok, became a symbol of the crisis, going to work as a tuxedo-clad waiter in a French restaurant in the Lotte Hotel, jewel of the Lotte group, one of the few chaebol that managed to hold down debts and minimize losses. "I Have to Live My Life My Way" ran the translation of the title of a book Suh wrote on his career switch.

The heir problem ranged from the lowest to the highest. One of the most noteworthy was that of Samsung. Chairman Lee Kun Hee, the son whom the late Lee Byung Chul had anointed as the most qualified to run the group, had been the model of a modern chaebol leader, if one could believe the laudatory articles published over the years in *Fortune, Forbes,* and *Business Week.* Lee's weakness, among others, was a lust for beautiful cars. He owned an entire fleet of them, and he could see no reason why he shouldn't build a car company rivaling Hyundai, Daewoo, and Kia. The plan seemed great in the go-go mid-1990s when Lee decided to invest a spare few billion in a motor vehicle plant with technology from Nissan, which was eager for a major opening into Korea. Lee, however, could not have chosen a worse time for his venture. He stubbornly persisted in a project that was doomed. In February 1998, Samsung invited journalists and others to a showing in Seoul. "Samsung has spent three years in preparing to break new ground in the Korean automobile industry," the invitation read. "You now have a chance to witness the unveiling of Samsung's first new car." The facing page proclaimed, "World-Class Automobiles from Samsung," and in parentheses were the words, "Quality, Safety and Comfort."[3]

Samsung was not alone in overexpanding in motor vehicles. Daewoo had done the same on a global scale. The Daewoo chairman, Kim Woo Choong, for years chafed at the reluctance of General Motors, which owned 50 percent of Daewoo Motor Company from 1978 into 1992, to agree to going into world markets. Finally Daewoo bought out GM's half of the company while Kim Woo Choong bought into state-owned plants in regions once "in the orbit," in the Cold War lingo, of the Soviet Union, including Poland, Rumania, Uzbekistan, and Ukraine. Plant expansion and globalization cost Daewoo about $5 billion. As if that were not enough, Daewoo then bought out Ssangyong Motor, the plaything of Korea's sixth-largest chaebol, the Ssangyong group, in danger of having to beg for court mediation.

The bottom line was that Daewoo Motor was now leveraged by at least five to one and had to go into negotiations with General Motors in hopes its old partner would come to the rescue. The bottom line for Samsung Motor was that it was now ready to negotiate with almost any international company for partnership.

Stories of the chaebol's anxiety for an infusion of foreign funds led to an overriding question: what were the chaebol doing about restructuring? The answer: beneath the pledges and resolutions, no one was sure. The word "transparency" became voguish in Seoul, with all the chaebol promising to open their books and come up with consolidated reports, as demanded by the IMF. The chaebol also promised specific plans for restructuring in keeping with the demands of the economic team of Kim Dae Jung, the perennial critic and dissident hero who narrowly won the presidency in his fourth attempt in December 1997. Kim's advisers were saying they should get rid of the money-losers and trim down to "core" industries. Through it all, there was much talk in the Korean media about "Big Deal." The phrase, uttered in English amid Korean newscasts and talk shows, stood for the exchange or "swap" of companies among the chaebol in order to trim the losers.

Everybody, both in government and in the chaebol, was now calling Big Deal a misnomer. The word harked back to the era of Park Chung Hee's examining the charts and graphs, telling chaebol leaders, you go into this industry, you go into that. Chun Doo Hwan sought to keep up the tradition of a Korean version of a communist command economy until the chaebol discovered how to win him over with bribes. The ultimate irony would be for Kim Dae Jung, after all those years in jail and under house arrest in the era of Park and Chun, to adopt their style of ordering the chaebol around. Kim's people all said, no, they didn't mean quite that, what they meant was to create the conditions under which unprofitable entities would naturally close.

They were attempting to foster such an environment in a number of ways. One was to eliminate entirely the system of cross-guarantees under which companies within a chaebol guaranteed loans to one another so, in the end, no one guaranteed anything. (If A guaranteed a loan to B and B guaranteed to C and C guaranteed to D and D guaranteed to A, who had the money to back up the loans?) They also were telling the chaebol owners to shut down the offices of planning and coordination that lay at the heart and soul of every group. These offices might be unique to Korea. Since none of the chaebol was actually owned by a single holding company, where did a group's senior executives sit to hold the reins over the empire? The answer: they had to work from whatever company was closest to the heart of the group chairman. In the case of Hyundai, they operated from Hyundai Engineering and Construction, founded by Chung Ju Yung in 1947, the "mother company" of the group. At Samsung, they operated from the Samsung Corporation, the group's trading company, and at Daewoo they operated from the Daewoo Corporation, Daewoo's trading and construction arm.

But how could the chaebol be forced to shut such offices? The government still had final power over bank credit—and planned to push for much greater rights for minority shareholders. Theoretically, they could stand up at meetings and say, "Why are you wasting the money in my company on group enterprises rather than plowing it into my company or paying higher dividends?" Shareholders under the new rules would even have the right to sue. For that matter, the administration had said it would push legislation permitting hostile mergers and acquisitions—something no one had ever heard of in Korea.

There were, however, several reasons why none of this was likely to come to pass right away. One was that the chaebol were fighting tooth and nail. They had already persuaded some of Kim Dae Jung's top advisers that, to offset hostile mergers and acquisitions, they should be able to invest a much higher proportion of their assets into companies within their groups. The chaebol had persuaded the emergency economic committee to endorse a new law that would do away with a law restricting how much a company could invest in other companies in the same group. That meant the chaebol could protect whatever company they wanted while making a show of downsizing by getting rid of some of the losers they didn't want. (The problem was that most companies did not have the money to buy other companies' losers, and what foreign company would be all that interested either?) Another factor, largely overlooked when the great plan to revise the M&A law was announced, was that only 26 percent of chaebol companies were listed on the stock exchange. There were numerous notable nonlisted companies. Daewoo Motor was one of them. True, one could buy a nonlisted company, but not having shares to sell publicly made a hostile M&A most difficult.

The chaebol were battling calls to do away with their planning and coordination offices by saying they needed them to plan and coordinate their restructuring programs. That was an excuse that was hard to dispute. This kind of debate would go on and on, but two factors would bring the Korean economic crisis to a turning point. One was that these companies would still have difficulty paying off their debts even with IMF bailout loans and rollovers of loans by foreign banks. Korea's total foreign debt exceeded $150 billion, and its domestic corporate debt was more than $600 billion.

The other great problem that might bring matters to a turning point—though not a solution—was that of labor. Kim Dae Jung thought he had a deal when a "tripartite commission" of government, labor, and business representatives agreed on a pact to allow mass dismissals of workers while the government set up an unemployment insurance fund of five trillion won. The Korean Confederation of Trade Unions called for renegotiating the deal before it got into law and promised a general strike of its 600,000 members. The total KCTU membership was half that of the Federation of Korean Trade Unions, which went along with the new law in the interests of economic realism, but the KCTU controlled the workers in the heavy industries. More than 10 percent of its members were from the Hyundai empire, most of them in the manufacturing center of Ulsan, where Hyundai produced its cars and ships, among other things. Some workers might be reluctant to walk out on economically strapped companies, but the KCTU appeared as a defender not just of its own members but of a workforce of more than 13 million people.

One question union leaders asked was why the workers should be the ones to sacrifice. Why didn't incompetent chaebol leaders step down—and put up their own billions of dollars for the sake of their companies? Lee Kun Hee of Samsung and others made a show of saying they would invest their "own money" for the sake of economic recovery. What was meant by "own money," however, was not clear. These men were multi-billionaires. They were talking about parting with millions—pocket money in comparison to the wealth they had accrued. While the chaebol system would endure, the weaker ones would fall off, swallowed by the stronger. That, however, was a long-term estimate. All the bankrupt chaebol were still in business through the first year of the economic crisis while awaiting court decisions on what to do about them and their assets. Hanbo still made steel. Kia made cars. Halla was even making ships. (Hyundai Heavy Industries, in the spirit of nepotism that often dominated chaebol thinking, relented and secretly extended loans to Halla Heavy Industries, Halla's shipbuilding unit.) Jinro was brewing soju and Dainong was producing textiles while Midopa stayed open down the street from the Westin Chosun Hotel, owned by the Shinsegae group, dominated by nearby Shinsegae Department Store, controlled by a younger sister of Lee Kun Hee.

The banks propped up these groups one way or another. In the end, reform of the banking system might have to come before reform of the chaebol. In the old days, chaebol chieftains could ask the banks for loans for this or that, and they would get what they wanted with no more than pro forma scrutiny of their books. Now the banks did not have the money and the chaebol were no longer able to borrow overseas. They had come to assume that foreign as well as Korean banks would roll over their loans almost forever, but they could not count on rollovers for more than a limited period. Ultimately a restructured banking system would be the means to keep the chaebol in line. How this story would play out, however, was far from clear as strikes broke out and thousands of small and medium-size companies, reliant upon the chaebol to purchase their products, dropped out of business. Somehow, though, it would be hard to believe that Korea's big five chaebol, Hyundai, Samsung, Daewoo, LG, and SK, would not survive.

The Asian crisis, it should be recalled, had erupted in 1997 to shrugs and the sense that it was far removed from the rest of the world, or from the parts of the world that mattered to most people outside Asia, namely the financial markets of the United States and Europe. The disintegration of the currency of Thailand in July, while worrisome to currency traders and Thai bankers, was hardly noted in the west. The spread of what came to be known as the Asian contagion to Indonesia and Malaysia did not seem all that troubling either, at least from afar. Wall Street was enjoying one of the best years in its history, and the sense was that Asia was suffering a severe cold that would go away after a few "corrections" and "adjustments."

Certainly the Asian crisis seemed far removed from South Korea, the Republic of Korea, where the economy in 1997 appeared to be riding the crest of the "Korean miracle" through another year of rising GNP and a prosperity that trickled down to the lowest levels of a highly stratified society. One might select any of several dates for when the Korean crisis tumbled into the abyss, but one that marked a clear dividing line was July 15, 1997. It was then, little more than a week or so after the Thai currency had begun its descent, that the Kia group, having delayed on some debt payments and failed to make others, acknowledged that it was bankrupt and asked the Seoul District Court for mediation. In other words, Kia would be given a reprieve from debt payment while the court tried to mediate a realistic schedule with the creditor banks, led by the government's Korea Development Bank.

Outside Korea, Kia's plight received if anything less notice than that of Thailand. It was, after all, just one group, one chaebol, the Korean word denoted by the Chinese characters for "fortune cluster." Kia's dilemma was a local story. It was about the comedown of a group whose flagship, Kia Motors, after having climbed to second place among Korean motor vehicle manufacturers behind Hyundai Motor, a core company of the mighty Hyundai group, the number one chaebol, had fallen to third. Although Ford Motor owned 9.4 percent of its stock and Mazda Motor of Japan, one-third owned by Ford, owned another 6.5 percent, Kia now ranked behind Daewoo Motor, powered by Daewoo Chairman Kim Woo Choong.

By itself, Kia's predicament was relatively inconsequential. Korea's motor vehicle industry was bloated by the expansionist urges of all its manufacturers. In retrospect, however, it was clear by July 1997, when Kia was going bankrupt, that large segments of the rest of the economy had begun their plunge. The slide was still largely unnoticed outside Korea until the government four months later, in November, had to approach the IMF for a package of loans needed to prevent the banks from defaulting on their debts to foreign creditors. Until the summer of 1997, the chaebol appeared to be growing rapidly despite halfhearted official efforts to reduce their size and power.

Government officials were caught between recognition of the role of the chaebol in Korea's economic advance and fear that the size of the chaebol was stifling real competition and discouraging smaller, more innovative companies. The question they asked was what they could possibly do to hold the chaebol in check without jeopardizing the entire economic structure. "The government cannot fully monitor and prevent them, there are so many different practices," said Cho Yoon Jae, senior counselor in the ministry of finance and economy. Besides, he added, "they are the major players in the economy, and we don't want to discourage them too much."[4]

A new version of South Korea's Fair Trade Act reflected the ambivalence of the government in attempting to curb the chaebol while also relaxing regulations that inhibited them. The Fair Trade Act, as of 1998, forbade any single company within any of the 30 largest chaebol from guaranteeing loans to companies in the same group totaling more than 100 percent of its assets. That provision represented a seemingly significant reduction from the 200 percent ceiling placed previously on intra-chaebol loan guarantees. The largest chaebol were growing so fast, however, that most of the top ten were big enough to be able to guarantee almost all the loans they wanted, according to Kang Hee Bo, a specialist on the Fair Trade Act on the staff of President Kim Young Sam.[5] Chaebol leaders, either the founders or the heirs to the founders of their groups, fought under the aegis of the Federation of Korean Industries, a powerful organization of owners, against tougher restriction.

"Originally the government suggested a very radical change," said Gong Byeong Ho, director of the Center for Free Enterprise of the Korea Economic Research Institute, an adjunct of the FKI. "The Fair Trade Commission suggested the ratio of cross-guarantees of credit be reduced to zero percent" by the year 2001, but the FKI protested so loudly that the law stuck to 100 percent.[6] The Fair Trade Commission, an agency with the status of a ministry, also suggested outlawing the crossholding system under which companies within a group, as well as a group's founder and heirs, invested in companies in the same group. The current ceiling on crossholdings was 25 percent, meaning that a company could invest no more than 25 percent of its assets in all the other companies within the same group.

Like the limit on debt guarantees, however, that provision had little meaning. Groups such as Hyundai and Samsung, the two largest chaebol in terms of both assets and sales, were so large that all the companies within each group still held enough shares to guarantee control by the founder and his family members, who also held significant numbers of individual shares. Hyundai represented the most tightly controlled among the top ten chaebol, with Chung Ju Yung and his family holding 15.4 percent of the invested capital while Hyundai companies through crossholdings owned another 46 percent, said the Fair Trade Commission. Samsung Chairman Lee Kun Hee and his family owned nearly 3 percent of their group, but Samsung companies controlled 46 percent through crossholdings. Daewoo Chairman Kim Woo Choong and his family owned nearly 6 percent of Daewoo, while Daewoo companies controlled 33.9 percent.[7]

Hyundai's assets by the end of 1996 were about $55 billion, and Samsung's were $51 billion. LG and Daewoo had assets of about $38 billion apiece—both more than double the assets of SK, the fifth-largest chaebol. Assets of the four largest chaebol represented half the top 30 chaebol's total assets of $367 billion. That figure, in turn, was up from $223 billion in 1992 when Kim Young Sam was elected president amid promises to cut the chaebol down to size. In terms of gross national product, the figures showed that the chaebol were also increasing in importance. Goods and services of the top 30 in 1995 accounted for 16.2 percent of Korea's GNP of $451.7 billion, 11th in the world, up from 14.2 percent in 1994, according to the Korea Economic Research Institute. The percentage of the chaebol's contribution to GNP was expected to increase somewhat in 1997, largely as a result of rising profits of the top five, while overall GNP approached $500 billion. The big five accounted for more than 60 percent of the contribution of the top 30.[8]

The Fair Trade Commission regularly criticized the chaebol while conceding it could do little to combat their rising strength. "It is true that large business groups have their merits, such as contribution to the rapid development of the Korean economy," said an FTC report, "but they are also the main culprits of economic concentration in Korea, raising the question of whether their wealth was accumulated in a just manner." Also, the report added, "they are the target of harsh criticism due to the income disparity stemming from excessive contribution of wealth." Despite policies "aimed at easing concentration" over the past decade, the report conceded, "the concentration is hardly being alleviated."[9] The chaebol, for their part, chafed under what they saw as excessive controls. The FKI had demanded that the government drop its ban on chaebol ownership of private banks, which the government said would otherwise function as "private safes" for the chaebol.

The chaebol, through the FKI, had also demanded that the government no longer reserve certain business lines for small and medium-sized industries—a measure seen as essential to preventing the chaebol from spreading control ever more widely over the fabric of the economy. As it was, the figures showed that literally all the top 30 chaebol were getting larger and the gap between the top five and the others was widening. The extended Hyundai group, including lesser groups controlled by Chung Ju Yung's younger brothers, chalked up "turnover" or sales in 1995 of about $80 billion, and Samsung, including two lesser groups, recorded sales of $83 billion. Sales at Daewoo and LG both exceeded $50 billion. Sales figures for 1996 for the big five were expected to go up by more than 10 percent.

"The rhetoric is far ahead of actual policy," observed SaKong Il, a former finance minister. The fact that SaKong had served two disgraced ex-presidents, Chun Doo Hwan and Roh Tae Woo, and had himself got a suspended sentence for his role in passing on bribes from industrial leaders did not diminish his appreciation of the paradox confronting Kim Young Sam. "In this globalized world, the situation is difficult," SaKong told me in mid-1997. "The paradigm has shifted in a way. The world is much more globalized" than it was years ago when cries were first raised for reducing the size of the chaebol. The question SaKong and others were asking was how the chaebol could downsize—break up into mini-chaebol—when they were the ones with the funds needed for investing on a global scale per the "globalization" policy. SaKong saw the transition of chaebol ownership from their founders to second- and third-generation heirs as a step on the way to breaking down monolithic structures amid modernization. "There will naturally be some separation. The old way of doing things now is changing."[10]

Always, the top groups provided the cutting edge for huge expansion both at home and overseas. The government, however, still had ultimate power. Thus the government refused in 1997 to approve Hyundai's bid to build a steel plant. "Hyundai steel is not acceptable," said Kang Hee Bo. "They wanted in-house sourcing for their products. That creates a vertical combination"—that is, a system under which products move up and down from one company to another in a group.[11] The government was caught in a vise: permitting Hyundai to finance the plant would mean that Hyundai could again expand significantly. Refusal, however, meant the government was going back on an oft-stated principle of "non-interference."

Observers saw a protracted off-again-on-again government-chaebol struggle—a fact of Korean life from the Park Chung Hee period. The Kim Young Sam government, without enunciating a pro-chaebol policy, might actually have been more pro-chaebol than the preceding regimes of Chun

Doo Hwan and Roh Tae Woo. Government policy now called for making it easier for the top ten chaebol to obtain credit as part of a program to encourage them to get into enormous national development projects, including sea, rail, and air facilities.

Kim Dae Jung, after having appeared critical of the IMF during his election campaign, seemed, as the foreigners put it, "to be saying the right things." One cold Sunday in January 1998, before his inauguration, he appealed again to the people, this time as a president-elect, forecasting "terrible hardships" if the country ever declared a moratorium on debt repayment. Kim issued the warning in a two-hour "conversation with citizens" broadcast live on the nation's four television networks—and watched in freezing temperatures by crowds in front of big-screen television sets at Seoul's central railroad station and in public squares in other large cities. Asked by a businessman in the live audience at the Korea Broadcasting System studio in Seoul if Korea's economy might collapse, Kim responded, "Not in a year but in a few days the country can go bankrupt unless we cope with the situation."[12]

It was Kim's first major public appearance before his people since he had been elected on December 18 with barely 40 percent of the votes, and it was the first ever town-meeting-type forum in Korean history. The 800 or so people in the KBS studio laughed, joked, applauded—and listened with rapt attention as the president-elect, alternately humorous and serious, called for unity and cooperation. "I have prepared for the presidency for many years," the 73-year-old Kim reminded his audience, his voice heavy with emotion. "Just trust me, I know I can do it, just trust me."[13] Kim spoke as a delegation of his top economic advisers was about to fly to New York for critical talks with Korea's creditor banks on rolling over more than $24 billion in debts. The purpose of the mission, he reminded countrymen who might have been hostile toward any attempt at kowtowing to the foreigners, was "to save the economy."

"DJ," as he was known in the Korean media, did not mention the technicalities of the talks, preferring to drive home a sense of pride among Koreans in what they had done to answer calls for help since the depth of the crisis had became clear in December. "I am very proud of the people's cooperation in turning in gold and dollars," he said. "Such efforts bring me to tears." He drew laughter when a woman in the audience asked about his health. "I've been campaigning for eight months, and you can see how I am," he answered. "People lied about my health. I think that made me healthier." He smiled when another woman asked whether he planned to cut his salary after he assumed office. "I don't even know how much I will be getting," he said amid more laughter. "After I start getting it, then I will let you know." He was applauded loudly when asked what he wanted people to think of him. "I don't want to be admired during my presidency," said Kim, limited to a single five-year term under Korea's constitution. "I would rather be admired after my term, and I would like to be missed by the students after my death."

The way Kim Dae Jung was behaving, he gave the impression that he had been ensconced as president for a year or two. He and his economic advisers were already getting the National Assembly to pass laws designed to bring about economic recovery. What was worrisome about Kim's enthusiasm was that every incoming Korean administration had taken over with big ideas that foundered on the realities of a system mired in tradition, authoritarianism—and corruption. Park Chung Hee, after seizing power in 1961, ruled with increasing dictatorial fury until his assassination by his own intelligence chief in 1979. Chun Doo Hwan, the general who took over soon afterward, resolved at once to investigate the owners of the chaebol.

Very quickly, Chun and his entire family were accepting bribes at a rate reminiscent of the gifts bestowed upon the Korean kings who ruled the country for centuries before succumbing to 35 years of Japanese colonial rule that lasted until the Japanese surrender at the end of World War II. Chun's Korea Military Academy classmate Roh Tae Woo, the general with whom he collaborated to suppress the bloody Kwangju revolt in May 1980, was less dictatorial but equally corrupt. Together they were jailed under Kim Young Sam for their roles in the Kwangju massacre as well as the enormous bribes they had accepted. Kim looked like a hero in his crusade against corruption—until his son was tried and jailed and the economy began falling apart.

Would Kim Dae Jung really do a lot better? He had suffered, as had no Korean president before him, for years in jail and under house arrest for defying first the authority of Park Chung Hee and then that of Chun Doo Hwan. Only in the era of democratic reform introduced under Roh Tae Woo, after riots that nearly toppled the government, was Kim able to emerge as a powerful democratic figure. After losing to Roh in 1987 and to Kim Young Sam in 1992, he defeated the government-backed candidate in December 1997. In the run-up to his inauguration on February 25, 1998, he managed to twist the arms of the leaders of the unions and persuade them to accept a draft bill that would legalize layoffs by major industries otherwise compelled to keep workers on the payroll "for life." The unions stopped issuing strike threats while waiting to see if the companies would challenge them by actually laying people off. He would not be content until the chaebol came up with plans for restructuring their empires.

The bottom-line question, though, was whether DJ would make good on his program once he became president. At 73, he ran the risk of running out of energy long before his five-year term was over. His power, moreover, rested on a flimsy coalition. While drawing popular support from the relatively poor southwest Cholla region, he had to unite with the conservative party of Kim Jong Pil, "JP" in the headlines. A prime minister under Park Chung Hee, JP was also remembered as founder of the Korean Central Intelligence Agency, which had once dogged DJ's footsteps and kept watch around his home. JP, in line to become DJ's prime minister, would make some of the right noises about reform, but his ties to the chaebol might be too tight to break.

The worst danger was that Kim Dae Jung, like his predecessors, might be lulled into his own compromises. There were differences between his party leaders and those of Kim Jong Pil. The chaebol were masters not only at exploiting them but also at coming up with great-sounding programs that offered very little substance. DJ's party had also accepted donations from the chaebol—less, certainly, than had his predecessors but enough to ensure that he was not out to destroy the basic system on which Korea's economy had rested since the Korean War. The showdown might come as Korea's long-term debt problem worsened, the chaebol laid off workers by the tens of thousands, and radicals, like DJ in the old days, defied the regime. In power, DJ would have to display all the stubbornness and resilience built up during his years in prison. How he responded would test the courage he had once displayed as Korea's most famous dissident. Ominously, he promised public hearings to fix responsibility for the crisis—an investigation that could embarrass if not ensnare numerous government officials and businessmen.

# Chapter 2

# IMF vs. Korea, Inc.

T he elderly women waited patiently in the narrow alleys twisting off the eight-lane avenue leading past Midopa Department Store and the central post office. Occasionally they chatted with passersby, then delved into large handbags or scurried into a room behind a nearby restaurant. "It's very difficult to gain much of a profit these days," one of the women, peering into a pocketbook containing rolls of $100 bills and 10,000-yen Japanese notes, told me. "You never know which way the money is going."[1] It had been about a month earlier, in November 1997, that the Korean won had begun its sharp downward slide. By now, on a cold morning in mid-December, business in the city's black market, festering in the back alleys and arcades winding from the Namdaemun market beside Namdaemun or South Gate to Myongdong, a high-fashion shopping district, had never been better.

"People need dollars or yen, and they can't get it at the bank," said a woman who had been changing money in the same back alley for more than 20 years. "They want it for overseas trips. Maybe they want to gamble in Las Vegas or buy clothes or presents when they go abroad." Then there were the clients who slipped into one of the alleys laden with hard currency—and high hopes of getting just a few won more to the dollar or yen than they would at a bank. "We can always do a little better than the banks," said a woman who ran her business from the entry to a nearby building where she had an office with a couple of phones.[2]

Now the government was telling everyone to change foreign currency and gold into won. Spurred by TV commercials showing Koreans in a tug-of-war against chaos, patriots were rushing to banks in a national campaign for hard currency, accepting won at terrible rates for their gold wedding rings and other gold jewelry. The government claimed nearly two million people had added $1.3 billion to the foreign reserves that way but failed to mention those who played the black market. "They think that now is the time for bargain-hunting, and they come to us," said a woman holding a pocketbook brimming with several types of currency. "They have too much to go to the bank. They are embarrassed to explain where they got it."[3]

These women were bag ladies only in the sense that they carried big bags. Otherwise, they appeared well-dressed and prosperous from a business that dated to the era when the government, after the Korean War, set the won at several times its real worth in hard currency. For many Koreans, bearing dollars earned or bartered or stolen in those depressed days from U.S. military bases and contractors, the only way to get full value for the won was to change illegally. The police for years had tolerated the black market, run from behind the scenes by gangsters reaping payoffs and protection money from nightclubs, hostess bars, and drug-dealing. "But it's dangerous, since we carry a lot of money," said one woman, working with her middle-aged daughter at the opening of an alley, both of them flashing fistfuls of $100 bills.

The fear of the won's losing ever more value was sending crowds of shoppers flooding nearby shops and supermarkets on a binge of panic-buying. Housewives placed top priority on sugar and flour, since Korea had to import all its sugar and most of its wheat. Another item high on the hoarding list: instant ramen, also made from wheat. With the won descending in value, merchants were raising prices or holding back on their stocks, waiting for the moment at which to sell at the highest prices. The government threatened to crack down on hoarders—difficult while banks and companies struggled to round up the hard currency needed to pay debts.

For the women changing money on the fringes of Myongdong, still bustling with affluent young people, the best bets were the entrepreneurs making money importing and exporting, going overseas to seal their deals and returning with attaché cases full of cash. A prosperous-looking man wearing a well-cut sports jacket stopped briefly near one of the women, smiled, and followed her down the alley and up a twisting staircase to her office. "We get customers like that once or twice a day," said the woman. The money also was flowing in from travelers arriving with large quantities of hard currency from China and Russia in search of quick bargains in textiles and clothes.

The women in the back alleys were about the only ones who seemed happy about the currency fluctuations, mostly, in the case of the won, in a downward direction. Against strong domestic opposition, South Korea's economic crisis was forcing the government into pressing the most sweeping economic reforms in recent memory. In the second week of November 1997 the government belatedly promised measures to stabilize both the stock market and the currency while the National Assembly voted on 13 long-delayed financial reform bills. "The government will announce measures on bad debts," said Kim Jun Il, counselor to Kang Kyung Shik, the deputy prime minister and finance minister responsible for the reforms. "We need to do something on the banks. These measures will provide an institutional set-up within which the government can act."[4]

Even as he spoke, however, several hundred Bank of Korea workers were demonstrating in front of the National Assembly building, protesting the government's plan to create a Financial Supervisory Commission dominated by the finance ministry. Bank of Korea workers charged that the plan was a scheme to destroy the bank's independence while enhancing that of the finance minister. They shouted the slogan, "We oppose banking reform," while police carrying shields and truncheons stood by. The police on November 14 briefly detained 176 of the protesters, who were joined by workers from the Securities Supervisory Board in an assault on the Assembly building reminiscent of the antigovernment rioting that had roiled Korean campuses for years.[5]

The move toward reform gained urgency as foreign rating agencies downgraded Korean banks. Moody's Investors Service on the same day indicated it might downgrade several more Korean banks after having downgraded four Korean banks two weeks earlier. Foreign bankers welcomed Korean moves to shore up its economy as long overdue. "They should have a financial reform bill which will address excessive investment and excessive borrowing," said Alain Bellisard, president of the European Chamber of Commerce and manager of the Société Générale branch in Seoul. "They have offered a reform bill which will show the right and fundamental issues are being addressed." Bellisard worried, however, that strong opposition might again deflect reform. "The next few days will tell," he said. "We know they have something in the oven."[6]

Would the government—and the business community with which it interacted—bite the bullet on the need for several banks to go under in view of debts from bad loans and bankruptcies conservatively estimated at the time at $26 billion? (The actual figure turned out to be many times that amount.) "They will have to merge some banks," responded

Cristoforo Rocco, branch manager of J. Henry Schroder. "It is not a modern system. Two or three banks should disappear to be competitive." He blamed Korea's troubles on the policy of government protection that had dominated the economy for decades. "The banking and security system are so protected, they cannot really compete with foreigners. History is coming too fast for them. They have to be quick." He wondered if the government comprehended the problem. "I'm not sure they are ready, but I am not sure they have a choice."[7]

The government, in a public relations blitz aimed at negative publicity abroad, insisted it would never have to go to the IMF for help, while IMF representatives in Washington argued that Korea needed foreign exchange reserves totaling three months' worth of imports, about $36 billion. The Bank of Korea claimed reserves of about $30 billion, but much of it was already committed to shoring up the currency, and reserves might be far less. Foreign business leaders emphasized that Korea had to engage in much more than public relations—and headline-grabbing reforms of little substance—to regain international confidence after a year in which eight of the country's top 50 chaebol had gone bankrupt. "There are different schools here," said Jean Jacques Grauhar, managing director of the European Chamber of Commerce. "The pessimists believe devaluation will happen and the government will have to call on the IMF." Moreover, "They will have to open wider the securities market," he said. "There are a lot of limitations. There's a big feeling that foreign companies will need to do more."[8]

The government was saying only that its reforms would streamline the banking industry. One way would be to raise the level of funds for the Korea Asset Management Corporation to take over nonperforming assets of banking institutions. "The plan will also set up funds to underpin mergers and acquisitions among the financial institutions," Yonhap, South Korea's semiofficial news agency, reported, and will "firm up measures for further widening inflows of foreign funds into the country and the bond market."[9]

The finance ministry had already made clear that some of the reforms would not go far enough to satisfy foreign securities firms. "South Korea is liberalizing its capital market in stages, centering around transactions of capital related to exports and investments," said a ministry statement. "However, we are withholding liberalization of short-term bond transactions as they are an easy target for speculation." The finance ministry was at pains to fend off comparisons of financial deterioration in Korea with that in Thailand or Mexico. "Korea will never become another Thailand or Mexico," said the statement, written in English. "It is hard to imagine the collapse of the Korean economy, and the assumption . . . is unrealistic." The statement promised the government would "tackle the structural problems, such as bad loans and weak capital structures of large corporations, by taking definitive measures in line with market principles, rather than employing a Band-Aid approach."[10]

In the marketplace, however, no one was listening. On November 17, as the currency crashed below 1,000 won per dollar shortly after 2 P.M., rumors spread that the government would have to call on the IMF. The market was suspended off and on until closing at 4:30 P.M. The won by then was pegged at 1,008.6 to the dollar—down nearly 20 percent since the start of the year.

Finance ministry sources blamed the won's descent on heavy borrowing by financial institutions overseas—and a report by *Chosun Ilbo,* South Korea's leading newspaper, that the economy needed IMF support. *Chosun Ilbo* front-paged the report on November 17 along with renewed government denials of any plan to go to the IMF for help. The article cited overseas borrowing and corporate bankruptcies as reasons for Korea's financial difficulties and the need to call on the IMF.[11] The plunge of the won triggered selling on the Korea Stock Exchange as the composite index fell 22.39 points to break through yet another barrier, the 500 level, closing at 496.98. The Bank of Korea gave up the battle to hold the won above the 1,000-to-one-dollar level after having intervened heavily all morning. Authorities at the stock exchange blamed the surrender on panicky demand for dollars on the part of institutions in a hurry to pay bills before the won lost more value.

Bankers and securities analysts viewed the fall of the won as significant but hardly unexpected. The government had artificially held the won to a level of 986 to the dollar at the end of the previous week by exercising firm controls in addition to bank intervention. "It's a managed currency," said Rocco at J. Henry Schroder. "The Bank of Korea and the ministry of finance and economy can control the level of activity on the domestic foreign exchange market." Analysts believed the won's downward slide was irreversible, at least until it hit between 1,100 and 1,200 to the dollar. "It will stop the day they call the IMF," said Grauhar at the European Chamber of Commerce. "I hope it won't reach 2,000."[12]

Bankers expected the won to go down further in value while the government battled to stabilize the economy. The National Assembly was to vote on November 18 for the 13 financial reform bills. Most important was a bill to establish a central body to supervise financial institutions, but the ruling New Korea Party said it might have to postpone voting until after the presidential election in December. The legislation had run into severe opposition from the Bank of Korea, whose executives and managers were denouncing it as "an evil monster" in daily demonstrations.

Contrary to all the finance ministry had been saying, the next day, November 19, the government had to confront the immediate need to ask the IMF for a bailout after the Assembly spurned most of its financial reform bills. The Bank of Korea, whose employees had threatened to strike if the Assembly passed the bill handing the bank's powers to a central supervisory agency, urged the finance ministry to consider an IMF bailout "as a last resort to overcome the currency problems." The bank also called on the government to adopt a stabilization program that would settle bad loans of about $20 billion by merchant banks, forcing a number of them to go out of business.[13]

South Korea's financial crisis was deepening by the hour as the dollar again passed the 1,000-won threshold. The dollar opened on November 19 at 990.6 won but rose in minutes to 1,012.8 before trading was suspended. Making matters worse, the ministry of finance appeared to be in conflict with the Bank of Korea over how to get out of the current crisis. Finance ministry officials again staunchly denied any notion of asking the IMF for help and insisted a new stabilization program would stop the won's slide. "We expect the stabilization package will alleviate the downward currency movement," said Kwak Sang Yang, with the finance ministry's currency division. He blamed the won's depreciation in part on reports that an IMF bailout was in the works and also on the downgrading of Korean financial institutions by foreign rating agencies.[14]

The purpose of the stabilization program was to open the domestic bond market to foreign investment and to set terms under which more than 30 merchant banks would either show they could pay their debts with an infusion of government funding or close. Many merchant banks were now unable to obtain credit internationally. After bickering all day, the finance committee of the Assembly in the evening approved just four parts of a 13-bill package. The most important was the financial deposit protection act providing for a nonperforming asset loan fund of 3.5 trillion won—nearly 3.5 billion dollars at the time—to help protect financial institutions. Another bill that made it through the committee was a financial deposit protection act aimed at protecting deposits in commercial banks. Finance Minister Kang Kyung Shik said he would resign if the Assembly failed to pass all the bills, then suddenly announced plans to speak the next day, November 20, at an unlikely forum, the Seoul Foreign Correspondents' Club.

By now all the measures—and the stabilization program in general—amounted to too little, too late. "They need to get out of their short-term lending crisis," said Daniel Harwood, regional director of ABN AMRO Asia. With more than half of Korea's $110 billion in debts to foreign creditors due for repayment within a year or so, Harwood said the government had to compel the chaebol "to cut unprofitable businesses" and "get efficient." Otherwise, he said, "Korea will go bankrupt" and "the economy is not going to survive" in its present form.[15]

Whatever happened, the government hoped not to have to ask the IMF for help until after the presidential election. For all the denials, however, IMF officials were already meeting with members of the finance ministry and the Bank of Korea. The bottom line was that the government did not have the money. "To stop won depreciation they have to sell dollars," said Jason Yu, banking analyst at Indosuez W. I. Carr. "They have only $20-30 billion in reserves"—well below the $36 billion, equal to three months' imports, recommended by the IMF. "There are so many defaults," said Yu. "If they reduce liquidity, it will raise more defaults. There are not many alternatives the government can try. Companies have aggressive investment plans. They want to do everything. First they have to get rid of unprofitable business. Overall expansion of the chaebol will have to slow down." As the central bank's assets decelerated, Yu came to an inevitable conclusion: "Eventually they have to ask the IMF for support."[16]

The crisis was building quickly. Kim Young Sam, a bright government critic in the era of Park Chung Hee and Chun Doo Hwan, had displayed a remarkable lack of understanding of economic issues since his election in December 1992. Touchy about criticism as the economy began to buckle, YS had dismissed a series of prime ministers as well as deputy prime ministers, who concurrently held the title of finance minister. Now he asked Finance Minister Kang, stonewalled by the Assembly, to resign and, on November 19, appointed his last finance minister, Lim Chang Yuel, former minister of trade, industry, and energy. The same day, the government adopted an emergency reform package to bail out the economy and stop the downward slide of both the stock market and the won. A veteran of seven years at the IMF in Washington, Lim said the program meant, "There is no need for financial help from the IMF."[17]

The package focused primarily on means to cover more than $28 billion acknowledged as bad loans of commercial banks. In the interest of what it described as "enhanced transparency," the package included a graph showing the skyrocketing loans, nearly double the amount at the end of 1996. As a first step, the program authorized the Korea Asset Management Corporation to spend the equivalent in Korean currency of up to $10 billion to "purchase bad loans from banks." That amount nearly tripled the fund's size. "The government is planning to write off 50 percent of all bad loans in this fashion by the end of the year and hopes to eliminate all bad loans within one or two years," said the announcement. In a bid to bring dollars into Korea, the package eased overseas financing by increasing the swap facilities of foreign banks from the current limit of $1.1 billion. At the same time, it encouraged "major public enterprises" to borrow overseas and authorized "selected business firms" to take out cash loans "without limit until the end of the year."[18]

The package addressed the currency crisis by broadening "the band in which the won is allowed to float" each day from 2.25 percent to 10 percent "to better reflect depreciation expectations in the market." More boldly, in a move long sought by foreign securities firms, the package sought to bring more dollars into Korea by opening the domestic bond market to foreign investors for the first time. Investors might now buy "guaranteed and nonguaranteed corporate bonds with maturities over three years," said the announcement, putting "the individual ceiling for foreign investors" at 10 percent and "the aggregate limit" at 30 percent. The package also encouraged the merger of weak financial institutions while protecting depositors to the full extent of their deposits through the newly formed Korea Deposit Insurance Corporation.

The program, however, did not seem at all likely to solve Korea's financial problems. The core problem was the government's foreign currency reserves, $20-30 billion by the most generous of estimates, were dwindling. On November 20, the beleaguered won declined almost immediately by the full 10 percent permitted under the new financial stabilization program. The sudden fall of the won to yet another record low—1,139 to the dollar—was followed by another decline in the stock market, which fell 14.18 points or 2.86 percent to 488.41 by the end of the day. The drops in both the currency and the stock market had the impact of a one-two punch on efforts to bail out financial institutions wallowing in debts totaling at least $110 billion—two thirds of it maturing within one year.

Finance Minister Lim, at the briefing arranged by his predecessor at the Seoul Foreign Correspondents' Club, blamed "the current economic crisis" on "psychological factors." Right away he picked an easy target: Korea's ancient enemy and former colonial ruler, Japan. Pleading for Japanese financial institutions "to provide safety nets" for Korean loans, he accused Japanese institutions of "calling bank loans rather than allowing rollovers." Menacingly, he reminded the Japanese that "instability combined with lack of cooperation will lead to instability in Japan" too. Currency reserves, he said comfortingly, totaled $30.5 billion as of the end of October. He responded sharply, however, when asked how large the reserves were after three weeks in which the Bank of Korea had spent several billion dollars supporting the won. "There is no government that announces its foreign reserve volume day by day," he replied. In any case, he said myopically, by widening "the band" under which currency could fluctuate from 2.25 percent to 10 percent a day, the Bank of Korea no longer had to defend the won.[19]

Lim still could not come to terms with the need for going to the IMF, making clear the government would rely on individual foreign banks and other institutions. "If we need IMF assistance, we should receive it," he said plaintively, but "we are already doing all those reform packages they are likely to request." He was open, however, to a regional bailout program in the form of an Asian monetary fund. "Asian countries are thinking of ways to support each other." He discounted reports that the United States opposed such a fund, but he already knew the Americans strongly believed it would undercut the power and resources to aid the region's hard-pressed economies. The same day, two officials, Timothy Geithner, assistant secretary of the treasury for international affairs, and Ted Truman, staff director of the international finance division of the Federal Reserve Bank, visiting Seoul, had recommended that South Korea go to the IMF if the reform package failed. "I do not discuss my private schedule," Lim huffed when asked about his meeting with them.[20]

The government, however, was anxious to convince Koreans it was taking the right steps without the face-losing need of asking the IMF for help before the presidential election. Lim said he would press for the Assembly to pass the economic reform bills spurned by the opposition but was likely to wait until after the election. He had an array of statistics to show the economy was basically in good shape. Predicting economic growth of 6 percent for 1997, he noted that consumer prices had gone up by less than 4.5 percent while the current account deficit had shrunk to less than $14 billion or 3 percent of the gross domestic product, down from $24 billion a year ago. "Korea's economic fundamentals are very much sound," he said, even though "the ominous effects of the currency crises in Southeast Asia" have "spilled over into Korea."[21]

The protestations of the finance minister, however, were about as hollow as the numbers issued by his ministry. All the cash the government counted on receiving from the IMF, the United States, and Japan would hardly resolve South Korea's financial crisis. The power of the IMF to set an agenda for financial reform might cause a number of the merchant banks that had sprung up in recent years to merge or close, but critics questioned whether the government was ready for drastic action. While the IMF might solve the short-term dollar problem, it would have more difficulty addressing the structural problem of overlending and overinvesting.

More daunting, the government faced what might be the impossible task of subduing an array of chaebol long accustomed to borrowing many times their value from banks equally accustomed to complying. By now, analysts were estimating the overall domestic corporate debt of Korean companies at the equivalent in won of about $200 billion. That figure did not count the latest estimate of more than $120 billion in foreign currency debts, half of it due by the end of the year. Without far more financial support than anyone expected the chaebol to receive, mergers, downsizing, and layoffs were inevitable over the next year or two. Korean companies, however, were far from resigned to this fate. An irony of the bankruptcies of eight of the chaebol was that they had not faded from view. Rather, they were supervised by government-appointed bankers and other directors while attempting either to climb out of bankruptcy or to find suitable buyers.

All bets were off if the economy continued its downward slide after January 1, 1998. Samsung, the second-largest chaebol, had made clear its interest in purchasing Kia, supposedly bankrupt but still producing cars for a glutted market. Hyundai considered Hanbo Steel, another bankrupt company, in hopes of fulfilling its dream of becoming a steel producer— not a likely possibility given the government's stake in Posco, Pohong Iron and Steel. The real question, however, was how much power the government would exercise simply in compelling companies to pay their debts and curtail projects. Only serious retrenchment would relieve the banks of the demands of the chaebol. The fear was that proud and stubborn leaders of the chaebol would fight to the end against any efforts to reduce their empires.

On November 26, while the government was still not quite ready to admit openly the need for IMF intervention, Samsung stole a march on its competitors by giving an impression of being the first to resign itself to the need for slashing new investment and focusing on core companies. In separate press conferences for foreign and local media, Samsung executives presented a plan whose ostensible purpose was to "bolster competitiveness in the face of South Korea's financial turmoil." The measures were the strongest response by any of the chaebol to the pressure created by enormous debts and nonperforming loans, all reflected in the country's depreciating currency and sagging stock market.[22]

Samsung, whose 1996 sales of $92.7 billion had led all Korean chaebol, said its "blueprint refocuses group-wide investments on core growth businesses, reorganizes management and cuts costs across the board." New investment at Samsung would go down 30 percent, from $8 billion to nearly $6 billion, said Samsung, citing memory and nonmemory semiconductors, telecommunications, and motor vehicles as "core growth businesses." The announcement said the group had already "divested 35 product lines, such as game machines," and "plans to divest an additional 34 product lines in 1998." Such divestment, Samsung claimed, would save more than $1 billion. The group also planned to save money by cutting salaries of top executives by 10 percent and pledged to base future increases "on merit rather than seniority"—a measure that Samsung saw as "a break from traditional Korean business practices designed to stimulate competitiveness." In "an effort to create a more frugal Samsung," by which the company hoped to save one trillion Korean won, nearly $1 billion, Samsung said it would "slash travel and entertainment expenditures in half and implement energy-saving measures." The result would be "to reduce consumption by 30 percent, saving the company annually over $150 million."[23]

Samsung issued its announcement as other chaebol were also responding, haltingly, to government demands for restructuring while finance ministry officials met secretly with representatives of the IMF. The government, asking the IMF for a $20 billion loan to help banks cover immediate short-term debts due in December, had promised to carry out a reform package that included restructuring of the financial industry. As a first step, the government said it had asked eight merchant banks to yield their foreign exchange operations to commercial banks.

A wave of dismissals seemed certain as key industries scaled back their plans. Halla Heavy Industries, the shipbuilding component of the Halla group, started the trend by saying it would lay off 3,000 of its 7,000 workers. That prospect raised the fear of widespread strikes. Samsung, however, would not dismiss any of its 260,000 workers, including 30,000 overseas, said spokesman Cho Jang Won. "The cultural background in Korea is quite different from western style," Cho explained. "We cannot fire workers so easily. We're going to set up combined organizations. We will be able to give workers other jobs."[24]

A larger question was whether Samsung should go through with its grandiose scheme to manufacture motor vehicles while the domestic market was shrinking. A sign of the difficulties facing the industry was that Hyundai Motor, the country's largest motor vehicle manufacturer, had 65,000 cars in stock, 15,000 more than normal. Hyundai Motor had begun eliminating overtime and cutting back on the regular hours of some of its 45,000 workers, had dismissed 35 of its senior executives, and planned to reduce its workforce from 45,000 to 40,000.

Negotiators from the IMF faced finance ministry officials in their first full day of fateful talks on November 27. That same day, planning officers from the top 30 chaebol called for a moratorium on debt repayments, asking the government to give them until January to repay more than $20 billion in debts falling due in December. They argued that the chaebol were finding it virtually impossible to obtain fresh credit from banks now calling in loans. The chaebol also pleaded for immediate suspension of the complicated procedure under which companies within the same group could merge with one another in order to increase efficiency and competitiveness. Observing that such mergers could take years for approval, planning officers called for a system whereby they got rid of borderline operations with a minimum of red tape. The planning officers demanded "immediate and strong action rather than step-by-step measures to prevent the collapse of the financial system."[25]

The government indicated it would try to comply with some of the demands—but not necessarily the plea for a moratorium on debt repayment. The ministry of commerce, industry, and energy promised to ease the procedure for mergers and acquisitions within the same group. The government, however, strongly resisted another chaebol demand—that it drop its insistence that investors use their real names on all accounts rather than be able to hide their assets under fake names. President Kim Young Sam had pressed for the real-name rule in order to prevent the chaebol and their owners from secretly amassing vast quantities of wealth. Chaebol leaders argued that the real-name system was forcing many account holders to hide their assets rather than use them to pay off debts. They did not confess their real fear—that they would face huge taxes and criminal charges if all their accounts ever surfaced.

By focusing on rule-bending to pay off loans, the chaebol were also resisting the substantial restructuring needed to turn Korea's economy around. While cutting back on investment plans, some chaebol leaders tried to view the crisis as short-term. Daewoo Chairman Kim Woo Choong said Korea had "the capability to cope with the current difficulties." He predicted the financial market would "regain calm by March next year at the earliest and in two to three years at the latest."[26]

Already, the IMF was an easy target. Worries about what the IMF would do or demand were partly responsible for a worsening stock market and currency. The won on November 27 closed at 1119.5 to the dollar after having closed the day before at 1,110.0. The government asked for at least a $20 billion credit line, to be supplemented by loans from Japan and the United States, amid bitter debate between the IMF and Korean teams. Hubert Neiss, the IMF's Asia-Pacific director, refused to comment, but the differences became clear four days later, on November 30. From the conference room in the basement of the Seoul Hilton, a Daewoo property up a steep garden slope from Daewoo group headquarters opposite Seoul Station, word leaked that negotiators differed sharply as they neared a deal. Finance Minister Lim, bargaining late into the night, said, "We are trying hard to end negotiations as soon as we can," but the government's Korea Broadcasting System reported "apparent disagreement" between Lim's team and its IMF counterparts on key issues.[27]

At the crux of the debate was whether Korea should force the closure of a dozen merchant banks and several commercial banks, as demanded by the IMF, or encourage mergers and acquisitions between weak and strong entities, as desired by the Korean government. Rather than shut down the weakest, most debt-ridden merchant banks, according to KBS, the government wanted to close only one and "gradually merge the rest" with merchant banks that were still relatively healthy. "The IMF wants the weakest 12 merchant banks to stop all new business immediately," said KBS. "In response, the Korean government says it will take care of the process not immediately but in six months."[28]

The government and IMF negotiators agreed, however, on the urgency with which Korea needed the first tranche of bailout funds to stop both the stock market and the currency from their precipitous decline. "The first rescue money of between $15 and $20 billion is likely to be here this week," said KBS. Finance Minister Lim, after having initially said Korea was requesting a credit line of $20 billion to help Korean companies pay about $80 billion in short-term debt, now said the government needed "far more." Most analysts believed Korea needed $50-60 billion by the end of the year to avoid mass defaults on loans by major companies. Japan and the United States had indicated they would lend billions of dollars provided Korea worked out the deal with the IMF.

The IMF was making other demands that Korean officials were reluctant to accept without a fight. The overarching theme of the talks was the need for a sharp cut in increase of a gross national product that had risen an average 8 percent a year since the Korean War. The IMF wanted the

GNP to go up by at most 2.8 percent in 1998, while the government still hoped for an increase of more than 3 percent. The GNP by the end of 1997 was expected to show an increase of 6 percent over 1996, when the GNP had risen by 7.1 percent from the year before. The credit crunch was blamed on the insistence of the chaebol on exporting as much as they could while borrowing heavily to cover manufacturing costs—and over-investing in facilities. IMF negotiators sought interest rates as high as 20 percent to stop heavy borrowing. Interest rates on November 29 fell to 15 percent, down from a high of 18.55 percent on November 26.

*Chosun Ilbo* on November 30 reported the IMF had made other demands for "tightening the economy" that might be difficult for Koreans to swallow. Besides calling for closure of merchant banks and a decrease in the growth of the economy, the IMF wanted Korea to permit more foreign banks and securities firms to open before the end of 1998. "Equal treatment for foreign and Korean financial institutions" was another point made by the IMF, said *Chosun Ilbo*. The paper said the IMF had also called for "flexibility of the labor market," meaning that Korean companies should have the power to lay off workers. Then too, the IMF was asking the government to increase the 10 percent band by which currency was permitted to go up or down every day. "They want to make it more free," said the paper. Thus, IMF negotiators were calling for complete privatization of the industrial sector. At the same time, "the government should guarantee transparency on business management," the paper reported IMF negotiators as demanding.[29]

Finance Minister Lim backed down on December 1 from his claim of an agreement with IMF negotiators. Somewhat chastened, he emerged from a bargaining session with leaders of the IMF team and acknowledged, "The agreement is not concluded." Rather, he said, "We have narrowed our differences," meaning he believed the sides would come to terms. By now word was that South Korean companies had to pay off $70 billion in debts by year's end to avoid bankruptcy. Negotiators were working out a package that included loans not only from Japan and the United States but also from international organizations, including the International Bank for Reconstruction and Development, otherwise known as the World Bank, and the Asian Development Bank.

Both Korean and IMF negotiators underestimated the calamity that was befalling the economy. Koreans feared that a significant fall in the growth rate would leave companies no choice but to lay off thousands of workers. Korean law forbade layoffs in most cases, and labor unions had threatened strikes and demonstrations if the law was changed. Lim was disappointed in a telephone conversation with Michel Camdessus, the IMF managing director, attending a meeting in the Malaysian capital of Kuala Lumpur of ministers of ASEAN, the Association of Southeast Asian Nations. "Until you have the last 't' crossed and the last 'i' dotted, the agreement is not there," Camdessus told reporters. Indicating the difference with Korea over growth rates, Camdessus said the country would need "one year, one year and a half" before the GNP could rise dramatically again as it had been doing ever since the Korean War.[30]

Lim, after facing Hubert Neiss across a table for several hours in a large room off the lower lobby a few doors away from the Hilton's blaring discotheque, was reluctant to face the South Korean reporters and cameramen surrounding him. Asked what was the problem, he responded, "No problem," in English. "We are still talking to working-level people," he went on in Korean, repeatedly saying, "I can't talk about any more details." Security guards fended off reporters, bundling Lim into an elevator that whisked him to his suite on the nineteenth floor down a long corridor from the rooms of members of the IMF team.[31] He and his aides had talked for hours during the day in the suite before confronting Neiss and other IMF negotiators. A cabinet meeting at which Lim was to have presented the IMF package was canceled. At the same time, the stock market fell 3.6 percent—enough to plunge it through the 400 mark to 393.16.

By December 2 the finance ministry was ready to meet the IMF demands partway by suspending most of the activities of nine merchant banks that the IMF had wanted Korea to shut down entirely. The suspension appeared as such a major step that the IMF team was cautiously optimistic about reaching agreement the next day. The sides, said Neiss, who had previously limited his comments to saying, "Nothing is certain," were "working out" their differences. Word that an agreement was in the offing, however, failed to stem the decline of both the stock market and the currency. The composite index fell another 16.29 to 376.87, sent downward after the bank suspensions.

Nonetheless, the sense of confidence prompted Lim to say that he and Camdessus would sign an agreement on December 3 for a bailout package under which South Korea would receive more than $50 billion. Korean officials still seemed reluctant to meet IMF demands for the closure of a dozen merchant banks and two or three commercial banks. Finance officials told Korean reporters the IMF had agreed to give the commercial banks until June 1998 to restructure and prove their viability, and Lim said the agreement did not actually require the closure of either merchant or commercial banks. "We won't let commercial banks close," said Finance Ministry spokesman Chung Eui Dong.[32] Resistance to IMF demands raised the possibility of complications. Camdessus, after the ASEAN sessions in Kuala Lumpur, said the talks were in their "last moments" despite "major difficulties."[33]

The Koreans, however, had to swallow their pride and yield to IMF demands if they wanted to get the initial tranche of at least $10 billion needed in the next week or so for companies to pay off immediate debts. Camdessus tried to be diplomatic in explaining why the Koreans had gone to the wire on critical points. "This takes time with the 11th economic power of the world, with a diversified and advanced economy," he said in Kuala Lumpur.[34] Besides negotiating on the banks, the IMF still believed it had to persuade Koreans of the need for a drastic slowdown in growth. No one suspected at the time how specious was this aspect of the debate. In the face of all efforts to prop them up, Korean companies were slowing down precipitously.

As proof of good faith, Lim said the merchant banks now barred from business would be liquidated if they failed to produce convincing plans for rehabilitation from their heavy debt burdens. An entity known as the Non-Bank Deposit Insurance Corporation would administer them, protecting depositors by issuing bonds. The issue of market opening was also critical, though not fully resolved. The finance ministry might permit foreign banks to open locally incorporated branches in Korea by 1998, said Yonhap.[35] The government also was said to have agreed to require the chaebol to release consolidated financial statements for all their firms and to require audits by international accounting firms of all major financial institutions. In addition, the agreement called for opening the market in short-term bonds, public bonds, and commercial paper to foreign investors before the end of the year.

The signing of the agreement on December 3 marked the opening of what Koreans called "the IMF era." From now on, it was said, Korea was under "IMF supervision" if not "IMF control." Stores advertised "IMF sales," and restaurants sold "IMF lunches," "IMF snacks," and "IMF drinks" from "IMF menus." The sense that a mysterious foreign entity, influenced by the United States and other western powers, was about to assume the levers of financial authority came as a shock. From black-suited executives in lavish suites to white-shirted salarymen in seas of desks on the floors below, the response was the same: how could a group of foreigners in town for just a few days challenge all the Koreans had fought to build since the Korean War?

Just what the IMF wanted Korea to do about the chaebol, however, was far from clear. As the agreement was reached, Kim Young Sop, senior presidential secretary for the economy, insisted that there was "no discussion of the dismantling of conglomerates or the forcible merger of banks." Such assurances did little to counter the widespread view around the headquarters of the 64 groups classified as "chaebol" that the IMF viewed them as the enemy in a battle to put Korea's house in order—and open its doors to foreigners. The Federation of Korean Industries spoke for the mercantile elite. "In Thailand, Mexico and other countries receiving IMF funds, has the IMF ever demanded the dissolution of conglomerates?" asked planning officers from the 30 largest chaebol. "We understand that the IMF apparently called for resolving the problem of chaebol's heavy borrowing practices."[36] The inference was that the IMF had failed to appreciate the courage of the chaebol, each founded by one man, almost all under family control, in daring to borrow four or five times their worth—and turn Korea from a pauper to a prince among nations.

"Without the chaebol, how can the country survive?" asked a director in the marketing department of Hyundai. He hardly had to know the terms of the agreement with the IMF to oppose it. "The IMF is intervening in the sovereignty of our country. Many people are afraid of what the IMF will do to the chaebol and to our country."[37]

All the IMF had done was to ask a few pointed questions about how deeply, exactly, were the chaebol in debt and what they could do about it. IMF and government negotiators agreed in principle that the loose payment guarantees among chaebol subsidiaries, which commonly bought and sold products within a group, was one reason some chaebol were in trouble. An obvious measure, long demanded by foreign bankers and investors, was the requirement for consolidated statements revealing financial realities of each company. The result might be to encourage the chaebol to get rid of costly entities, many of them run by sons and other relatives of chaebol chieftains. "Inefficient companies must go," said Park Ho Won of LG Securities, an offshoot of the LG group. "That is the lesson we are learning at the expense of national pride."[38]

To outward appearances, the chaebol were responding by cost-cutting measures to show they realized they could no longer amass such debts and hope to stay in business. Not to be outdone by Samsung, which had said it was selling off unprofitable entities and docking executives' salaries, Daewoo on December 3 announced a 15 percent salary cut for executives, a 10 percent cut for managers, and a freeze on pay for the rest of its

200,000 employees. "Everyone will share the pain," said Daewoo spokesman Lee Jeong Seung, anticipating "a hard time under the IMF guideline."[39] While issuing such pronouncements, however, the chaebol were also likely to fight a rearguard battle against what many of their leaders saw as an IMF effort to destroy them.

No chaebol displayed more old-time drive, along with less desire to change, than did Hyundai. At a rally in the Hyundai headquarters on December 8, five days after the IMF agreement was announced, 1,000 senior executives, ranging from members of the owning family to presidents and directors, sang the company song, waved clenched fists, and shouted slogans. Then they listened to Chung Mong Ku, the founder's oldest surviving son, now cochairman of the group, fire them up for the "crisis" confronting Korean business and industry. "Hyundai has to raise the flag high to contribute to the development of the Korean economy," said Chung, raising his own fist before his executives. "Many big companies are bankrupt. The management crisis has become more serious."[40]

Chung, however, saw good emerging from bad. "If we make every effort to discard our loose thinking, then we can make this crisis a chance to gallop." He linked the sloganeering and speechmaking to a practical message. Hyundai, he said, was cutting investment for research and development as well as facilities by 30 percent in 1998 while slashing the two-month bonus paid to executives. At the same time, he said that Hyundai would increase profits from business overseas from $12.1 billion in 1997 to $17 billion for 1998, up 40 percent. The group's total sales would reach about $90 billion in 1998, up 14 percent from the year before, thanks largely to sales abroad. Hyundai's announcement of a cut in spending had already made it the third of the top five chaebol to announce reform measures since the IMF agreement was announced.

The rally epitomized the manner in which many Korean groups and companies were psyching up their workers as they fought to pay off debts and survive. The appeal was to company loyalty and old-fashioned nationalism—the same forces that had driven Korea to economic greatness under Park Chung Hee in the 1960s and 1970s. Beneath the rhetoric ran a subtext, unspoken publicly but repeated to me by a Hyundai manager. "In Korea, every company is in trouble," he said. "That's the intention of the IMF. The IMF and America are trying to break up the chaebol." He warmed to the sense of national shame as he talked on. "The Korean financial institutions are in trouble. The Korean system is in trouble. How can Hyundai not be in trouble?"[41]

The recriminations were not isolated. The largesse of the IMF inspired an extraordinary reaction. On television and in the press, Koreans complained that the IMF has been guilty of egregious imperialistic meddling. To hear Finance Minister Lim tell it, the country had gone down to defeat in a hard-fought war. President Kim Young Sam, asking his compatriots to accept "bone-carving pain," told the nation he was "truly sorry."[42] From the street to the executive suite, the crisis-speak generated a sense of doom that tapped into Korean xenophobia and humiliation at having to beg. The next step might be for the doors to close on imports—a defensive measure that might protect industry in the short term but would also invite retaliation and insulate it from the reality check it needed of foreign competition. In the moment of worst crisis, the government, the elite, and the newly rich chaebol class, as well as the media, provided no leadership in dealing with a sentiment that might turn dangerous.

The chaebol leaders, titans of the miracle, crawled into their shells. First they came up with defensive claims that they were reforming—not true, on close examination. While both Hyundai and Samsung said they were cutting back on new investment by 30 percent, that was out of necessity. They did not have the unlimited credit on which they had once counted. Samsung made a show of announcing it was reducing the number of "entities" but refused to specify which ones—except for the audio division, transferred, as a Samsung executive reluctantly admitted, to Saehan, a lesser chaebol controlled by a brother of Samsung Chairman Lee Kun Hee. Daewoo, like Hyundai, claimed to be plunging ahead with motor vehicle expansion worldwide, to mixed reviews. Some analysts believed that Daewoo companies abroad were getting all the credit they needed. Others wondered where they hoped to sell all those cars produced on plants in the former Soviet bloc.

The willingness of the chaebol to issue consolidated statements of the sort that the IMF had demanded—and that bankers needed to assess credit-worthiness—would be a great test of their sincerity. The chaebol now talked about "transparency" the way they talked about "globalization" and "restructuring"—words to throw around but not to carry out except on their own terms. The suspicion was the chaebol would obfuscate to the fullest on providing a complete, open, honest chance for outside accounting firms to evaluate profits, losses, and assets. The chaebol leaders were also saying there was no need to "break up" the chaebol. All that might be true, but the chaebol were overlooking their own disgraceful record of pressuring their friends in the banks to overextend loans for many years—so many that disaster was inevitable. Relations between chaebol leaders and bankers remained extremely close, as did those between the hacks in the National Assembly whom the chaebol called on to talk to friends in the banks.

All too often, illicit money had greased the way to comfortable deal making, with little collateral to back up the loans. Reading the sweeping promises of "reform" by this or that chaebol, one saw scant evidence that chaebol leaders were determined to reject the old ways. None said a thing about the merits of accepting far smaller loans so the banks would have enough for medium-size and small firms that found the going toughest. Ideally, the chaebol could fight to straighten up their own extremely confused, obfuscatory accounts. The temptation would be to find ways to get around IMF stipulations, particularly on transparency.

In considering the significance of the bailout, the least important stipulation was the one requiring that Korea hold its annual growth rate to 3 percent in 1998 rather than the usual 6 to 8 percent. Korea's economy was now so weakened that the GNP had to go down in 1998. In berating the foreigners for interference in Korean affairs, the politicians and the media forgot that the economic miracle had incurred a tremendous downside risk that was now catching up with the country. IMF meddling, far from penalizing Korea, might help to minimize the suffering. The country, like the chaebol, had to escape the narrow confines of an inward-looking nationalism. That was a lesson that the chaebol were reluctant to accept after years of growth on the backs of policy loans, guaranteed markets, and legislated isolation.

Koreans preferred to attribute their troubles to foreigners. Finance ministry officials and chaebol executives alike blamed the United States for influencing the IMF to negotiate for reduction of chaebol power—and for foreign interests to be able to own up to 50 percent of stock in Korean companies. "The United States may have played some role behind the IMF call with the aim of holding Korean firms in check in the world market," said a senior executive at a leading chaebol. That view went deep into society. "We oppose freedom for foreign investment in financial markets," said the Korean Confederation of Trade Unions. "Opening of our financial markets won't help solve the problem. Rather, it will bring us into slavery." The fear among the workers, as among the chaebol chairmen, was that during the IMF era Korea, Inc. would fall under foreign control.[43]

The immediate impact, however, was quite different. Rather than go into bankruptcy, much less slavery, Korea, Inc. got a new lease on life. The agreement called immediately for the IMF to guarantee at least $58.5 billion to cover short-term debts in return for opening the country wide to foreign investment and imports. Finance Minister Lim and IMF Managing Director Camdessus signed letters of intent on the evening of December 3 after a few last hours of frantic negotiations over details. The bottom line was that the IMF, after receiving the approval of its executive board, would provide a three-year credit line of $21 billion while the World Bank provided another $10 billion "in support of specific structural reform programs," said Camdessus. At the same time, the president of the Asian Development Bank recommended to his board another $4 billion for "policy and institutional reform." Camdessus expected "a number of countries, including Australia, Canada, France, Germany, Japan, the United Kingdom, and the United States," to contribute "in excess of $20 billion." Japan promised $10 billion of this amount and the United States, the major force behind the IMF, $5-6 billion.[44]

Camdessus smiled happily for photographers as he met Kim Young Sam and then appeared with Lim to announce the deal. His appearance contrasted with that of Lim, who looked glum before the cameras and, an hour and a half after the announcement, came close to apologizing on national television for what he had done. "I am here to speak to you with utmost regretful feeling," said Lim, who had been leading the resistance to key points of the package. On Korea's four major television networks, Lim said that his government had been "trying hard to regain public trust" but that "our economic strategy has failed and we regret it very much." Now, he explained, "our government is going to build up a new strategy." Henceforth, Korea's highly leveraged companies "had to be discreet in deciding on their investments and should conduct transparent business deals." As if he were surrendering to a bitter enemy, Lim appealed to Koreans to "understand our suffering."[45]

The deal marked a historic turn for a country and a people that had traditionally resisted foreign influence and inroads into the economy and culture. Among its most important elements was a provision under which the government agreed to lift the ceiling on foreign investment in Korean firms listed on the stock market from 26 percent to 50 percent in 1997— and entirely in 1998. Foreigners would theoretically be able to buy up Korean industries, banks, and securities companies, off-limits previously to significant foreign penetration, and could also set up their own branch firms and companies. The Korean side, moreover, was largely unsuccessful in attempting to persuade the IMF to soften its demand for closure of a dozen merchant banks and several commercial banks. The Koreans agreed to push through financial reform bills, stalled for months, calling for merger and acquisition of commercial banks and liquidation by stages of commercial as well as merchant banks. The net effect of the IMF agreement, if the terms were fully carried out, might be not only to rescue but to open an economy that ranked among the most notoriously closed of any capitalist nation's.

Undoubtedly the most foolish aspect of the talks was the emphasis placed by negotiators on both sides on the proper rate of economic growth. Debate on this point showed how unrealistic were the calculations of the most experienced minds, both Korean and foreign. The IMF also seemed short-sighted in its insistence on a commitment to reducing the national budget, meaning deep cuts in public works projects. By the autumn of 1998, after foreign exchange reserves had risen from about $3 billion, the actual amount on hand in December 1997, to more than $40 billion, and the current accounts surplus had risen from a deficit to $37 billion, IMF economists would be singing a different tune. They would be telling the government not to be afraid of a budget deficit, and they would applaud the 5 percent deficit incorporated into the 1999 budget announced in September 1998. At the height of the crisis in late 1997, however, economists on both sides panicked as they counted how much the government was spending on seemingly extravagant schemes for expressways, railroads, and other ventures.

Such lack of foresight seemed all the more surprising considering the IMF team's insistence on dismissal of superfluous workers as companies downsized and merged. One of the most controversial areas of the agreement called for the government to revise the law banning companies with more than 100 workers from laying anyone off. The IMF viewed dismissals not only as inevitable but as a healthy response as the growth rate declined and unemployment, now 3 percent, seemed likely to reach 10 percent, the highest level since the period of social upheaval after the Korean War. IMF negotiators talked about a "social safety net" in the form of expanded unemployment benefits, unheard of in Korea, but no one imagined that next year they would extol the advantages of a budget deficit and public works spending to provide jobs as part of the net.

Yet another area that seemed questionable, not only at the time but over the next year, was the requirement for keeping interest rates high in order to discourage more borrowing—and spending. IMF negotiators were about the only ones who did not seem disturbed by the phenomenon of rapidly rising interest rates. As far as they were concerned, the high rates would impose discipline on companies that might otherwise borrow much more than they could ever repay just to keep their companies alive. The IMF position on interest rates was an understandable response to the uncontrolled borrowing of the past, but the danger was that excessively high rates would deprive companies of much if not all the credit they had to have to import the goods needed to produce for export. As interest rates soared above 30 percent by New Year's 1998, the richest chaebol and the banks were the beneficiaries. The leading chaebol got most of the credit, while their banks made up for some of their nonperforming loans by profiting from the high rates.

One of the greatest challenges confronting the government would be how to control the chaebol that controlled most of the economy through their own companies and thousands of smaller companies with which they did business. The government, under the IMF agreement, was committed to cracking down on loans between companies within a chaebol and also to compelling the chaebol to issue consolidated statements. It was unlikely, however, that such an agreement could curb the excesses of the largest chaebol. The pressure for mergers, for getting rid of the losers, would promote a contest of survival of the fittest, in which the rich got much richer while many of the rest fell by the wayside. The IMF, in its zeal to bring about discipline needed to stave off bankruptcy, encouraged disaster for small and medium-sized enterprises. Many of the lesser chaebol would survive as also-rans, picking up the leftovers of credit, while the largest positioned themselves to dominate the economy more than ever.

For better or worse, however, December 3, 1997, was a critical date in Asian history. It was like so many of those dates in previous centuries in which foreign interlopers had first got to see the Korean king, threatened Japan with their black ships, and made their way in gunboats up the rivers of China. This time they had come in dark business suits, packing computer files jammed with facts and stats, demanding more facts and stats from the dark-suited mandarins who greeted them with tense, tight smiles and anxious stares. There were 17 of them from the IMF in Seoul; they stayed for a week to ten days, and they left having pried open a system that had refused to do what it had to do of its own free will. The need for the biggest bailout in IMF history to pay off debts that everyone had known for years were piling out of control compelled Koreans finally to yield to serious foreign investment.

The implications were frightening to a system, a society, and a culture attuned to resisting and rejecting foreigners until finally there was no other choice. The idea that foreign companies might own controlling stakes in more than a token few major Korean firms was anathema. The thought of foreign cars arriving in Korea at almost competitive prices was out of the question in a market already glutted by its own excess production. The notion of foreigners owning and running Korean banks was about as far-out, for Koreans, as had been the specter of those first missionaries implanting their crosses on the soil of the hermit kingdom. Now that the unthinkable might happen, under IMF duress, the phenomenon of Korean xenophobia would not disappear. Rather, it would reflect the strength as well as the weakness of Korean society.

At the highest levels of business and commerce, the doyens of the chaebol would be as resistant to serious transparency as were the finance ministry bureaucrats to coming to terms with the IMF in the first place. As far as the chaebol were concerned, releasing full accounting of how well or badly their firms were performing was like opening personal mail for the world to read. They lived in their own secret world in which corporate interactions, like private homes and private lives, were closed to all but close relatives. Chaebol leaders revealed themselves at opportune moments to favored journalists for specific purposes. Interviews meant publicity, which meant sales. "Reforms" and "streamlining" when announced by a chaebol were more likely to portend window dressing than a new model.

Nonetheless, the deal the IMF had wrested from Korea's finance ministry bureaucrats marked a pivotal first step. The aggrieved protests of chaebol leaders against intrusions on sovereignty were battle cries signaling a hard fight. They expressed the paranoid fear that Korea would become a colony again—an outpost of foreign concerns driving the chaebol elite into secondary, subservient status. Thus the IMF mission unleashed a storm of complexes, of ill-defined frights and fears and nightmares of goblins in the night. It was one thing to pay off a few debts, but another to sell out to the wicked Satan. There would be a great deal of talk like this over the next few months. The challenge for both Koreans and foreigners was to try to put the IMF reforms in place without inspiring a counterreaction that might undo the program. Regardless, South Korea, as of December 3, had entered a new period in a restless, unpredictable struggle for survival in a world of foreigners—barbarians, in the deepest recesses of the Korean mind, who were out to get them in the end.

# Chapter 3

# "DeeJay, DeeJay"

In the midst of the new city of Pundang, a model of high-rise apartment blocks, glittering shopping plazas, and carefully planned parks and playgrounds in the urban sprawl east of Seoul, real estate agent Ahn Hyo Hee waited one day in mid-December 1997 for calls that never came. "My last sale was in November," said Ahn, whose clients included mainly young couples looking for safe and comfortable homes within commuting distance of Seoul. "From the time they announce the IMF, they don't sell." What had stifled the mood to buy and spend in a community that epitomized Korea's newly affluent middle class? "It is the IMF influence" was all she could say, relaxing in an armchair beneath a detailed map of the city, whose population had soared to about 400,000 in five years. "People are not comfortable with the IMF."[1]

Ahn, whose husband owned a motor vehicle repair company, could hardly imagine how low the economic downturn would descend. She believed, however, the worst was yet to come—and soon. "Oil prices will go up," she said. "Everybody will have to pay much more. Nobody will be able to buy anything." In three or four months, she feared, she would have to close her office, while her husband's garage scraped by on almost no business. "We have two girls, aged six and four. People are not happy. We don't expect any future."

A stroll along the wide streets beside the park revealed a vista of gleaming shopping centers and office towers dominated by Samsung Plaza, an architectural wonder of balustrades and atriums and glass-and-marble arches, all providing the setting for stores and shops bearing fashion brands from around the world. Just a month earlier, Samsung Plaza had opened in a blaze of speechmaking and ribbon-cutting and exploding firecrackers. Clerks in neat uniforms smiled from behind the counters, but few were buying. Christmas gift certificates lay in neat stacks under glass, ready for giftwrapping in ribbons and bows. Trouble was, the numbers on the certificates—50,000 won, 70,000 won, 100,000 won—were losing allure to customers who either had no money or viewed them as bad investments.

"It's all because of the IMF," said Choi Yeun Hee, an advertising designer. "People don't have money. A lot of people are window-shopping." Her husband was an engineer for Samsung Electronics, flagship of the chaebol that owned Samsung Plaza. Recently, they had moved into a two-bedroom apartment a few minutes from Samsung Plaza, from which a new extension of one of Seoul's subway lines moved thousands of commuters into the city every day. "Many young people are living here," she said. "It is a place for the young"—as evidenced by mothers with babies and children playing in the nearby park as well as a twinkling district of bars and restaurants a few blocks away. "We all have the same problem. The economy is going down, down, down. People are saving up on sugar and flour before the prices go up." She had a suspicion of whom to blame. "Maybe the IMF and our president did not do a good job."[2]

Anticipating the backlash, the IMF had obtained the signatures of all three candidates in the presidential campaign as a precondition for any agreement. Kim Dae Jung of the National Congress for New Politics, Lee Hoi Chang, the candidate of the ruling New Korea Party, and Rhee In Je, who had broken away from Lee to run on his own, were all on paper as supporting it. That detail was overlooked as the two front-runners, on December 4, vied to unleash the strongest criticism during the three-week campaign permitted by law. After his party declared December 3 "national shame day," DJ himself said he would demand new talks, a new agreement, with the IMF. A spokesman sought to soften the tone, saying that DJ had told President Kim Young Sam he would respect the agreement "in principle if elected" but would "enter into further talks with the IMF."[3]

The official three-week campaign leading to the presidential election on December 18 was potentially the most divisive since mass demonstrations had ushered in a new constitution and an era of quasi-democracy ten years before. As YS wound up his term, DJ saw his last and best opportunity for power. He claimed he was in perfect health except for a somewhat shuffling gait and uncertain arm—the result of the "accident" in the only presidential election held under the dictator Park Chung Hee in 1971. Kim, injured when his car tumbled off a road to avoid an onrushing truck, had embarrassed Park by winning 46 percent of the votes. Now he believed he had the brass ring almost in his grasp. This time the establishment forces as represented in the government and the New Korea Party were hopelessly divided.

The good news about Lee Hoi Chang, a former supreme court justice, was that he had appeared fairly incorrupt while carrying out Kim Young Sam's anticorruption campaign as chairman of the Board of Audit and Inspection before YS appointed him prime minister. Strangely imperious, hypersensitive to criticism or even mild disagreement, YS dumped Lee as prime minister two years later in a clash of egos—and then had to persuade him to run for president to get a big name on the slate. Still, the machinations were far from over. The former mayor of Seoul, Cho Soon, dropped his own campaign and threw his weight behind Lee, who began overtaking DJ. The shifting alliances made it seem as though the election were bereft of issues—an exercise in names in which one member of an elite supplanted another. Just to show the cynicism that dominated the campaign, Lee professed to reject the support of YS, who had made a show of quitting the ruling party, renamed the Grand National Party. Scarred by revelations that his two sons had avoided military service by saying they were "underweight," Lee threatened to steamroller his way into the Blue House in a photo finish.

Ordinarily, the alliance of government and chaebol would have ensured victory for the ruling party. In December 1997, however, the campaigning rolled on against the backdrop of a drama that seemed much closer to home, much more real—the economic crisis. YS lacked the economic sense to understand what was going on, to undertake reforms that were urgently needed. His presidency had been severely compromised, moreover, when one of his own sons was found guilty and jailed for accepting payoffs during the 1992 presidential campaign. With YS on the sidelines, business and politics were now linked in different fashion as the candidates blamed YS for not having moved nearly fast and hard enough to blockade all that easy credit for the chaebol.

These issues touched the deepest nationalist sensitivities. It was difficult to identify them with any one candidate, however, since each viewed the IMF presence as humiliating infringement on sovereignty. To Koreans, the IMF represented quintessential economic imperialism—a group of economists sent out from Washington to tell them how to run their country. The sermon of "liberalization" ran directly against Korea's own brand of nationalist conservatism. Koreans had never wanted foreigners owning their institutions. The country had accepted them as a matter of expediency. Only the threat of losing markets had compelled Korea to recognize the need to accommodate the rest of the world within its institutions.

Would Kim Dae Jung, the champion of Cholla, be ready to call out the police and the troops and order the volleys of tear gas needed to quell demonstrations? Would Lee Hoi Chang, with no real governing experience, have any idea how to respond—other than to relinquish authority to the generals? While spouting criticism and invective, none of the candidates made clear his answers. In the game of business and politics, the economic issue was paramount but elusive. All that was clear was that Korea faced a prolonged crisis that would surely not get better until it had gotten much worse—and the next president managed to come up with the solutions that had evaded his predecessor.

To be sure of victory, DJ had to pick up significant votes outside his native Cholla region; voters were moving toward Lee just to deny DJ the presidency. At the top of DJ's agenda was an absurd concern about revising the forecast for growth, held to 3 percent under the IMF agreement. Such low growth would "lead to mass layoffs and many corporate failures," said DJ's spokesman. Rather, "the growth rate must be adjusted upwards" from the 6 percent increase it was expected to post for 1997. The statement called on President Kim and his cabinet to "kneel down before the people and apologize." Lee too was critical of YS. Having broken away from the increasingly unpopular YS as a campaign ploy, Lee's Grand National Party now denounced the IMF demands as "rude acts that encroach upon the autonomy of a sovereign state."[4]

More talks with the IMF were likely every time the country was about to receive another tranche of the promised money. The IMF deposited the first tranche of about $5.5 billion on December 4, the day both DJ and Lee Hoi Chang were attacking the agreement, and promised another $3.6 billion after the election. For all the criticism, the two daily statistics that counted the most still showed the deal had inspired confidence. The composite stock market index shot up 26.50 points, a gain of 6.99 percent, while the won closed at 1,170 to the dollar after closing at 1,196 the day before.

On December 5, 1997, Korean authorities weighed criminal charges against those responsible for the "economic debacle." DJ targeted the country's top leaders for questioning. Recriminations and trials loomed even as Finance Minister Lim, proud of the numbers if not the conditions, falsely claimed that the IMF bailout fund was soaring above $60 billion and promised to close "ailing banks" and announce details on opening Korea's opaque financial world to foreigners. A "comprehensive screen-

ing process" was now under way to fix the blame "for the present economic crisis," said Park Soon Yong, chief investigator for the supreme
public prosecutor's office, citing "public opinion that someone has to take
responsibility for the economic debacle."[5]

Kim Dae Jung seized on the "national disgrace" of the IMF bailout to
call on Kim Young Sam to "take the witness stand" at a public hearing
after the election. Touring the market around Namdaemun, Seoul's South
Gate, he was a little more emphatic. Kim Young Sam "should be
impeached," he told Korean reporters, unless he promulgated an "emergency decree" forcing economic reforms. DJ's National Congress for
New Politics named YS as well as Lee Hoi Chang and three other senior
officials as "enemies responsible for the economic problem." Lee, unre-
pentant, also blamed the president for the country's economic woes but
did not suggest calling him as a witness. "Now is not the time to find a
scapegoat," he said virtuously.[6]

The Blue House appeared extremely defensive in view of YS's own
record of having insisted on trials of the two presidents who had preceded
him, Chun Doo Hwan and Roh Tae Woo. "It is illogical to hold government officials criminally responsible for economic policy mistakes," said
the Blue House. Rather, government ministries might "conduct a self-
examination of those responsible." Investigators, however, were already
questioning past and present economic officials. The search for culprits
appeared likely to gain momentum even as confidence spread by the IMF
bailout package radiated through the market. Share prices rose a record
28.31 points on December 5, closing at 434.12. The percentage increase
of 6.98 was second only to the 6.99 percent increase the day before when
the market soared back above the 400-point benchmark. Downside, however, the won depreciated to 1,230 to the dollar—significantly below the
closing rate of 1,170 on December 4 as merchant banks frantically sought
hard currency to pay debts.[7]

The presidential candidates did not underestimate the vote-winning
value of attacks on the IMF agreement. Posters emblazoned with the letters "IMF'd" represented more than a fringe element who knew enough
English to pun on the initials. Koreans were driven by what all seemed to
feel was the humiliation of begging. The Korean government—and an
establishment of elite business and media interests—failed in those first
few weeks to provide real leadership in dealing with this complex. After
advising Koreans to endure "bone-carving pain," Kim Young Sam did not
attempt to combat the reaction to the foreign "intervention" and "interference" that was sweeping society. One might forgive the reluctance of

madly campaigning politicians to pounce on the agreement for political purposes. YS, however, had nothing to lose. No matter what, he was stepping down at the end of his five-year term in February. He did not, however, advise his people to stop the recriminations.

As much as Kim Young Sam, the chaebol leaders refused to face up to their guilt. Instead they left the demeaning details of dealing with the public to managers, directors, and vice presidents, who came up with defensive claims of reform. Finally, the Korean media, invoking the specter of national disgrace, left the impression that the foreigners from the IMF were plotting a takeover of the country. Would it have been too much for the papers to suggest that retreating behind a wall of antiforeign shame and anger would not solve the problem? The best hope was that hysteria would prove only a temporary phenomenon and that a leader would emerge with the ability and vision to hold this reaction in check. Once elected, the victor would have to take advantage of whatever the IMF could do to help Korea regain its place as an economic power.

Kim Eun Sang, branch manager at Morgan Stanley, summarized the conflicting sentiments. "As a professional I am very happy," he said, "but as one of the people of this country, I am hating the U.S. as a gangster that did too much to this small country." The United States, "actually the negotiator instead of the IMF," he claimed, was responsible for the strict terms under which Korea had to open its markets and bring about the closure of banks and financial institutions that were hopelessly in debt. Nonetheless, he agreed that Korea's financial condition was far worse than the government had indicated. He estimated foreign currency reserves at $24 billion but said all but $5-6 billion of that amount was already committed.[8]

Even with the bailout funds arriving, analysts predicted a series of bankruptcies. One of the first on the chopping block was the Halla group, founded and still chaired by the next-younger brother of the Hyundai founder, Chung Ju Yung. Halla, the twelfth-largest chaebol, almost went into bankruptcy the same day the IMF and the government announced their agreement. On December 3, the morning of the IMF agreement, creditor banks agreed to lend almost $400 million to the group after Halla Engineering and Construction admitted it could not meet a deadline on a loan and would have to ask for court mediation. Without Hyundai behind it, Halla could not have found the extra funds needed to get it through the year. It was the kind of life-saving deal that would keep the losers afloat against all the advice of the IMF when they should have long since drowned in their own red ink.

Three days later, on December 6, having defaulted on $220 million in loans, Halla announced it was applying for court receivership for Halla Engineering and Halla Merchant Marine. At the same time, Halla sought court protection for rescheduling debts of Halla Engineering and two other companies, Mando Machinery and Halla Cement Manufacturing. Another Halla company, Halla Pulp and Paper, was also due to seek court receivership or court protection. Halla was going under despite a transfusion from Hyundai, which had secretly pumped in nearly $1 billion after the group had amassed debts from construction of the shipbuilding unit at Halla Heavy Industries.[9]

Halla, by the time of its collapse, had racked up debts of about $6.3 billion, mostly from merchant banks that were unable to grant a moratorium while trying to pay off their own debts. Halla's solidest enterprise was Mando Machinery, which made components such as the motors for windshield wipers and windows on motor vehicles. Hyundai Motor was Mando's biggest customer, but Mando also produced components for other motor vehicle manufacturers. Hyundai did not contemplate taking over Halla and did not "have the means to continue providing financing," said a Hyundai official.[10] Halla had acquired orders for more than $1 billion worth of ships in 1996 and 1997 but had failed to get the order it wanted to produce a liquefied natural gas (LNG) carrier.

Finance Minister Lim, on December 5, sought to buoy confidence both in Korea's determination to live up to IMF demands and his equal desire to placate banks and securities firms worried that they would suddenly have to close. He insisted the government had not agreed specifically to shut banks but "to deal with troubled banks after assessing their asset and liability conditions." Institutions could survive if they came up with plans "deemed viable" for self-rescue while "ailing financial institutions deemed impossible to recover" should merge with other competitors or shut down entirely. He admitted the government had moved too slowly after bankruptcies earlier in the year had shaken the economy. "We should have acted more decisively to deal with banks' snowballing nonperforming loans," said Lim. "Another serious problem was that we had paid little attention to banks' heavy short-term debts."[11]

Lim did not comment on a KBS report that more security firms and ten merchant banks were expected to go bankrupt. For Lim, pressured into an agreement that he hated, the best policy was to appear as a foe of the whole deal. The strategy made excellent sense for a man who knew that his brief term as finance minister would expire as soon as Kim Young Sam stepped down on February 25. Besides, Lim was already running for governor of Kyonggi Province, the heavily populated, industrialized region surrounding Seoul, and he was well aware that any realistic admission of Korea's needs would win him few votes.

A spike in the three-year corporate bond rate from 21 to 24.5 percent on December 8, however, reflected the inability of companies to sell their bonds. The fear was that a number of merchant banks and commercial banks would face bankruptcy as they failed to obtain fresh loans to cover their immediate debts. The government was pressuring commercial banks to rush to the rescue of nine merchant banks on the verge of default. The financial market was paralyzed amid fast-declining confidence in efforts at persuading bank presidents to extend emergency loans to merchant banks and securities firms. The decline in the stock market and the currency fueled fears of a long-term downward spiral. "Depreciation of the won indicates that foreigners haven't come into the bond market," observed Cristoforo Rocco at J. Henry Schroder. "You have a currency weakening rapidly and interest rates soaring."[12]

What hurt was that the IMF prescription, after an initial show of confidence, was not working. As of December 9, the economy was plunging perilously toward depression despite efforts to buck up support among foreign bankers. The won closed at 1,460 to the dollar, down by the maximum 10 percent permitted by the government, while the stock index again crashed through the 400 barrier, falling 26.83 points or 6.46 percent to close at 388.

In a bid to encourage foreign investors, Lim promised "greatly expanded opportunities for foreign financial institutions." Under terms of the IMF package, "Mergers and acquisitions in a friendly manner and on equal principles will be allowed by foreign bankers."[13] Foreign banks could now buy equity in Korean banks beyond the current limit of 4 percent, and foreigners could enter the corporate bond market with no restrictions. Securities firms were heartened by the promise of transparency of corporate records, guaranteed by annual audits by international accounting firms. "It's what investors are looking for," said Daniel Harwood at ABN AMRO Asia. "If you can trust them, you stick with them."[14]

Such reassurances, however, did little to stop the decline of key indicators and raised fears that companies within some of the largest chaebol were in trouble. Analysts cited reports that Korea's short-term debt was far more than the sums raised by the IMF, which deposited the first tranche of $5.6 billion on December 5 as the market was falling. Short-term debt was now estimated at $110 billion, nearly twice previous estimates, while foreign exchange reserves were about $6 billion. The question was when Korea's swift financial deterioration would affect daily lifestyles in a brightly lit city bustling with Christmas shopping.

By December 10, 1997, almost no one was willing to sell dollars in exchange for fast-depreciating local currency. The value of the won rapidly sank to the 10 percent daily limit, hitting an all-time low of 1565.90 to the dollar in a panicky desire on the part of nearly bankrupt companies to pay off dollar debts. On the bright side, the stock price index almost returned to 400, closing at 399.85, an increase of 3.05 percent, buoyed by the news that foreigners could soon buy up to 50 percent of stock in a listed company, up from 26 percent.[15] The government had set December 15 as the date to raise the limit on foreign holdings and also to permit foreign direct investment in the corporate bond market but advanced it to December 11 in a bid to bring in foreign currency.

In the meantime, fear swept the markets that more large companies were on the verge of failure. Lee Eung Baek, chief of the foreign exchange division of the Bank of Korea, cited "psychological panic over the possible bankruptcy of more corporations and the collapse of the financial system" as the cause for the won's decline. Lim, announcing a new stabilization package, said the Bank of Korea would supply short-term loans to commercial banks as well as liquidity for securities firms and investment trusts in order to "protect depositors in the event of failure at financial institutions and prop up the restructuring of the financial sector."[16] With the finance ministry admitting that short-term loans totaled well over $100 billion, the debt crisis reverberated through the economy. The Korea Construction Association said that 89 construction firms had gone bankrupt in November, with electrical and civil engineering companies suffering first because of a drop in orders from general contractors. Petrochemical firms had to cut the output of oil products amid rising oil prices.[17]

One Korean response appeared immutable: export or die. The bright side to the precipitous decline of the won was that the price of Korean goods was falling on world markets. "Now it's about 1,700 to the dollar," said Lim. "Can you imagine the impact of this exchange rate on competitiveness?" Thus, he said, "I am quite optimistic that our exports can make it by allowing the free floating of the won."[18]

The influence of Park Chung Hee lived on in the structure of the chaebol and the banks—and an instinctive drive to export at any price. Koreans in both government and chaebol seemed oblivious to the implication in the IMF agreement that the government should deemphasize exports because they often required massive infusions in plant investment. Nor did they appear aware of a renewal of complaints that Korea might arouse resentment from trading partners for dumping low-priced goods into foreign markets while importing very little in return. One area of dispute was that of foreign motor vehicles. Since the won's sudden descent, foreign car sales had plummeted. Mercedes-Benz, Ford, and Chrysler sold ten cars apiece in Korea in December, while BMW sold none.[19]

Foreign manufacturers' associations, embroiled in bitter disputes with Koreans before the current crisis, renewed the offensive through the IMF. The Korea Foreign Trade Association predicted that exports for 1997 would reach $136.8 billion, up from $129.7 billion in 1996. A sign of lack of foreign exchange or access to credit was that Korea's trade deficit was steadily declining from 1996, when imports reached $150.2 billion, to 1997, with imports likely to wind up at $146.5 billion, and would turn into a $40 billion surplus in 1998.[20]

While emphasizing reductions in new investment and cuts in bonuses and salaries, Korean manufacturers talked up hopes and plans for increasing exports. Daewoo Motor, for instance, planned to enter the American market in 1999 as part of a group plan to increase overall exports from $15 billion in 1998 to $17 billion in 1999. "It is part of our globalization strategy," said Lee Jeong Seung, a Daewoo spokesman. Lee cited three reasons why he believed Korea now faced economic difficulties—"short-term foreign exchange, noncompetitiveness, and the trade deficit." In response, he said, "We have to export more."[21]

Daewoo buttressed its export hopes on December 8 by acquiring controlling interest in Ssangyong Motor after agreeing to take on about $2 billion of the company's crippling debt. If Ssangyong's venture into motor vehicles illustrated the failure of overinvestment, Daewoo's move demonstrated a determination not to heed the lesson. After selling 53.5 percent of the stock in its motor company to Daewoo, Ssangyong laid off 83 of its 273 top managers on December 10 and cut the wages of the rest by 30 percent. Other employees suffered 15 percent pay cuts. "Our economy is at the crossroads between growth and decline," said Chey Jong Hyun, chairman of the SK group. "We overcome the crisis by sharing the burden and through sweeping restructuring."[22]

Details might differ, but the philosophy was the same at all the major chaebol. Far more than IMF intervention, sagging stocks and bankruptcies would be needed to affect attitudes inbred since the drive to export was generated by the repression of 35 years of Japanese colonialism—and then the tragedy of the Korean War. "We'll do our utmost to increase our exports," said Park Hong Kyu, a manager with the Hyundai Corporation, the Hyundai group's trading arm. Although the government no longer told banks how much to lend to which company, he said, "the government encourages exports because that's the only way to get over our problem and because of the shortage of foreign currency."[23]

There were, however, severe obstacles. Hyundai Heavy Industries was still doing well, but Hyundai Motor and Hyundai Electronics were not. "Motor vehicles are getting difficult to export," said Park. "Other countries are having a hard time economically, and total demand is down." As for electronics, both Hyundai and its competitors, notably Samsung, were suffering from a worldwide oversupply. "Everybody knows the price of semiconductors is sharply reduced," said Cho Jang Won, a manager in Samsung group headquarters. "We have to diversify our exports. We are emphasizing semiconductors, shipbuilding, consumer electronics, and chemicals."[24]

But how could Korea produce for export without the money? Cho Yoon Jae, a Sogang University economics professor now retired from the finance ministry, suggested a way around the dilemma—but one that might not appeal to the IMF. "We should do something to release the constraint on financing," he said. "The government can increase the capital base. It can issue government securities." Lee Jung Tae, an economist with the Korea Institute for Industrial Economics and Trade, saw the problem as one of "confidence about our economic restructuring for the bankers and investors." The chaebol, "to make good investment," he advised, "should sell existing factories and companies and increase their own capital." Lee saw more shocks ahead before confidence returned. "I have heard rumors about Samsung motor vehicles and electronics," he said, but he was hopeful about semiconductors. "Next year memory chips change from 16 DRAM to 64 DRAM, and there will be strong overseas demand again." No matter what, exports were key. "For next year, exports is the only way the government can promote economic growth."[25]

For now, however, Korea was approaching meltdown. On Black Friday, December 12, the stock market and the currency descended to new lows as the government fought both to stop the freefall and to head off anti-western emotions. With Koreans blaming the United States and other western powers, the government seemed as concerned about the rise in antiforeignism as about the economy. Ralliers at Pagoda Park, the staging ground in central Seoul for a short-lived revolt against Japanese rule in March 1919, denounced "U.S. imperialism" and western powers in general for imposing a new form of imperial rule on Korea through the IMF.

"Western countries are conspiring to rule Korea" was one of the cries heard at Pagoda Park. "Western corporate hunters are exploiting our difficulties." President Clinton was blamed as "an enemy" who had conspired to foment the crisis. "The Japanese want to colonize Korea again through the IMF" was another placard-borne slogan. The Blue House, sensitive to the negative impact, called on groups such as the National Alliance for the Unification of the Fatherland to tone down their language for fear of undermining efforts to bring in foreign money. Such "inflammatory" verbiage, said Ban Ki Moon, presidential assistant for international affairs, would disturb "foreign lenders and investors who may want to help."[26]

As demonstrators gathered at Pagoda Park, the stock market fell 24.79 points within minutes after opening on December 12 and closed at 349.49, down 26.69 points, 7.07 percent, from December 11. The dollar rose almost immediately to a new high of 1,891.4 won and appeared well on the way to 2,000 won to the dollar when the Bank of Korea intervened, selling $200 million. The quick fix sank the dollar to 1,710 won at the end of the day, down slightly from the closing of 1,719.8 the day before.

The government, however, had no clear strategy for reversing the overall downward pattern as two widely anticipated efforts at market opening met extremely disappointing results. First, foreign investors purchased only about $191.2 million worth of stock in Korean companies on the 11th, the first full day on which the ceiling on foreign ownership of listed Korean companies was raised from 26 percent to 50 percent. Second, foreign investors showed remarkably little interest in purchasing South Korean benchmark three-year corporate bonds on the 12th, the first day on which the market was open to foreigners. Investors were deterred by yields of nearly 25 percent—and then hung back as Dongsuh Securities, the fourth-largest brokerage, suspended operations for a month while seeking court receivership. Dongsuh, with 82 branches and 1,450 employees, declared itself insolvent with loans of about $15 million falling due.[27]

As a sense of despair set in over the financial community, President Kim Young Sam called a meeting on December 13 of the three major candidates in the presidential election. The purpose, said Shin Woo Jae, a Blue House spokesman, was to convince them that "pan-national cooperation is necessary to overcome the economic difficulties, rebuild a healthy economy, and boost international credibility." YS was concerned by statements by Kim Dae Jung, now ahead in the polls, that he wanted to renegotiate the IMF agreement. DJ, who had advertised his disagreement in newspapers, said that he wanted only "supplementary" talks. He claimed that local newspapers had "misquoted" him in demanding renegotiation of the entire agreement.[28]

The Blue House also said that Kim Young Sam might send envoys to

Washington, Tokyo, and other capitals in hopes of drumming up quick loans from government officials as well as bankers. U.S. Treasury Secretary Robert Rubin had earlier rebuffed Seoul's plea for money, calling on the government to carry out all the IMF requirements for closing debt-ridden banks and opening markets. Signs of the crunch were rapidly gathering strength. The ministry of commerce, industry, and energy said local refineries had only 25 days of crude oil, while private companies had enough oil and gasoline to last for 33 days. Authorities still hoped the United States could advance funds quickly. The government said it had asked Washington to extend $1.6 billion in credit for agricultural exports. Feed producers in Korea were close to defaulting on some of their payments. The government also sent Kim Mahn Je, chairman of Pohang Iron and Steel Company, to the United States in a further effort at persuading Washington to hurry up with loans.

At the same time, the government sought to bring in more dollars by permitting companies to borrow money from abroad and to issue foreign-currency bonds starting December 15. Finance Minister Lim noted that the IMF had called for lifting all restrictions on borrowing overseas as one of a number of liberalization measures. Companies, "private and public alike, should go overseas to borrow hard currency," said a finance ministry spokesman. The Bank of Korea was also issuing about $6.5 billion in emergency loans to stave off more bankruptcies.[29]

Finance ministry officials employed elementary calculations to show that Korean companies would not default on $16.3 billion in short-term debts due the end of the month. Putting foreign exchange reserves at about $10 billion, Chung Tok Ku, assistant finance minister, said that Korea was to obtain another $7.5 billion under the IMF package in the next two weeks. That figure included $3.5 billion from the IMF, $2 billion from the World Bank, and $2 billion from the Asian Development Bank. Thus, said Chung, "there would be no substantial problem in the foreign exchange demand and supply."[30] The arithmetic, however, failed to convince observers that Korea would not have to approach the IMF for a new bailout. Stephen Marvin, then research chief at Ssangyong Securities, estimated that Korean companies would owe about $85 billion within a year. Overall, the country's short-term debt still exceeded $100 billion.

Loan-givers were holding off somewhat while awaiting not only the results of the presidential election but also the response of the winner to the IMF agreement. The answer came soon enough as the election results became clear on the night of December 18. Although Kim Dae Jung had once led in the polls by as much as 5 or 6 percent, Lee Hoi Chang had been gnawing at his lead. As the returns rolled in, however, DJ assumed an edge he never lost. Aside from his skill as a populist orator, DJ had two other factors going for him. One was that Rhee In Je, running separately after failing to win the ruling party's nomination, picked up 4,925,391 votes, 18 percent of the total, most of which would have gone to Lee. The other was Kim Jong Pil's decision to back out of the race after DJ promised to name him prime minister. Aided by popular anger over the crisis, DJ received 10,326,275 votes, 40.3 percent, 1.7 percent more than the 9,935,718 ballots cast for Lee.

In the Cholla region, confidence in Kim, born near the fishing port of Mokpo in South Cholla Province, was unwavering—he won 754,159 or 97.3 percent of the votes in Kwangju. The tally in surrounding South Cholla was 1,231,726, 94.6 percent, for Kim, and 1,078,957, 92.3 percent, of North Cholla's voters went for him as well. The overwhelming support for DJ in Kwangju and the Cholla provinces reflected a view dating deep into Korean imperial history. "Our infrastructure, our factories, our industries, everything is so far behind," said Chung Soon Joon, an official in the office of the governor of Kwangju, an independent city of 1.3 million surrounded by South Cholla Province. "We don't overly expect Kim Dae Jung will do many things for us, but probably deep in our hearts we expect more development."[31]

From the viewpoint of fulfilling their dreams, DJ could probably not have been elected at a worse time. Difficulties were sure to worsen as the price of oil went up along with sugar and wheat. In Cholla, however, the fact that DJ, who had won the hearts of the region with a populism that marked him as a foe of oppression, could now sit in Seoul as leader of the country was enough to counter the mood of depression.

"DeeJay, DeeJay," young people shouted as they walked around the large circle in front of the governor's office, repeating the initials in English. It was at that very circle that government troops, on May 27, 1980, had fired into crowds of student rebels, killing scores of them and finally driving them from the government building. DJ was in Seoul at the time, but the president, Chun Doo Hwan, a former general, accusing him of fomenting the riot, had had him arrested, tried, and sentenced to death. Then, in 1983, under pressure from Washington, Chun freed him to go to the United States, where he spent 26 months before being permitted to return to house arrest in Korea in 1985.

One reason for Kim's popularity was that his worst foes had all been from the neighboring Kyongsang provinces, dominated by the independent cities of Taegu and Pusan. Park Chung Hee, Chun Doo Hwan, and Roh Tae

Woo, all generals, were leaders of what the Korean media called "the TK Mafia"—TK standing for Taegu and the Kyongsang provinces, North and South. "They got all the industry and the infrastructure," said Yun Young Ju, another staff member at Kwangju city hall. "They think the Cholla provinces are only for agriculture. The road between Kwangju and Mokpo is not even a highway. We are far behind them."[32] DJ long had played on these sensitivities, winning his first election to the National Assembly in May 1961, two days before Park Chung Hee seized power. The firebrand Kim was soon arrested, banned from politics, then charged with plotting to overthrow the government—and elected again to the Assembly in 1963 and 1967.

Whatever the special circumstances, DJ's victory on December 18, 1997, ranked along with the ascent of Nelson Mandela in South Africa and Lech Wałesa in Poland among the great political reversals of our time. Permanently injured in the campaign motor vehicle "accident" of 1971, kidnapped from a hotel in Tokyo two years later, tried for treason on trumped-up charges of fomenting the Kwangju revolt in 1980, sentenced to death, exiled to America, Kim was one of the most durable figures of the twentieth century. The story, however, was only beginning. What could he do to conquer South Korea's economic problems, reconcile with North Korea, and hold his people together with only 40.1 percent of the votes? He might be the darling of American liberals, but his bedrock was Cholla. Now he had to show that he could transcend the regional support that had propelled him to power. He had to graduate from politician to statesman.

At the least, the election proved that democracy could function in a most unlikely setting, solidifying the democratic process in a society that remained authoritarian in structure and style. Even during the first two elections under the new constitution, the supremacy of the candidate of the governing party was never in doubt. Roh Tae Woo had won in 1987 after splitting the opposition between YS and DJ. YS, by allying with Roh, had had no trouble defeating DJ in 1992. This time, as the polls closed, few people had been willing to pick a winner between two very different front-runners.

One common criticism of this celebration of democracy was that, actually, the candidates did not differ much on issues. Korea's economic duress, however, emerged as the focal point of everyone's concern—even if, superficially, the candidates appeared to get behind the need to abide by the reforms demanded by the IMF. DJ would have liked to stake out an independent, critical position in a ploy for votes but was seriously embarrassed by the response. The final irony was that DJ wound up accusing Lee Hoi Chang of slandering him by claiming that he, Kim, if elected, would subvert IMF efforts.

Shows of support for the IMF deal, however, would count for little if the incoming administration forgot the fine print on transparency and market-opening and fought to salvage debt-crippled banks and companies that should fall by the wayside. Korean groups and companies were psyching up their workers as they struggled to pay off debts and survive. The basic appeal was to company loyalty and old-fashioned national-ism—the same forces that had driven Korea to economic greatness under Park in the 1960s and 1970s. One result of the election would be to mute some of the criticism of the IMF. DJ's camp quickly forgot the campaign cant about "National Shame Day."

With the election now just another date in tumultuous Korean history, the new administration could get on with reform. The risks were enor-mous. Kim Young Sam had another two months before stepping down. The first thing his administration had to do was to guarantee an independent Bank of Korea, free of finance-ministry (read political) interference plus other reform measures recommended by the IMF. The government also had to deal with labor to keep cash-strapped companies afloat—that is, those that deserved to remain afloat.

Korean history, under a new administration and "IMF supervision," as Koreans called it, had entered a new phase. The prospect that demonstra-tions would turn to violence was real. The protest issue was complicated by the process of four-party talks involving the two Koreas, the United States, and China. All the while, the administration had to counter the regionalism that had much to do with how Koreans voted. It would take imagination and toughness to deal effectively with disparate trends. The campaign and election of 1997 would have been a success if that type of leadership emerged.

Having won the prize, DJ showed his eagerness to cooperate with the advice of both the IMF and the United States. On December 22, the begin-ning of the first week after the election, he dared to say what would have been unthinkable during the campaign, that layoffs would have to be per-missible under certain circumstances. As the financial crisis worsened in the face of concerns that Korea did not have enough reserves to pay off debts by the end of the year, those circumstances were clear. Government officials said there would be no "debt moratorium," but the reality was that the foreign exchange reserves and funds already received through the IMF totaled no more than $20 billion.

As if to confirm Korea's dilemma, Moody's Investors Service and Standard & Poor's both downgraded the ceiling for bonds and bank deposits for Korea along with those of Southeast Asian countries. The state-invested Pohang Iron and Steel was among nine companies and 20 banks whose ratings were cut by Moody's. The others were all major entities: Hyundai Motor, LG-Caltex Oil, Yukong Oil, Korea Electric Power, Korea Telecom, Hyundai Semiconductor America, Samsung Electronics, and SK Telecom. The cut reflected the declining value of Korean currency along with soaring interest rates and fast-disappearing credit.[33] Standard & Poor's, slashing Korea's long-term foreign currency rating, estimated Korea's foreign exchange reserves at $4.7 billion—5 percent of the country's short-term foreign debt. The S&P report also struck at Korea's largest chaebol, cutting the rating of such giants as the Daewoo Corporation, Samsung Electronics, Hyundai Motor, and LG-Caltex Oil.[34]

The ratings carried disastrous implications. "The banks cannot open letters of credit to oil, electronics, and motor companies," said Shin Seung Yong of Indosuez W. I. Carr. "All such companies have to import and then export. Korea is an export-driven country. Without raw materials, we cannot make money. The crisis is getting worse and worse." The problem was most acute when it came to crude oil, all of which was imported. "We have to pay cash to buy it," said Shin. So far only one refinery, Hanwha Energy, part of the Hanwha group, on the brink of bankruptcy, had cut production, but the four others were considering such measures. One result was that gasoline would rise from the current 1,083 won per liter to at least 1,400 won. "If the gasoline price increases, then the price of everything goes up 50 percent," Shin said.[35]

The downgrading of Korea's credit rating to junk bond status precluded many foreign investors, such as managers of pension funds, from putting dollars into Korea. Another problem was a report spread by aides of DJ that actually the government's total foreign debt was $300 billion. A vice minister of finance, Kang Man Soo, estimated Korea's foreign debt at more than $200 billion. Some analysts believed that Korea's reserves might be somewhat more than $6 billion, but the math remained the same: Korea needed to repay about $15 billion within a week. The government believed that it could obtain a total of $13 billion in loans from the IMF, the World Bank, and the Asian Development Bank to supplement its reserves but would encounter worse problems in January when another $10 billion fell due. One sign of the deteriorating economy was that the benchmark yield on three-year corporate bonds rose to 29.95 percent. The government expanded the interest rate ceiling on bonds from 25 to 40 percent.

For now, nothing seemed to be working. On Tuesday, December 23, five days after the election, two days before the Christmas holiday, the economy was still careening toward meltdown as the government fought to stave off fears it would have to declare a debt moratorium. A rapid decline in the value of the Korean won precipitated a sense of panic in a financial community desperate for hard foreign currency to pay off debts due this week and next. "All bets are off, the bottom's fallen out," said Richard Samuelson, director of Warburg, Dillon Read Securities in Seoul, watching the won drop to 1,962 to the dollar by the time the banks closed.[36] The won almost reached the 2,000 mark during the day, reaching an all-time low of 1,995 before recovering slightly. The search for dollars had the same impact on the stock market as share prices slipped 7.5 percent, 29.7 points, amid growing fears that more companies would go bankrupt without the credit or investment needed to produce, export, and stay afloat. The yield on three-year corporate bonds also soared above 30 percent, finally closing at 31.11 percent.[37]

The government appealed to foreign bankers to extend $20 billion in loans falling due by mid-January as possibly the only way to avoid default. A finance ministry spokesman estimated that Korean institutions would be able to pay off $10 billion in debts provided foreign banks "roll over half of their Korean loans." Foreign bankers, however, were divided. Some did not want to refinance at all. Others only wanted to roll over loans for a month instead of the normal three to six months. Japanese and American banks, with huge interests in Korea, were leaning toward extending loans. Major foreign banks, including Citibank and Chase Manhattan, might form a consortium to drum up another $10 billion in credit.[38]

The Koreans sought more help in meetings with top American troubleshooter David Lipton, undersecretary of the treasury and a key figure behind the IMF bailout package. Lipton, however, countered with four demands. First, he called for deregulation of the foreign exchange market, tightly controlled by the government even though the won as of the previous week was floating freely. Second, he wanted the government to authorize Korean companies to lay off unneeded workers as soon as possible rather than delay the process under pressure from labor unions and outmoded labor laws. Third, he asked the government to protect minority shareholders, who had no power to express their views within Korean companies. Fourth, he demanded compliance with the terms of membership in the Organization for Economic Cooperation and Development when it came to transparency and opening of markets.[39]

What Korean officials feared most was a debt moratorium, which would be a political and economic disgrace of far greater proportions than an IMF bailout. Under a moratorium, Korean companies, which relied on imports of raw materials, would have to pay cash on delivery for everything. Finance Minister Lim, fighting to counter the mood of pervasive pessimism, assured DJ that the country would not default and would "lift legal restrictions" on the foreign exchange market. Complying with the demands of the IMF as well as the World Bank, the government on December 23 loosened control over the monetary system. Having removed the band in which the won could move up or down each day, Korea now permitted investors to buy and sell won as they would any truly free currency.

Koreans wondered if they could count on the United States to come to the rescue if nothing worked. American diplomats stopped a carefully phrased step shy of pledging never to permit Korea to go into default but refused to rule out the possibility. Confident that Korean institutions would manage to pay off more than $15 billion due by December 31, a senior American diplomat promised, "We'll give full consideration to whatever they are asking."[40]

That pledge alone had persuaded many Koreans that the United States would come to the rescue when the chips were down. DJ's aides said the United States would announce emergency aid for Korea—in addition to the IMF package—when Kim visited Washington in early January. Not so fast, was the conflicting signal the Americans were putting out in keeping with U.S. Treasury Secretary Robert Rubin's rejection of additional funding until Korea had fulfilled the IMF requirements. "Nobody wants them to default," said the American diplomat, warning "it would be a mistake" for Koreans to conclude the United States was ready with an eleventh-hour handout.[41]

American officials were ambivalent about the Korean record in sticking to the terms of the IMF bailout. "The program in the first three weeks hasn't worked out," said the diplomat, noting the rapid descent of both the won and the Korean stock market in December. At the same time, U.S. officials put the best interpretation they could on Korean efforts to abide by the terms of an agreement that called for the almost complete opening of a financial and mercantile world that was still largely closed to foreigners. "What Korea has done is quite remarkable," he said. "They have complied fully with commitments on the monetary side and on restructuring. The government is pushing enactment of foreign reform bills."

Through such remarks, however, ran an element of wishful thinking, as evidenced by the IMF's hammering out a second set of requirements for Korea on December 24 and also by the World Bank's drafting detailed stipulations in return for rushing a $3 billion loan as part of the package. An underlying point that all the foreign experts were trying to get across was that the Koreans, products of a culture that prized secrecy, would be better off letting the world know the real facts about their economy. Public disclosure was one of the IMF's central messages. The World Bank, meanwhile, insisted on specific structural reforms to make it easier for foreign financial entities to operate in Korea.

Yet another point of emphasis was interest rates. The fact that the three-year corporate bond rate had gone above 30 percent still did not satisfy the IMF. It wanted all restrictions off, permitting interest to go to any limit in a truly free, market-driven economy. Western economists saw free interest rates as key to credibility and stability in the marketplace. Investors should find a milieu in which they could not only invest but take over troubled companies. "Foreign direct investment is the best kind of investment," said an American diplomat. "This is a very attractive economy. They've got a tremendous human resource, modern equipment. They could very badly use foreign investment."[42]

The future now rested in large part on Kim Dae Jung, but a legacy of regional prejudice ensured nonstop opposition. People from most regions scorned those from Cholla as backward, poor, and uneducated. DJ had learned to disguise his heavy Cholla accent when on the hustings elsewhere around Korea, but in the region of his birth he talked in tones and words that appealed to the heart and the mind. "We don't expect any special favors for Kwangju," said a garrulous woman behind the desk of a small inn near the railroad station. "We just want to be treated the same as everybody else." For the first time in memory, people in Kwangju believed they were going to get that kind of break—treatment equal to that accorded the more affluent regions to the north and east.[43]

Although Kim Dae Jung now spent most of his time in Seoul, his oratorical style, along with his seeming support for labor and the poor against big government and big business, ensured his popularity in Kwangju. "We have not so much business here," said Kim Tuk Han, a small merchant in a shopping district near city hall. "Kim Dae Jung may be able to influence other factories to build here."[44] Or at least he might prevent industries that were there from closing. Halla Heavy Industries in South Cholla Province had gone into bankruptcy, announcing 6,000 layoffs, while Asia Motors in Kwangju, a subsidiary of the bankrupt Kia group with 6,000 more workers, was in receivership.

DJ's identification with the suffering of the people in the Kwangju massacre accounted in large part for his continued popularity not only in Cholla but also elsewhere, especially among antigovernment critics in Seoul, many from Cholla. "He came here and mourned and cried a lot," said Park Jung Koo, a former local official, at the heroes' monument soaring over the cemetery with the graves of more than 200 people killed in the massacre. "Park Chung Hee when he was president never came here. He killed many people even before the massacre."[45] Now DJ had to identify with the suffering of all Koreans, a majority of whom were dead set against him as president.

# Chapter 4

# Criminals on Parade

L
ike the spring winds blowing dessert sand in from Mongolia, dis-
gust with politics and politicians was chilling and polluting the
atmosphere in 1992, five years after Chun Doo Hwan had staved
off riots by having his ally Roh Tae Woo announce a new era of democracy on June 29, 1987. National Assembly elections in March 1992
revealed disillusionment with Roh, who had won a five-year term in
December 1987 in the first presidential election after his declaration. With
Roh constitutionally barred from succeeding himself, there was hardly a
charismatic figure capable of inspiring widespread support in the presidential election coming up in December 1992.

The ruling Democratic Liberals clung to a bare Assembly majority
after persuading one or two "independents" to join them, but they were
also divided against each other as they tried to select a presidential candidate. Much of the disillusionment focused on Kim Young Sam, party
cochairman. Once hailed as an idealist with the courage to stand up
against Park Chung Hee, Kim had run for president against Roh in 1987.
When he merged his party with that of Roh in early 1990, he seemed to
many followers to have betrayed their trust. Kim's strategy was clear—he
would support Roh in return for Roh's assurance that he would become
the new party's presidential candidate.

Roh could not run again for president under terms of the democratic
system that he himself had demanded in 1987 after some of the biggest
demonstrations ever seen in Seoul. Still, he might select the party standard-
bearer—the one he hoped would succeed him in February 1993. Ever
since Park Chung Hee had thrown out a corrupt and inept civilian regime

in 1961, the generals had held the ultimate power. Roh had come to appear as an apostle of democracy as he called for an end to the heavy-handed rule of Chun Doo Hwan, but he too was a retired general. His division had backed up Chun, then commander of the capital military district, when Chun staged his "12/12 mini-coup" on December 12, six weeks after Park's assassination in October 1979.

No one forgot this legacy as YS pinned his hopes on a party convention in May 1992 at which Democratic Liberals would choose a candidate. Kim claimed he had Roh's support, but critics charged that Roh differed little from Chun, that his police were as tough as ever on union organizers, that the National Security Planning Agency tortured dissidents as it had done under Park when it was called the Korean Central Intelligence Agency. The level of such harassment had fallen in recent years, but the real test was yet to come. Roh had to resist pressure from military allies as students hit the streets in the usual springtime outburst of rock-throwing rage. The temptation to resort to martial law in the tradition of Park and Chun would rise if the riots appeared more violent than those of 1991, when several demonstrators had burned themselves to death.

The nature of the opposition added to the turmoil. Kim Dae Jung in 1992 was stronger than ever. His party, already the leading opposition group in the Assembly since YS had gone over to Roh's side two years earlier, now held nearly a third of the seats. It seemed unlikely that DJ could win a presidential election, but he could paralyze legislation while playing on human rights and corruption. Another potential spoiler in 1992 was the billionaire tycoon Chung Ju Young, founder of the Hyundai empire. Chung, who had built his industrial machine with easy credit provided under Park and Chun, had clashed head-on with Roh's bureaucrats, who stuck him with a tax bill of $180 million after he divvied up his holdings among his sons.

Chung had got his revenge by founding his own party early in 1992 and taking more than 10 percent of the Assembly seats. Chung blamed Roh for plunging the country into "economic crisis" while ignoring the strikes and unrest at his own companies. YS cited government "mismanagement" but remained a leader of the government party. Thus business and politics were inextricably linked, not only by Chung's candidacy but by relationships between the chaebol and the other candidates. Flawed though it was, however, the advent of some semblance of democracy in South Korea was one of the region's more startling phenomena. Was it really true? Just a decade before the 1992 campaign, the country had been under the thumb of the corrupt Chun Doo Hwan. A decade before that, Park Chung Hee was tightening the screws of his own dictatorship. The difference between Park and Chun was that Park, in nearly 19 years in power, had demonstrated some real capacity to mold and lead his country into the modern industrial age.

The 1992 election provided the chance for the gadflies of dictatorship, Kim Young Sam and Kim Dae Jung, the front-runners, as well as Chung Ju Yung, a gadfly of Roh Tae Woo, to show that South Korea could sustain a democratic miracle paralleling the economic one. The winner would be Korea's first civilian president in 32 years. YS, armed by a military-backed establishment, won by a wide margin, picking up 44 percent of the votes to 36 percent for DJ and most of the rest for Chung. Immediately after his election YS began to pursue bureaucrats and military officers and petty politicos with a vim that surprised the mass of Koreans who had elected him by such a comfortable margin. He promised to keep up the drive in the face of serious threats; the generals and intelligence operatives who had propped up Park, Chun, and Roh still controlled hundreds of thousands of troops. The leaders of the chaebol had the resources to pay off just about anyone.

Chung Ju Yung, who had regularly denounced Kim Young Sam as "a stonehead idiot" during the campaign, got a suspended sentence the next year for diverting more than $60 million from his companies to finance his foray into politics.[1] Chung then paid obeisance to his erstwhile foe by praising him to the skies, withdrawing from politics, and making a pretense of retiring from the Hyundai group. The bitterness between YS and Chung lingered on, with Hyundai executives complaining about difficulties in getting all the credit they needed.

Whoever sat in the Blue House, Koreans had a feeling that bureaucrats, politicians, and chieftains of business exchanged bribes and presents and favors for almost anything—contracts, permissions, entrance to universities, articles in newspapers. "We are a cash society," a Korean journalist explained to me during the 1992 presidential campaign. "It is part of our system." Thus he rationalized the practice of *chonji,* literally "envelope," brimming with cash, for reporters on a beat or even an individual story. He had received "a couple of envelopes" containing "between 100,000 and 200,000 won"—between $125 and $250—while temporarily covering Kim Young Sam's campaign. Colleagues on regular assignment were "entitled" to much more.[2]

There was a special irony to the handouts from Kim's headquarters. YS was shaking up the political and bureaucratic establishment by insisting on "full disclosure" of the assets of legislators, prosecutors, and vice ministers and going after top military officers charged with selling promotions. Some of them were wealthy indeed. Real estate holdings, including lavish "second" and "third" homes and commercial buildings whose landlords' identities were secrets to their occupants, not to mention outsized bank accounts, all raised questions. How did these people acquire such riches on their salaries as National Assembly members and bureaucrats working their way up through the system? Who was paying off whom? Top officials, initially thrilled to be serving under yet another Democratic Liberal Party administration, were dropping like cherry blossoms in April. First to lose his job was a mayor of Seoul—an appointive position in a society where skin-deep "democracy" only penetrated down to election of the president and National Assembly. He reported to his posh office in city hall for less than a week before resigning amid questions about a large home and garden intruding upon "protected" land on the capital's fringes.

The mayor's demise was just the beginning of a series of resignations, apologies, and hand-wringing over corruption. How long, though, could YS keep up the pressure before slowly yielding to other forms of pressure? Kim's predecessors had also mouthed high principles before yielding to temptation. After Chun Doo Hwan seized power in 1980, he and his subordinates in the armed forces and bureaucracy also swore they would stop the rot. Roh Tae Woo gladly accepted enormous handouts and left his post an extremely rich man.

Kim Young Sam seemed more determined than the generals before him to live up to his moral precepts. He talked as if he meant it when he demanded full revelation of assets. Would the targets of his inquest submit to public embarrassment before they turned against him, opposing his programs, stonewalling on real reforms, undermining his presidency? No one believed the revelations of assets were truly complete. There was much talk of "secret" ownership of shares and accounts held under false names. There was not much inclination to ram through a full disclosure requirement that might reveal still more hidden assets—in the hands not only of legislators and bureaucrats but of thousands of business people who did not believe in the public's "right to know" much of anything. The brouhaha was a test of Kim's power over potential foes—generals caught up in the investigation, business leaders accustomed to paying off whoever was in power, and civilian officials who believed "the system" owed them much more than their low salaries.

Corruption threatened not only the Korean miracle but also the strength of the South in whatever contest might erupt with the North, increasingly impoverished and dangerously frustrated. The society, not just the ruling establishment, could fall apart, as had happened in the late Chosun dynasty a century earlier as the Japanese exploited Korea's social weaknesses, if Kim's campaign rebounded against him and his civilian presidency.

The collapse of the Sampoong Department Store near the end of June 1995 provided a metaphor for the much greater crisis ahead, exposing the pattern that created and masked an underlying evil. I watched as rescue workers, looking like extras on a phantasmagoric movie set, dug through the night under glaring spotlights for hundreds of bodies trapped in the rubble of the country's worst construction disaster. Crowds behind police barricades and on neighboring rooftops cheered every time someone emerged alive as the workers pulled away debris covering the crushed basements of the five-story building that had crashed to the ground at the height of the evening rush on June 29.

Prosecutors, questioning the owner and managers of the store and the construction companies, promised indictments. They said the store president, Lee Han Sang, had met four hours before the collapse with his managers but had decided only to close the fourth and fifth floors temporarily. The top executives had all left the building by the time of the collapse at 6 P.M. The majority of victims were women—store clerks or housewives. The store had crashed down floor by floor from the fifth floor all the way to the third subbasement 20 yards below ground. Now Koreans feared more disasters in the fast-growing capital, dominated by soaring skyscrapers, office blocks, and apartment buildings thrown up at record speeds over the past decade. South Korea's construction companies had long been notorious for their rough-and-ready tactics. Less than a year earlier, in October 1994, a span of a bridge across the Han River, half a mile from the store, had plunged into the river, killing 32 people. Then in April 1995, 101 people had died in a subway explosion blamed on the carelessness of construction workers who had cut a gas line while digging a tunnel in the south-central city of Taegu.

In the aftermath of the Sampoong accident, the question was how an extremely rich family could have gotten by for so long with such a terrible building, covering up its efforts at cutting corners with contractors and customers. One factor was the custom of passing on top jobs to sons of the founder regardless of qualifications. A court meted out stiff sentences for the owner, his son, and their top managers as well as the building inspector whom they had bribed. Prosecutors might also have gone after a legion of others to whom corner-cutting was an habitual practice. Standards were improving—but not fast enough for the more than 600 people who died in the Sampoong Department Store, once an eyesore in pink, now an eyesore in collapsed steel, glass, and concrete.

Ten months later, North Korea's decision to turn up the pressure along the demilitarized zone dividing the two Koreas obliterated corruption as a number one issue in campaigns for all 299 National Assembly seats. Fears of a second Korean War provided a fresh boost for YS just five days before the Assembly elections of April 11, 1996. How serious a threat was North Korea posing by its decision on April 5 to abandon its duties along the DMZ as stipulated in the truce that had ended the Korean War? For the next few nights, North Korea sent anywhere from 120 to more than 200 troops into the Joint Security Area in Panmunjom. Television networks broke into regular broadcasting on April 8 to report the latest incursion at 8:05 P.M., then broke in again to say the North Koreans had pulled out at 10:30 P.M. They were hefting heavy weapons, including mortars and machine guns.[3]

Kim Young Sam owed the North Koreans a debt of gratitude when his party did better than expected in the Assembly elections. Although the party fell short of an absolute majority, it was able to piece together a razor-thin edge with the help of independents and minority party members. Without the insecurity engendered by the North Koreans, YS might have suffered serious humiliation. While pressing corruption charges against both Chun and Roh, he was severely embarrassed by the arrest in March of an aide charged with having accepted nearly three quarters of a million dollars in bribes from companies in return for favors. His popularity also suffered when a college student died in a demonstration the week before against the government. An autopsy showed that he had suffered a heart attack, but radicals blamed government policies and police.

In the aftermath of the April 1996 elections, YS had the confidence to press the cases of corruption, treason, and mutiny—and responsibility for the Kwangju massacre—more zealously than ever. He interpreted the results not only as endorsement of his rule but also as a vote of confidence for his campaign against corruption, as epitomized by the cases against Chun and Roh. While losing popularity in the TK region, Taegu, and North and South Kyongsong provinces, the government hoped to gain in Cholla by winning convictions on the Kwangju massacre charges. The intensity of Cholla sentiment against rule by Seoul mandated stiff sentences for Kwangju—an issue of far more importance there than corruption.

YS was more interested in going after Chun and Roh than any chaebol leader. By having his prosecutors bring them to trial, he saw political popularity as well as vengeance—even though he might have been grateful to Roh for backing him in 1992. Through the summer of 1996, the two stared stony-faced and glum from TV screens and newspapers. Prosecutors wanted to keep them behind bars into the next century. Chun, recovering from a 30-day hunger strike, was charged with having accepted $273.35 million in bribes in return for contracts in construction, aerospace, and other key industries in six years as president. He already faced charges stemming from the 12/12 mini-coup as well as the slaughter of more than 200 dissidents rounded up after the Kwangju revolt. Prosecutors said that Chun's slush fund exceeded $1.2 billion. That figure placed him well ahead of Roh, languishing in the same prison as Chun on charges of piling up about $650 million.

"Transparency," the ability to penetrate the mysteries of the payoffs, was the code word, taken from efforts by foreigners for full disclosure of financial records of companies, for revelations in pursuit of Chun and Roh. The ex-leaders admitted that they had accepted hundreds of millions of dollars in the time-honored custom in which Korean kings accepted gifts from their subjects. Anyway, they claimed, they had spread most of the loot among their aides. There was no doubt they had skimmed several billion dollars more in bribes and donations from chaebol leaders and others who needed their patronage to win the loans, the contracts, and the perks and privileges to build up their empires.

Park Jin, an Oxford-educated presidential assistant who wrote some of Kim Young Sam's speeches and interpreted for him for English-speaking audiences, urged me to "see the woods, not the trees," in judging the arrests of the two ex-presidents. Asked how YS could have allied with Roh's party in his quest for the presidency, Park had a ready reply. "It was a political venture into the tiger den. He caught the tiger. Now he has his own party"—the New Korea Party, formerly Roh's Democratic Liberal Party. "At the beginning of this presidency, complete reforms were introduced," said Park. "All nine four-star generals are new appointees. All the three-star and most of the two-star generals are new. We have a new elite. We have five years of presidential tenure in which to pursue reform." He conceded the investigation "could have some harmful effect" but called it "a painful blessing."[4]

Those who gave the bribes were not viewed so sternly. The government in April 1996 evinced its confidence in Samsung and Daewoo as business giants and tools of diplomacy by approving new deals for them with North Korea even while pressing charges against their chairmen, Lee Kun Hee and Kim Woo Choong. Samsung Electronics was authorized to invest $7 million in a telephone switching system in the free trade area of Rajin and Sonbong in the northeast corner of North Korea. Daewoo won approval for a $6.4 million deal to produce household electrical goods ranging from color TV sets to microwave ovens in the west coast port of Nampo in collaboration with a North Korean state company. Both deals were linked to the government's desire to bring North Korea into four-party talks proposed by President Clinton and Kim Young Sam in April.[5]

"It is really business as usual, even the more the better," said Samsung spokesman Cho Jang Won. "Chairman Lee Kun Hee is going free. He is in a very free position. He is quite the same."[6] True, the Samsung chairman had already faced three hearings on charges of having given about $13 million to Roh Tae Woo, but he was still pouring billions into new factories. "Those who are indicted but not in jail have no problem in doing business," said prosecutor Moon Young Ho, in charge of all the cases springing from the trials of Chun and Roh. Moon's curt written response to questions that he had asked me to submit in writing reflected the government's attitude toward the nine "tycoons," as Koreans called them, using the English word, whose cases he was diligently prosecuting.[7]

"The general view of the people is it did not hurt the production," said Chang Suk Hwan, director-general for planning and management at the ministry of commerce, industry, and energy. "The business operation of the business groups is going perfect." That might have been an overstatement, but the record showed the chaebol as setting the pace for a country whose gross national product in 1996 exceeded that of the year before by 9 percent. Samsung, the leading chaebol in 1995 with group sales of nearly $84 billion, $3 billion more than those of Hyundai, set its sights on sales in 1996 of $105 billion. Daewoo ranked third in 1995 with sales of $57 billion and targeted more than $70 billion in 1996.

The most peripatetic of chaebol leaders, Daewoo's Kim Woo Choong, got back to Seoul in April to face charges after overseeing the opening of new motor vehicle plants in Poland and Rumania and checking on expanded sales operations in England. It was all part of a $5 billion investment scheme for catching up with Hyundai as Korea's largest motor vehicle manufacturer by 2000. Kim, accused of passing along $20 million in payoffs to Roh, appeared more concerned about bolstering Daewoo's car sales from about 450,000 in 1995, third behind Hyundai and Kia, to 2.2 million by 2000. "We want to be in the top ten worldwide," said Kim's assistant, Lee Sung Bong. "It's a promising business."[8]

Daewoo officials said proudly that Kim's return flew in the face of claims that he was traveling overseas to avoid the case against him. They denied that another top Daewoo executive, Lee Kyung Hoon, had fled to the United States—and a new posting as chairman of Daewoo USA—to avoid charges of money laundering and hiding assets under false names. "He will come back," said a Daewoo spokesman, Lee Jeong Seung, even though Lee Kyung Hoon had resigned as chairman of the Daewoo Corporation "because of the court cases." J. S. Lee offered the customary rationale for the gift-giving—and for why hardly anyone believed the accused businessmen would serve a day of the jail terms from one and a half to four years asked by the prosecutor. "The circumstances in Korea are that many people know about the former government of Mr. Roh and Mr. Chun," he said. "There is no one who denies or refuses the president's request. Everyone knows, they understand the chiefs of big companies cannot but accept."[9]

So sure were Koreans that none of the tycoons would receive anything other than suspended sentences and fines that routine hearings in their cases went unnoticed and unreported. That was in contrast to the sensational trials of Chun and Roh, lavishly reported every time either of them was escorted under heavy guard, wearing light blue prison uniforms, their hands bound behind their backs, from their prison cells for another session in court. Far more important than the bribery, which the public took for granted, was the evidence that they had masterminded not only the takeover of power in late 1979 and 1980 but also the Kwangju massacre of May 1980. Thus the government hoped to defuse antigovernment campaigns led by dissidents from Cholla in the footsteps of their hero, Kim Dae Jung.

Daewoo's overriding concern was not the trial but whether Kim Woo Choong could succeed in his drive to advance Daewoo production from twenty-third in the world as of 1997. Daewoo ran two plants in Poland, including FSO, Fabryka Samochodov Osobowych, Daewoo Motor Polska, and was opening a plant in Uzbekistan in July. "In eastern Europe we are making cars," said Lee Sung Bong. "In western Europe we are selling them." Daewoo also hoped to get back in the North American market in 1997, four years after buying out its partner, General Motors, for which it had produced the Pontiac LeMans, last shipped to the United States in 1993. Daewoo's $5 billion investment in motor vehicle expansion included $1.2 billion for a new plant in Korea capable of producing another 300,000 cars a year.[10]

The other top tycoon on trial, Lee Kun Hee, was pressing just as hard on a similar mission—to get Samsung into the motor vehicle business with engines and design licensed by Nissan. Samsung, pouring $5.5 billion into "short-term" investment needed to open its plant in Pusan in time to produce 80,000 cars in 1998, anticipated "long-term investment" of $13 billion by the year 2010. To outward appearances, Lee Kun Hee, like Kim Woo Choong, was unfazed by the case hanging over his head. He had recently inspected Samsung Tijuana Park, an "integrated electronics complex" in Tijuana, Mexico, that had cost $200 million and would cost another $580 million to expand. He had also dropped by the site of a DRAM fabrication facility under construction in Austin, Texas, for a price of $1.3 billion.

The leaders of lesser chaebol caught up in the prosecutor's net behaved with equally supreme confidence. Dong Ah Chairman Choi Won Suk, charged with having bequeathed about $20 million to Roh, appeared on the front pages of the newspapers beside Libyan President Moammar Gadhafi. He was in Tripoli as Gadhafi publicly asked Dong Ah to build the third and fourth phases of "the Great Man-Made River Project"—a $10 billion deal in addition to the irrigation and water system Dong Ah was already building there. None of the reports mentioned Choi's legal difficulties at home—or, for that matter, Gadhafi's own reputation for harboring and encouraging terrorists.

Among the most notorious of the accused, Chung Tae Soo, chairman of the Hanbo group, ranked fourteenth among the chaebol, was held for 20 days before he was finally released on bond—and now faced charges of giving about $13 million in bribes. "No problem," said Song Jin Myong, general manager in the planning department. He predicted the group would double sales, mainly in steel products and construction, from about $4.05 billion in 1995 to about $8.12 billion for 1996. "We believe not only one businessman did it but many, many Korean heads do it," said Song. "It's all up to politics. It has nothing to do with business." Song preferred to talk about the group's overseas projects—irrigation in Jordan, a hydroelectric dam in the Philippines—rather than his boss's travail.[11]

Several months later, in August, after Chun was sentenced to death and Roh given 22 years and 6 months, Kim Woo Choong and three other chaebol chieftains who bribed them got jail terms. Lee Kun Hee was among another five whose sentences were suspended.[12] It was inconceivable, however, that any tycoon would go to jail. The appeals court in December 1996 suspended all the prison sentences while commuting Chun's death sentence to life and cutting Roh's term to 17 years. All the accused chaebol leaders were forgiven their sins. In a hierarchical society where the top leader demanded obedience in exchange for favors, the cases gave YS the power to obtain fealty by compassion. His mercy may have been a mistake. For also corrupting the miracle, the chaebol chieftains were as guilty as anyone.

How much was likely to change as a result of the trials of 1996? To what extent did the sentencing of a few chaebol chieftains signal a shift from the old ways—or was it merely an instinctive effort to bring the chaebol into line behind Kim Young Sam? There may not have been an explicit understanding between government and chaebol leaders on the corruption cases, but the government showed it had no desire to penalize their companies—or jeopardize the nation's vibrant economy. A display of forgiveness, however, would not mean the end of the struggle between big government and big business.

Through all the scandals, the chaebol were steaming on as "engines of the miracle" through the mid-1990s despite persistent talk about shedding holdings, diversifying, or even "breaking up" into mini-chaebol. How could the chaebol downsize when they were the entities with the funds for investment on a global scale in keeping with Kim Young Sam's "globalization" policy? No matter what, the top groups provided the cutting edge for huge expansion both at home and overseas. In 1996, for instance, Samsung Aerospace was manufacturing F-16s for the Korean air force on license from Lockheed Martin while Hyundai Space & Aircraft pursued plans for light planes and Daewoo was developing a propeller-driven trainer for the Korean air force. Bureaucrats might dwell from time to time on the need to encourage small business and frustrate the designs of some of the chaebol, but they were growing even faster than the overall economy. The New Industry Management Academy in Seoul reported that the 30 largest chaebol had actually increased in value by about $72 billion in 1995. That meant that the overall worth of the top 30 had gone up by 30 percent from 1994.[13]

YS and his bureaucrats, however, sought to give an impression of making good on their 1992 campaign pledges to curb overweening chaebol expansion. An example of the battle between big government and big business revolved around the longtime effort of the Hyundai group to build a steel mill near Pusan. Hyundai believed it was edging closer to fulfilling the group's dream of going into steel, for years adamantly opposed by a government bureaucracy that held the final say. "Government officials are moving not to oppose the project strongly," said a Hyundai spokesman in August 1996, claiming victory from seminars that had "wiped out the fear of a potential oversupply in a precise analysis."[14] The government, however, was caught in a contradiction. If the government were to refuse permission, critics would say the government had betrayed its oft-stated and oft-forgotten, principle of noninterference.

The Hyundai steel case had two other dimensions that revealed its complexity. First, the government had a stake in protecting Pohang Iron and Steel, Posco, the nation's showcase steel plant. The government owned the largest percentage of shares in Posco, and Posco executives fervently deplored the Hyundai scheme. Second, YS had not gotten over his distaste for Chung Ju Yung. Hyundai and the government also clashed over Hyundai's efforts to get into aircraft production with McDonnell Douglas and to gain control, with partners, of an investment banking firm in violation of government policy against chaebol taking over that field.

"The chaebol, you can't live with them, you can't live without them," said Korea analyst Nicholas Eberstadt. "The Kim Young Sam government can't figure a way to restructure the economy in any way that the chaebol don't have an enormous role. It's one thing to push a couple of your enemies around and another thing to come up with a new system."[15] The Federation of Korean Industries, however, did not believe the bureaucrats were so kind. The FKI complained that the government wanted not only to curb overseas investment for fear it would compromise expansion at home but also to block mergers and payments among companies within a group—key to chaebol prosperity.

If the criminal transgressions of the chaebol leaders were history, the ordeal of Kim Young Sam was just beginning. Following an oft-repeated pattern, the anticorruption drive rebounded to undercut his presidency long before his incompetence in economic matters was fully exposed. Chants of "kick out the president" echoed across university campuses in May 1997 as students demanded Kim's resignation for having refused to reveal the source of funding for his 1992 presidential campaign. Radicals seized on corruption as an easy target after the arrest of Kim's second son, Kim Hyun Chul, charged with having accepted funds from Hanbo for both himself and his father's campaign.

A declaration by YS that he had "no records" of campaign spending and no clear idea how much he had spent ignited the riots. "The son is just an example of corruption in government," said Noh Kyung Hwe, a senior at Hanyang University, the focal point of a showdown between rampaging students armed with steel poles, rocks, and bricks and policemen holding them off with peppery tear gas. "The president is the most corrupt person. We want a real democracy. Korean society is corrupt. We need to get rid of all the bad leaders."[16] At the height of the 1997 spring riot season, some questioned whether YS would even be able to serve out his term. As it was, in his last year in office he lacked the authority to govern effectively.

Kim Young Sam's troubles worsened when Chung Tae Soo, ex-chairman of the Hanbo group, was sentenced in June 1997 to 15 years in prison for bribery and embezzlement just as Kim Hyun Chul was going on trial. Chung's son, Chung Bo Keun, by now Hanbo chairman, got three years for embezzlement. They were found guilty of bribing eight others, including a former home minister, four assemblymen, and three ex-chiefs of banks, all of whom also got prison terms. What made "Hanbogate," as it was known in the Korean media, so different from that of other cases involving bribery by tycoons was that the money trail led straight to the 1992 campaign. Kim Dae Jung, eyeing the 1997 election, pledged during the spring riots to "go after corruption" and demanded the president make a full accounting. DJ claimed to have "evidence and witnesses" to prove that YS had far exceeded the spending limit in 1992.

It was a measure of DJ's willingness to compromise, however, that he went along with the decision of YS to free Chun and Roh after getting elected as president. There was no doubt that YS planned to grant them amnesty, but DJ's assent was disillusioning to those who had suffered through the worst of their rule. On the bright, chilly morning of December 21, 1997, less than two years after Chun and Roh had gone to prison, the news spread around the May 18 Cemetery, named for the first day of the revolt that had ended in the Kwangju massacre of May 27, 1980. The mourners arrived at the cemetery, outside Kwangju, praying before the heroes' monument, then approaching the gravestones dug in mounds piled in rows on the slope of a hill.

The mourners, most of them women, widows and mothers of those who were gunned down or stabbed or beaten to death in the massacre, sat, stared, or knelt in silence, leaving behind flowers. They had just heard that Chun and Roh, convicted of ordering troops into the city and firing into mobs, were leaving prison the next day. Sitting in front of the grave of her son, Kim Soon Hee could hardly believe the news. "Those who committed crimes against our people should be punished most severely," she said. If Chun "had not once been president he would have been executed." Kim Kil Ja, who had lost her high school son, demanded that Chun and Roh "kneel down and apologize before the graves of all those they killed." Only the fact that Kim Dae Jung had approved the amnesty somewhat mollified the women, some alone, others standing or sitting in twos and threes, scattered through the cemetery.[17]

If DJ had to endorse amnesty for Chun and Roh as a matter of expediency, he had more reason than most to heap scorn on the guilty—and to go after still more villains from the old days. For his inauguration on February 25, 1998, under cobalt-blue skies, before 40,000 people massed on the plaza in front of the domed National Assembly building, the former presidents were all gathered for the first time. DJ, having been prosecuted and jailed by Chun, shook hands with Chun, Roh, and YS in a show of national purpose and "reconciliation," but the display of president and former presidents did more to accentuate differences than unity. All three sat expressionless as DJ took the oath as the first opposition leader ever elected president of Korea. That done, he launched into a speech blaming his predecessors for the current turmoil.[18]

At his oratorical best, DJ said he could not "help but feel limitless pain and anger when I think of you, the innocent citizens, who are bearing the brunt of the suffering over the consequences of the wrongdoing committed by those in leadership." The country, he said, would not have been in trouble "unless the political, economic and financial leaders of this country were tainted by a collusive link between politics and business." Reconciliation forgotten, DJ hinted at revenge, meaning investigation and trials of the perpetrators of the crisis. "We must calmly and squarely look back to find out how we have arrived at this state of affairs," he said with more than a trace of menace. The setting alone dramatized the fissures that threatened efforts at reform. Behind the dais loomed the entrance to the massive Assembly building where the Grand National Party, Kim Young Sam's political organization, still held a majority.

The new administration soon began searching for villains from the previous government with a familiar zeal. This time, the multiple investigations reflected a widespread desire to find just how South Korea, the land of the miracle, had plunged into an abyss from which it showed few signs of speedy recovery. "A lot of people suspect that revenge is the motive," said Hahn Chai Bong, a politics professor at Yonsei University, "but the public at large feels there were errors somewhere, and they want to know what happened."[19] Could the search for the truth behind Seoul's economic problems lead all the way to Kim Young Sam? Prosecutors on April 20 confirmed that they planned to submit questions to YS in writing to see if he could explain why his government, in the months before going to the IMF for rescue, had failed to heed clear warnings that the economy was in trouble. "They will really try their best not to get Kim Young Sam," said Hahn, noting that DJ had said he did not seek revenge against members of previous administrations.

Would DJ, however, try to curb the enthusiasm of members of his own administration for going after their predecessors? "Every single administration coming in has conducted an anticorruption campaign of one sort or another both to justify itself and to undermine the previous regime," noted Michael Breen, longtime business consultant in Seoul.[20] Now prosecutors were bringing new charges—first for the economic crisis, second for smearing DJ during his campaign, and third for bribes by major companies eager for licenses for their cellular phone operations.

Latest to receive a summons from the prosecutor's office was Lim Chang Yuel, former deputy prime minister and finance minister, who had signed the agreement with the IMF on December 3. The prosecutor wanted to ask him what instructions he had received from YS before finally agreeing to terms set by the IMF. Lim joined three other former deputy prime ministers/finance ministers on the prosecutor's list. Prosecutors had already questioned Lim's predecessor, Kang Kyung Shik, who had insisted shortly before he resigned in November that the country would not have to go to the IMF. One question was whether bribes were involved in such horrendous misjudgment. Prosecutors charged that major companies had paid off officials from slush funds just as they had done in the 1980s and early 1990s under Chun and Roh. Prosecutors banned four managers of the Hansol group, a chaebol led by a sister of Samsung Chairman Lee Kun Hee, from leaving the country. They were questioning them as well as two senior finance ministry officials about bribes passed in exchange for licenses.

When I saw Kang Kyung Shik in prison three months later, on July 18, 1998, he was suffering from high blood pressure and losing sight in one eye and had no desire to read the fine print of newspaper stories about Korea's financial crisis. "Economic problems give me a headache," said Kang. "Most of the time, I look forward to meeting my family and looking at books."[21] The 62-year-old Kang spoke to me from behind the wire mesh and glass of a prisoner's interview room in Seoul Detention House about 20 miles from downtown. The prison was slightly to the southeast of the capital, in Kyonggi Province, now the fiefdom of his successor, Lim Chang Yuel, elected governor of the same province when DJ was elected president. Lim had skillfully jumped onto the DJ bandwagon, then helped to land Kang in jail by saying that Kang had briefed him improperly on the gravity of the crisis.

Beside Kang as he spoke to me was a prison official, scribbling notes as Kang talked in English about his arrest and trial on charges of "negligence of duty" and "abuse of power." Kang, garbed in prison blue, identified on his shirt as prisoner number 1199, floor 2, cell 3, talked about the implications of his arrest and that of one other former senior official, Kim

In Ho, held on the same charges, for others now in charge of economic policy. "I very much worry," he said, "if I am treated this way, our economic management cannot function properly." Those now responsible for navigating the country through economic turmoil, he suggested, might fear that someday they too would be jailed, awaiting trial on equally vague charges. Held without bail since May 18, Kang and Kim were to go on trial on July 24 in Seoul District Court. Until the court finally granted their applications for bail, 110 days after they were jailed, they remained in separate cells, each seven feet by five feet, with no running water and a hole in the floor for a toilet. The highlight of every day was a ten-minute meeting with a visitor, usually a family member.

Kang believed that he and Kim, former chief economic secretary to Kim Young Sam, had been singled out for a crisis they did not create and had done their best to solve. "Some people think that I'm a scapegoat," said Kang, when asked if he would agree with the use of that word. "I am trying to figure out why I'm here." Many others were asking that question. Investigators had failed to discover ill-got wealth, and he was not charged with conflict of interest. "In no advanced country or developing country that I know of can we find such a case," said Lee Han Dong, a former prosecutor and judge who was now acting president of the opposition Grand National Party. "I only hope the ruling of the court will be fair and it will be a ruling that we as a nation will not be ashamed of."[22]

Kang, who retained his National Assembly seat as an independent but had been a member of the ruling party of Kim Young Sam, avoided mentioning Kim Dae Jung by name. "Since most people are suffering from economic disaster," he said with deliberate vagueness, "somebody thinks we need somebody to blame." The indictments against Kang and Kim In Ho accused both of them of having failed to report clearly on the country's slide. Kang was also charged with usurping control over the exchange rate from the central bank, while Kim was charged with pressing the bank to lend funds to a chaebol that had gone bankrupt. Then the indictment added that Kang had hindered negotiations with the IMF by not telling Lim Chang Yuel what he had negotiated before resigning.

Kang said he had had to resign so quickly that he never had a chance to brief Lim and defended his record. "Our economic situation was not good, but we never thought we were in a position to ask the IMF for assistance," he said. "The situation changed so rapidly in November. There was a lot of speculation on the exchange, and all these foreign papers reported it. Our government already had lost a lot of confidence in the international market and community. I had in mind to show we were moving into action." A buzzer sounded; the guard said "time over" for the interview. "They accuse me as a criminal," said Kang. "We will try our best to make this a nonsense case. Anyhow, I have to survive."[23]

The investigation into business and economic issues paralleled the "North wind" investigation in which prosecutors indicted the former head of the National Security Planning Agency, Kwon Young Hae, for an attempt to get voters to believe that North Korea supported Kim Dae Jung for president. Prosecutors in March 1998 questioned NSPA officials about a letter, purportedly written by a South Korean religious leader who had defected to North Korea, suggesting that DJ was in touch with the North Korean regime. Why had the NSPA publicized the letter during the campaign? The answer was that it was a forgery to make DJ appear as a pawn of the North. "This type of action might help the ruling party to dismantle the opposition," said Park Kyong San, a political consultant. Asked how much the prosecution was tinged by a desire for vengeance, he responded, "About half and half."[24]

Fittingly, the once omnipotent National Security Planning Agency, which had tortured political prisoners and nearly killed Kim Dae Jung, was now the target of a top-to-bottom investigation. The justice ministry on March 12 banned Kwon Young Hae and 30 other past and present NSPA officials from leaving the country. Warrants were issued for three other top officials, in addition to two already under arrest. All faced charges of libeling and slandering DJ by spreading rumors that North Korean leader Kim Jong Il was contributing to his campaign.[25] The case, known as *bukpoong,* "wind from the North," threatened to blow up into a pervasive inquiry going deep into the history of the NSPA, called the Korean Central Intelligence Agency until 1981. (The agency in 1999 was renamed the National Intelligence Service, again for the sake of the image.)

Publicly, DJ insisted he had no desire for vengeance. Behind the scenes, however, Kim's aides were determined to press charges against those responsible for arranging three news conferences, in Seoul, Tokyo, and Beijing, in which a Korean-American businessman had said he had evidence that North Korea was bribing DJ. The businessman, Yoon Hong Joon, also was under arrest. Prosecutors charged that NSPA officials, led by Lee Dae Song, chief of the agency's international department, had paid Yoon $19,000 to state that DJ was a communist sympathizer who depended on donations from the North. The news conferences, according to prosecutors, were part of a broader plot masterminded by Kwon. They planned to question Kwon about a letter that the NSPA had disseminated in which a South Korean defector to North Korea had supposedly expressed his support for DJ.[26]

The NSPA revealed its guilt by destroying records of all activities linked to Kim Dae Jung's election campaign. The Korea Broadcasting System said the reports were destroyed over a one-week period after the election. DJ, after his inauguration, imposed his own authority over the agency by appointing his campaign manager, Lee Jong Chan, who had once been a KCIA official, as director and naming three vice directors whom he believed he could trust. Lee promised to divorce the agency from politics and focus on its "proper function" as an intelligence-gathering agency. He had less success, however, in persuading political activists, many of them allied with DJ, not to press a campaign of revenge for the agency's past acts of cruelty.

Kim was under equal pressure to show compassion for the NSPA's worst victims. He had said he would release some of more than 400 "prisoners of conscience" by Friday, March 13, but activists wanted more, including repeal of the national security law under which thousands had been arrested. Chun Jae Soon, mother of two sons jailed for their involvement in a political party linked to North Korea in 1992, said they were given electric shock treatments and smothered with water-soaked towels while in pretrial NSPA custody. She herself was held for four months when she inquired about them.[27] Nam Kyu Sun, secretary-general of Mingahyup ("Democratic Family Committee"), representing families of political prisoners, demanded "prosecution of those responsible, compensation of victims and full disclosure of the truth."[28]

DJ the president displayed a rather different attitude toward political prisoners from that of DJ the dissident. He wanted to appear ready to extend mercy to several hundred people who had been held for political crimes, some for many years. He did not, however, wish to seem overly lenient in the eyes of probably a majority of South Koreans who had little sympathy for those identified with the North. He particularly did not want to offend an establishment of business and political leaders, military officers, and government officials, among others, whose support he needed to buttress his own power and programs. The first prisoner release, celebrating his inauguration, was a compromise. From among more than 400 "prisoners of conscience," only 74 were freed. Most of them were either very old or were due for release in the next few months anyway. Their stories reflected the suffering of a divided country and a people uncertain how to respond to the same threats that had undermined the country since the end of Japanese colonial rule.

Yoon Yong Ki wept as he described the greatest agony of his 39 years in South Korean jails. "My wife died of cancer four months ago," said Yoon, 73, the country's longest-serving prisoner of conscience among those released. "Right before she died, she came to see me in prison. She said, 'They say you can come out if you just sign a statement of confession.' She had no other wish. She said, 'I just wish I could live with you before I die.'" As he had done countless other times since his capture on a spy mission for North Korea after the Korean War, Yoon refused to sign the requisite statement. "More than the beatings or torture or threats, what my wife said was hardest to deal with. I thought she may have believed I had no love for her. It hurts me to think she passed away thinking that I did not love her, just for not signing a statement."[29]

Yoon turned from tears to anger, however, as he talked about the suffering of the political prisoners and DJ's failure to add more than 400 others to the total of 2,304 prisoners included in the general amnesty. "I had no guilt when I heard the verdict on my case on January 29, 1960," said Yoon, captured and charged as a North Korean spy just two days after crossing the DMZ. "I did nothing wrong. I was very proud even when wearing handcuffs." Yoon "felt guilty of a crime," however, when he learned he was leaving the prison in Taejon, south of Seoul, while his close friend Woo Yong Gak, the country's longest-serving political prisoner, and others remained in their tiny cells.

The difference between the cases of Yoon and Woo was simple: although both had steadfastly refused to sign confessions, Yoon was released along with five other septuagenarians just because they were more than 70 years old. Woo, still 69, was to have his seventieth birthday six months later, and might be a candidate for release at that time. As long as so many of his friends remained in prison, Yoon was reluctant to talk about the pain they endured. "Just because I'm free doesn't mean I'm free to talk about whatever goes on inside," said Yoon. "You don't know what kind of consequences there may be. I don't want to inflict that pain on them. They have suffered enough."

Still, he offered glimmerings into life in prison as he talked about the pressure to confess. "You can't be in prison for following your conscience, but we were imprisoned so long because of our conscience," he said. "Once inside, they told us to sell our souls, to denounce our beliefs, everything we believed in. As a human being I could not." For the first two or three decades of his imprisonment, physical punishment was almost a daily routine. "They would shove statements in front of you and subject you to all sorts of torture," he said. "Nowadays, they leave it up to you."

Not all "prisoners of conscience" succeeded in resisting. Some yielded and signed. Yoon, however, did not blame them. "These people who signed the statements were by no means selling their conscience. They were at a crossroads between life and death. There were many who refused to sign and ended up losing their lives. All the people inside were only held there because they refused to give up their conscience. There is really no hazard from these people, yet they keep them in prison."[30]

The next amnesty, on August 15, the independence day marking freedom from Japanese colonial rule in 1945, would also highlight celebrations marking the 50th anniversary of the founding of the Republic of Korea in 1948 on the third anniversary of the Japanese surrender. The clemency was to include pardons for at least 1,650 prisoners, including 100 prisoners of conscience, and parole from prison for another 2,100. Among those who were likely to leave jail were former military officers who had helped Chun Doo Hwan seize power in 1980 and then had had Kim Dae Jung jailed for treason.

None of the 17 longest-serving political prisoners, however, was among them. The reason again was that all of them, led by 69-year-old Woo Yong Gak, had refused to sign the pledge. "Prisoners of conscience think it's the same as the old system when they were asked to sign statements of 'conversion' from communism," said Oh Wan Ho, Amnesty International director in Korea. "This system does not meet international human rights standards."[31] Defying the government as usual, Woo, sent from North to South Korea after the Korean War and later arrested as a spy, led the 17 nonsigners on a two-day hunger strike the weekend before the release. In solitary confinement in the jail in Taejon, Woo had almost no visitors but had probably spread word of the strike during daily exercise periods.

Those who signed the pledge were hardly traitors to the cause. The criminal record of Korean poet and labor activist Park No Hae read like a personal history of Korea's transition from dictatorship to democracy. In the first few months after his arrest in 1991, said Park the day after his release, "I was tortured and beaten like an animal," until finally he was sentenced to death for violating the controversial national security law through his activities as a radical labor leader. Next, the death sentence was commuted to life in prison. "Then somehow it was shortened to seven and a half years," and on August 15 he walked out of jail in the southeastern city of Kyongju.[32]

Park was perhaps the best-known of the 103 "prisoners of conscience" who received amnesty that day. Like all the others, he had signed the pledge to "abide by Korean laws," only to be greeted by chanting demonstrators waving banners saying, "No to the law-abiding oath." All told, 2,071 prisoners were freed while 13 others got reduced sentences and civil rights were restored for 4,820 people on parole. Also among those let go was Kim Seong Man, who had spent "13 years and 2 months" in 6 different jails. A former student at Western Illinois University in McComb, he was convicted of violating the national security law by meeting North Korean officials in Budapest. "I was arrested a month after getting back to Korea," Kim told me. "At first I was treated badly. They would tie my hands together and put me in a room with no sunlight for some unreasonable reason."[33]

Kim Nak Jong, released after serving six years for meeting North Korean propaganda agents, said the decision of Woo and the others not to sign the pledge reflected fears for their families. "They said their families are living in North Korea and, if they sign, North Korea will make it difficult for them," he said. As for about 100 others who refused to sign, he said they were student radicals who only had a few more months to serve. More than 200 others who made a show of refusing to sign the pledge were "still awaiting trial and were not yet eligible."[34]

As Kim Dae Jung approached the end of his first year in office, the government again tried to temporize with the critics. This time the justice ministry, on February 22, 1999, promised unconditional amnesty to the 17 long-termers in a deal that authorities hoped would promote an exchange for 300 South Koreans held captive by North Korea. Justice Minister Park Sang Cheon said the government was considering "extraordinary measures" under which the 17, including Woo Yong Gak, now in his forty-second year in prison, might return to North Korea provided the North freed the South Koreans, some of them held there since the Korean War. It was unlikely, however, that any of the long-term prisoners would want to return to North Korea—or if the North would consider an exchange. A dozen of the prisoners being freed had originally come from South Korea but had enlisted in the North Korean army during the Korean War and were captured in the 1960s while conducting espionage in the South.

The 17, each of whom had served at least 29 years in jail, were among 1,508 prisoners going free as Kim again sought to display "national harmony" among disparate political and regional groups as well as between the two Koreas. The amnesty covered 8,800 people, including another 7,292 previously paroled from prison, given suspended sentences, or ordered to pay fines for offenses ranging from participating in illegal strikes and demonstrations to breaking traffic laws.[35] The amnesty, the largest ever, was intended to placate the critics who had demanded that the hard-liners go free in the interest of human rights. The release marked a sharp departure from the previous policy of releasing prisoners only after they had signed the infamous pledge.

Although the amnesty was far more sweeping than the two others authorized by Kim during his presidency, Mingahyup said it included only a tiny minority of those held for political crimes. Among those still in jail were 18 labor union leaders whom the justice ministry charged with leading illegal strikes. About 30 members of Mingahyup, shouting, "We demand the release of all the prisoners," demonstrated in front of Seoul Station, displaying photographs of more than 300 prisoners of conscience. Nam Kyu Sun said the amnesty included only 41 political prisoners, most of them either sick or old or both, while the government still demanded that most political prisoners sign the pledge. "Korea is the only country in the world demanding such a pledge," she said.[36]

The zealots in DJ's entourage were less interested, however, in salvaging the remnants of the lives of old radical prisoners than in pursuing members of the former ruling party. In their passion for revenge, they dug up crimes not only from the most recent political campaigns but from the distant past. There was even evidence, they said, that the brother of the defeated presidential candidate, Lee Hoi Chang, and two others had gone to Beijing before the 1997 presidential election for the purpose of bribing North Koreans at the embassy there into staging an attack across the DMZ. A few shots across the DMZ, it was said, were needed to frighten voters in the South away from DJ, always portrayed on the side of radicalism. If true, the story showed the depths of corruption of the previous regime. If false, it revealed the paranoia of the current regime, witch-hunting for victims in a familiar cycle of retribution.

The atmosphere of witch-hunt pervaded an investigation by an Assembly committee seeking, in January 1999, to pin the blame for the economic crisis on the previous government. The committee on January 22 ordered Kim Young Sam to testify. Calling YS "a key figure responsible for the crisis," Chang Che Shik, chairman of the committee, said he "must inevitably come to the Assembly to provide his own account" of the crisis and threatened "legal action" if the former president tried to avoid the summons. The committee also summoned Kim Young Sam's son, Hyun Chul, convicted of accepting bribes from the bankrupt Hanbo group but no longer in jail. Hanbo's former chairman, convicted in the same case and still in prison, would also testify, as would the former chairman and former president of the Kia group, both convicted of such misuse of funds as to bankrupt Kia.

Lawmakers from the ruling coalition promised to present evidence showing that bribery was much more widespread than previously realized. The ex-chairman of Kia, Kim Sun Hong, handed over bribes of approximately $10 million to YS in three separate meetings, according to testimony on January 22. YS had extended dinner invitations to businessmen in exchange for the equivalent in Korean won of more than $1 million, said one witness, while lesser donors were invited for tea. As long as the coalition stuck by its plan to summon YS, members of the Grand National Party would boycott the investigation. Lawmakers evoked memories of the arrest and trial of Chun and Roh as they expressed determination to put YS on the stand. "No exceptions," said Jang Sung Won, a member of the coalition, planning to summon more than 90 people to the investigation before it ended in February. "We must get to the bottom of it."[37]

The government insisted on the investigation while citing a series of statistics showing the economy had recovered faster than appeared possible a year earlier. The Bank of Korea now forecast a 1 percent growth rate for the first half of 1999 and a 4 percent growth rate for the second half, for 3.2 percent overall growth for the year.[38] "Even though our economy is a little better now, we should identify what were the real causes of the financial crisis," said Kim Tae Dong, senior secretary for policy and planning on DJ's staff. Koreans must know "what went wrong, whether there was corruption which affected policy decisions." He denied the government was motivated by a desire for revenge. "This is a historical public hearing because their mismanagement of the economy resulted in our crisis," he maintained.[39]

The hearings turned into a round of recriminations and rationalizations as Korea's four television networks carried much of the goings-on live. YS and his son both ignored the summons, but the public got the chance to size up a parade of bureaucrats from the previous administration, most of whom had sunk into anonymity. Opposition leader Lee Hoi Chang charged the ruling coalition with "oppression of the opposition"—the kind of charge that DJ, in opposition, had made for years.

Lee himself, however, was not altogether above reproach. His two sons, it emerged during the 1997 campaign, had both avoided the draft. One of them, when it was revealed he had gotten out on the pretext of being underweight, did penance by working in a leprosarium. Their cases typified the pervasive corruption of a selective service system in which Korean men by law had to serve two years and two months in the army before they were 30. About 12 percent managed to fail the physical—a figure viewed as suspiciously high.

The images of young men in white T-shirts and green shorts flashed across television screens in May 1999 as investigators asked why military doctors pronounced so many of them not as fit as they looked. The scenes were induction centers where the men were undergoing physical examinations. Prosecutors arrested more than 200 people, including parents, doctors, and brokers accused of passing along bribes for the sons of the rich and powerful to dodge the draft. Draft-dodging had been a social and political problem for years, but the prosecutor's office opened its investigation in keeping with promises to go after corruption on all levels. One quarter of National Assembly members had never served in the armed forces, and 60 percent of the young men from Kangnam, the district of wealthy apartment and office buildings south of the Han River in Seoul, were somehow also adjudged physical misfits.

It was "no surprise that this type of thing was going on," said a career military officer. "That's the way it was in Korea."[40] The surprise was the crackdown—and a new law designed to embarrass some of the draft-dodgers and their families by requiring that National Assembly members and senior government officials reveal not only their own service records but those of their children and grandchildren. The purpose was to provide "transparency for these officeholders as well as their families," said Lee Dong Bak, who helped to write the bill as a member of the defense committee of the Assembly. "We want to encourage anyone aspiring for public office to make sure they have done military service. We have been harassed by one revelation after another involving people with relatives in high places, of high social class and wealthy backgrounds, who are turned down for the service."[41]

The larger question was whether the pattern of draft-dodging had undermined the morale of South Korea's army of 600,000 men. "Everybody knows these rich people get out of the army," said Chang Yong Jin, an office worker who had done his time in service. "It doesn't matter so much to the rest of us." What mattered more was another problem—that of an attitude in the armed forces of total subservience to authority. One curious manifestation was that wives of young officers had to do maid work for senior officers to ensure promotions for their husbands. Officers also were known to pay off superiors for good assignments and promotions. The impact of such practices had been to encourage the rise of a senior officer corps promoted only partly on the basis of merit.

More than anything else, avoidance of the draft fostered and reflected a class system that separated the well-educated and well-to-do from the rest of society. "This is a Korean social problem and should be corrected," said Brigadier General Cha Yung Koo, spokesman for the ministry of national defense. "We see in the rich people's areas, the sons of the rich have violated the law. That kind of trend is not healthy."[42] One typical way to evade the draft—and prosecution—was to remain out of the country as a student until the age of 30, the cutoff for compulsory service. More common, however, was the pattern of bribery for which prosecutors were pursuing athletes and TV personalities as well as sons of politicians and businessmen. A member of the coaching staff of the Hyundai Unicorns, Hyundai's professional baseball team, for instance, was arrested for arranging for a bribe on behalf of a player.

Bribes ranged between 10 million and 20 million won, $8,500 to $17,000, split between broker and doctor, but could go higher or lower. The easiest way to avoid induction was to fail the eye examination. Another favorite pretext was a back problem. Investigators complained that doctors often suffered only minor penalties if caught taking bribes. One investigator told Korea Broadcasting System that the investigation so far had been superficial. "When we started this investigation, we vowed to root out all crimes," said the investigator, his face deliberately blurred in the television report. "Then the investigation was stopped. It certainly came from higher up." Such remarks, said the KBS report, "indicate big involvement of higher-level politicians and officials," usually at the heart of what was sometimes known as "the Korean disease."[43]

No one, it seemed, was immune. On June 2, Bae Chung Sook, wife of Kang In Duk, who had just stepped down as unification minister, was charged with having attempted to extort the equivalent of $20,000 from Lee Hyung Ja, wife of Choi Soon Yong, a former chaebol chief charged with bribing government officials. Lee said Bae had told her she would give the money to Yon Chung Hee, wife of Kim Tae Joung, former prosecutor-general, newly named justice minister, to keep Choi out of jail. The case became known as "the mink coat scandal"—or "furgate" or "minkgate"—amid reports that Lee had ordered a mink for Yon. DJ accused the media of a "witch-hunt" but then, embarrassed, was forced to dismiss Yon's husband as justice minister just two weeks after appointing him. Everyone, DJ tut-tutted, "must take a lesson from this case and make efforts to enhance morality."

DJ's foes, however, planned to turn the corruption issue into a major platform in an effort to defeat his allies in crucial National Assembly elections in April 2000. "This regime began its term as a reformist regime," said Lee Bu Young, a key member of the opposition Grand National Party in the assembly. "Revolution is easier than reform." Lee spoke from an unusual perspective. Like DJ, he was jailed under previous corrupt administrations for his dissident views, and he had joined DJ's party before switching to the Grand National Party in a mood of disillusionment. "This reformist government turned out as corrupt as the previous administration," said Lee. "Widespread corrupt behavior is rooted in every leading group—politicians, bureaucrats, managers of big business."[44]

Such criticism struck a responsive chord as evidenced in comments one often heard when Shin Chang Won, 32, a convicted murderer and bandit who had eluded police for two and a half years after escaping from prison, was captured in an apartment with a girlfriend in July 1999. Police were especially upset by a Reuters report that labeled Shin the "Robin Hood bandit" on the basis of his claim that he had donated to an orphanage. The police said he had stolen the equivalent in Korean currency of more than $400,000 as well as numerous cars, pieces of jewelry, license plates, and identification cards needed to help him slip through police dragnets, and also charged him with raping one of his girlfriends. More disturbing, however, was the revelation that Shin had walked into a police station while on the run and bribed an officer to drop battery charges against the brother of a girlfriend. All told, 20 officers were charged with misconduct in the pursuit of Shin, whose case raised the issue of who was worse, a petty bandit such as he or the kind of powerful official who routinely accepted bribes.[45]

The question seemed pertinent in view of the arrest around the same time of one of the country's most active political couples. Lim Chang Yuel, the governor of Kyonggi Province, who had negotiated the IMF deal while serving as finance minister, and his wife were charged with having accepted more than $400,000 from an executive who hoped to keep the government from closing his bank. They said the gifts were donations, all returned after the bank was shut down anyway, but the affair was a special annoyance to DJ, who had supported Lim for governor. Lim's position, however, saved him. Deciding he should return to duty, a court in Inchon freed him on October 5, suspending a one-year term and fining him $83,000, the amount he personally was said to have accepted. His wife, Joo Hae Ran, a physician known as "the Hillary" of Kyonggi province for her social prowess and her campaign against AIDS near a U.S. base, was not so lucky. Resented for her high-profile socializing, an affront in a male-dominated society, she got a year and a half in prison, not suspended, for having received 400 million won, $333,000, despite her claim that the money was for fighting AIDS.

"Below the top guys, it's the same group of bureaucrats and politicians," said Peter Bartholomew, managing director of Industrial Research and Consulting, a firm that he had set up in Seoul in the early 1980s. "DJ's cabinet may be better than the earlier ones. He intends to clean it up. The latest revelations were a bit of a shock. Koreans were hoping the new government would be better, but look at the faces of the bureaucrats. They haven't changed."[46]

The government sought to ward off criticism by issuing guidelines forbidding senior bureaucrats from receiving any kind of favor. One rule banned them from announcing dates of weddings and funerals—opportunities for tycoons and politicians to woo influential officials with bribes disguised as gifts. Another rule banned "any gathering of wives of high-ranking officials"—a measure to break up "wives' clubs" notorious as conduits for influence-peddling and illicit gift-giving. The guidelines, however, were not likely to have much effect. "There was some resistance to them," admitted an official who did not want his name used. "Some people thought they were not reasonable, probably not enforceable. People have been donating for a long time. It's a general tradition."[47]

The scandals among those close to DJ inspired a counterreaction against the vendetta to ensnare members of the previous government. On August 15, 1999, independence day, the date on which the government often granted amnesty to prisoners of all sorts, DJ relieved Kim Hyun Chul, the son of Kim Young Sam, of any fears about having to serve the rest of his jail term on charges of bribery and tax evasion. Kim Hyun Chul, who was already free on bail while appealing his conviction, still had to pay a fine of well over $1 million. Nor did he regain the civil rights that his conviction had cost him. Clearly, however, DJ was anxious to avoid the claims of Kim Young Sam and other foes that he was persecuting, not prosecuting, Kim Hyun Chul in his zeal to punish his enemies.

Five days later, on August 20, the Seoul District Court cleared Kang Kyung Shik, Lim Chang Yuel's unfortunate predecessor as finance minister, and Kim In Ho, the former chief presidential secretary for economic affairs, of having failed in their responsibility to prevent the economic crisis. There was poetic justice here, for Lim had testified that Kang had failed to brief him on the seriousness of the situation when he became finance minister several days before appealing to the IMF. Kang and Kim In Ho "could be blamed for failing to appeal" to the IMF, said the court, but prosecutors had not proved that they had "attempted to cover up or downplay the seriousness of the market situation." The court found both men guilty, however, of having abused their positions to persuade banks to extend loans to the Jinro and Haitai groups, among the eight that collapsed as the country entered the IMF era. Indicative of a widespread desire to relegate the entire affair to history, the court suspended sentences.⁴⁸

In a culture of vengeance, it was the turn of DJ's people to face claims of malfeasance. In June, while reeling under "furgate," as the Korean press called the mink coat affair, DJ dismissed You Jong Keun, governor of North Cholla Province, as a top economic adviser after a burglar stole about $100,000 from his lavish home. The question was how the money had gotten there in the first place. For foreigners, the case was compelling since You had been among their most trusted seers on the topic of Korea's economic dilemma.

One never knew who would be next on whose list. The government in October pursued a seemingly unlikely target—Hong Seok Hyun, publisher of *Joongang Ilbo,* one of the country's largest daily newspapers, was indicted on October 18 on charges of having evaded more than $2 million in taxes and receiving a kickback of more than $500,000 from a construction company. Hong hardly fit the role of a persecuted crusader. He was the brother-in-law of Samsung Chairman Lee Kun Hee, whose group had founded the newspaper in 1965. More than half the money on which Hong had purportedly not paid taxes was income inherited from his mother, Lee Kun Hee's mother-in-law. In an appearance of downsizing, Samsung in 1998 spun off the newspaper as a separate entity, Lee's wife, Hong's older sister, ran a museum and foundation in the imposing building where *Joongang Ilbo* leased office space—and printed editions of *The Asian Wall Street Journal* and *Newsweek.*

Technically, the charges revolved around Hong's role as major shareholder in the Bokwang group, whose holdings included a large resort area southeast of Seoul. The question, however, was why prosecutors did not simultaneously pursue hundreds of other business leaders whose tax records were open to question. Government officials zealously denied that the charges were in retaliation for the paper's having sided with the opposition in the 1997 election—or favoring opposition candidates in the National Assembly elections coming up in April 2000. Had not the prosecutors also gone after DJ's allies, Lim Chang Yuel and You Jong Keun? In the game of charge and countercharge, no one was beneath suspicion. Whoever won at the polls, one could be sure of more arrests in a parade of criminals whose stories showed the risks as well as the riches shared by those vying for power along with wealth in modern South Korea.

# Chapter 5

# Policy and Politics

In the honeymoon before his inauguration in February 1998, DJ's advisers counted on Koreans to rally behind him but feared a backlash. "Our aim is to frontload our toughest economic policies early on while the president enjoys his highest popularity," said You Jong Keun, economic adviser and governor of North Cholla Province. "By the time the pain of reform starts to bite, we hope the reform process will have progressed to the extent it cannot be reversed."[1]

The strategy for overhauling the Korean economic system illustrated the delicacy of Kim's position as a one-time dissident about to take charge of a coalition. Kim vowed to combat the chaebol but counted on his alliance with conservatives to combat the majority party of the outgoing government in the National Assembly. Calling on the chaebol to submit restructuring plans, he berated them for extending their power far beyond their means. Typically, he charged the chaebol with smothering competitors, overpricing products, and speculating in real estate. His goal was to prevent them from advancing beyond certain core industries. Kim's advisers acknowledged, however, that persuading the National Assembly to reform the economy would be an uphill battle. "If we fail to implement many of the reforms, if we go on piecemeal," said You, economic problems "could go on three to four to five years."[2]

The mood in the winter of 1998 was one of growing confidence in Korea's ability to deal with the crisis after a negotiating team, including You, persuaded representatives of international creditor banks in January to roll over $24 billion in short-term debt into bonds maturing in one to three years. Kim's advisers pointed out, however, that the country had to pay more than $10 billion in debts falling due in March, while both domestic and foreign banks were reluctant to extend fresh credit. They estimated Korea's foreign exchange reserves at $15 billion, half of what the IMF stipulated as the minimum for fiscal security.

The FKI indicated the fight ahead, accusing DJ of ignoring "economic realities" by criticizing the chaebol for having too many subsidiaries. Instead, the FKI urged the government to rely on the market—a plea for the chaebol to be able to carry on as before. While reluctant to sell off any of their major companies, the chaebol showed signs of cutting back on their own. Hyundai Electronics sold Symbios Logic in Fort Collins, Colorado, for $775 million. Hyundai, which had bought the company from AT&T for $340 million in 1995, planned to invest the proceeds in its semiconductor plant in Eugene, Oregon.[3]

Such measures, however, were not the sweeping reforms that DJ and his team had in mind. "In the short term things will get worse," said consultant Peter Bartholomew. "Many companies are down to four, three, two, or even one day of work a week. The inventory of materials will be depleted by the end of March. All they're importing are critical materials like oil, iron ore, cooking coal and medicines." Bartholomew foresaw a pattern in which "bits of pieces" of failure were "growing into larger chunks and the food chain of the economy is slowly being shut down."[4]

The incoming administration wanted to look as if it were attacking problems before taking over the Blue House. The high-speed railroad from Seoul to the southern port city of Pusan made an easy target for budget cuts that were sure to slow it down if not throw it off track. A team from the Anglo-French company GEC-Alsthom was going ahead on the project along with a dozen Korean firms, but You said budget considerations dictated that "the speed of implementation has to be adjusted." He indicated that the new government would also consider charges that the high-speed railroad had been a political extravagance that the country could no longer afford. "The incumbent president"—Kim Young Sam—had wanted to push the project "to show the results," he said.

One factor was that YS was from Pusan, his political base and the TGV's final destination if it ever got that far. "There has been too much waste," said economist You of the project, for which the government had signed a $2.1 billion contract in 1994.[5] Under the contract, GEC-Alsthom was to lead a consortium in providing the technology and rolling stock for the *train à grande vitesse,* modeled after TGV projects in France. GEC-Alsthom confirmed that a government commission named by Kim Dae Jung had asked the Korea High Speed Rail Construction Authority to consider "postponing the construction of part of the dedicated high-speed line."[6]

You Jong Keun said that "everything must be reduced"—including projects for a huge airport serving the Seoul-Inchon metropolitan region as well as facilities to turn Pusan into one of Asia's leading ports, competitive with Hong Kong and Shanghai. The country faced a much more severe problem, however, than commitments to build new railroads, airports, and highways and other schemes for bringing the infrastructure to a level befitting an advanced industrial society. Prestige projects were needed to sop up newly unemployed workers and pump money into a floundering economy. By the time the railroad was completed, perhaps "the IMF crisis" would be history—and the new line needed to help fuel a renascent economy. The government's attack on the project symbolized its desire to get at much deeper, more intractable problems. Foremost among them was that of the chaebol.

Eager to show they were doing the right thing without doing much of anything, the top leaders of four of the top five chaebol, two weeks into 1998, pledged sweeping reforms of the system. The four, including the chairmen of the Hyundai, Samsung, LG, and SK groups, theoretically committed themselves to an overhaul of their groups on January 13 in a 90-minute session over breakfast with the president-elect in the Assembly building. DJ and the chairmen wound up with a statement declaring the reforms would "help stabilize the financial markets and the national economy." The chaebol leaders agreed on an accord that included some of the major conditions set by the IMF.[7]

Point one was a pledge to release consolidated financial statements—critical for bankers and investors, who had never been certain of the financial status of many companies, notably major entities not listed on the stock exchange. The chaebol chairmen also conceded the inevitability of layoffs of thousands of workers on the payrolls of companies burdened by debt and poor sales. The accord, however, said nothing about breaking up the chaebol, whose power over the economy had increased steadily since Park Chung Hee had established the chaebol system. Instead, as point three, the chaebol leaders agreed to stop cross-guarantees of loans among chaebol entities. They also promised to reduce debt-equity ratios.

The speed with which Kim Dae Jung acted to support the IMF demands came as a relief to Michel Camdessus, IMF managing director, who, on January 13, at the end of a two-day visit, said that he had found "almost unanimous support" for the IMF program despite "a difficult beginning." Korea, he said, had "demonstrated to the international community that these reforms, revolutionary though they might appear, will be implemented steadily." The sense was that Korea might work its way out of crisis. The stock index, which had fallen well below the 400 level in December 1997, closed at 460 points, up 1.65 percent despite a brief downturn triggered by the collapse of Peregrine Investments in Hong Kong. "You shouldn't cry victory prematurely," said Camdessus, clearly elated. "You have won the first battle. You haven't won the war."[8]

Camdessus was sensitive to criticism of the IMF for demanding both high interest rates and layoffs. On interest rates, he said, "There is no other way of stabilizing an economy and reestablishing the exchange rate at the proper level." Although benchmark rates had dropped from 30 percent to 22 percent in the first 13 days of 1998, he argued that "high interest rates must be there as long as confidence is not properly reestablished." As for layoffs, he preached the IMF mantra that "the government distribute equitably the framework of adjustment" and shareholders and managers "share the burden" with workers.[9]

Amid great bursts of publicity, the largest chaebol on January 20 revealed plans for scaling back their empires in the interests of efficiency—and survival. Hyundai, the largest in terms of assets, abandoned its cherished dream of building its own steel plant, suspended work on a semiconductor plant in Scotland and a motor vehicle plant in Indonesia, and promised to get rid of unprofitable subsidiaries. LG, the fourth-largest chaebol, said it too would give up the losers and modernize its corporate structure, while Samsung said it was closing regional headquarters in London, Singapore, and Ridgefield Park, New Jersey. The plans for cutbacks reflected the anxiety of owners to appear to fulfill the demands of the IMF.

Hyundai promised measures designed to bring its 58 companies into the modern age of business and industry. The group's octogenarian founder and honorary chairman, Chung Ju Yung, in "retirement," would exercise power through a strategic planning committee in place of the group's central planning office. Two of Chung's sons, Chung Mong Ku, the eldest surviving son and successor as chairman, and Chung Mong Hun, the fifth son, newly promoted from deputy chairman to cochairman on an equal status with his older brother, would remain in overall charge of 14 other companies. Four other surviving sons and relatives remained in control of the rest of the group.[10] Critics doubted Hyundai's commitment to change. "Hyundai's announcement lacks substance," said a Korea Broadcasting System commentary. "There were very few specifics. It's not up to people's expectations."[11]

A faithful hired hand, Park Se Yong, chief of the strategic planning committee, not Chung Ju Yung or any of his sons, announced the program. Chung Mong Ku and Chung Mong Hun were notably unavailable. Like other chaebol chieftains, they left annoying details of dealing with the press to underlings. Park, jailed for several months in 1992 for his role in siphoning off Hyundai funds to finance Chung Ju Yung's presidential campaign, later rewarded with top executive posts, said, "All available assets owned by founder Chung Ju Yung and his family members will be used to improve the financial statements." While selling off assets, the group would "block possible hostile efforts at merger and acquisition" of Hyundai companies.[12]

Both Samsung and LG announced cutbacks on January 20, though they were not as far-reaching as those at Hyundai. LG said it hoped to reduce its ratio of debt to equity to less than two to one over the next four years by unloading about 90 enterprises valued at 2.4 trillion won, about $1.6 billion, deemed as borderline and unprofitable. An LG spokesman also said that companies within the group would cease to guarantee each other's debts while the owners transferred money from personal funds and new outside directors ensured "transparency" and protection of minority shareholders.[13]

Samsung Chairman Lee Kun Hee said he too would go into his fortune, promising to sell 128 billion won worth of "personal property" and invest it in cash-strapped subsidiaries. Lee planned to put 90 percent of his annual income of 8-10 billion won in an employee welfare fund as long as the country was under the "IMF-sanctioned program." He would donate 10 billion won "from his savings and shareholdings to a fund to help retrain employees and assist with their relocation." The chairman's pledge amounted to more than $100 million, assuming Korea was bound to the IMF for another two years.[14] The significance of such pledges by owners was minimal, however, since their fortunes were in the billions—of dollars.

Like Hyundai and LG, Samsung promised "transparency of corporate accounts," including "consolidated financial statements," and also promised to "abolish group-wide cross-debt payment guarantees" under which companies within a group guaranteed each other's debts. At the same time, Samsung reiterated a previous pledge to unload "unprofitable units" and focus on "three to four major industries." Samsung said its most profitable company, Samsung Electronics, would be listed on the New York Stock Exchange by 2002, but the group canceled plans for a 102-story headquarters "in order to free more money in technology development and payment of debts."[15]

Battling to get through the crisis, increasing numbers of chairmen and owners assumed the posts of chief executive officer. Daewoo Chairman Kim Woo Choong in February became chief director of the group's three most important companies: Daewoo Motor, Korea's second-largest motor vehicle manufacturer; Daewoo Heavy Industries, Korea's second-largest shipbuilder; and the Daewoo Corporation, the group's trading arm and construction company. The chaebol viewed assumption by their leaders of clear-cut company titles as evidence of their desire to restructure. Most of the chaebol, including Daewoo, pledged hands-on control by owners in deals signed with their creditor banks.

The incoming government, however, was far from satisfied. Displeasure in DJ's camp set the stage for a showdown between government and chaebol. The bottom line, as far as DJ was concerned, was that Korea's tycoons had to implement the agreements to which they had sanctimoniously committed themselves. The announcements by the chaebol inspired skepticism among many foreign observers as well. "The right things are being talked about," said David Young, in charge of the Boston Consulting Group's Seoul office, "but there's some distance between announcements and actions." One problem was how to find companies with enough money to buy off marginal operations. "It's pretty complicated how divestitures will work when everybody's short of cash," said Young. "You don't see a long line of buyers."[16]

Figures released by the National Statistics Office added urgency. Imports of machinery, which the chaebol had to have to maintain production, fell by 47.3 percent in January 1998 from January 1997, while industrial output dropped 10.3 percent from the previous January. More than 3,000 small and medium-size companies went bankrupt in January, and industries in general were operating at only 65 percent of total capacity. Unemployment had nearly doubled, rising from 2.3 percent to 4.1 percent, the worst rate in more than a decade.[17]

Pressured by companies of all sizes, Hubert Neiss of the IMF cautioned against lowering interest rates prematurely in an effort to jump-start the economy. "As long as the exchange rate is undervalued and markets are not stabilized, it's difficult to lower interest," he told me after getting back to Seoul on February 1. The economy was "moving and that, of course, gives hope," he said, but the chaebol still had to restructure.[18] Korean officials, meanwhile, spread the word that the country was ready for a sharp drop in interest rates in view of restoration of confidence shown by foreign creditor banks. Finance Minister Lim felt "very sorry for companies suffering from recently soaring interest rates under the IMF program."[19]

Not yet president, Kim Dae Jung behaved as if he were, keeping up a drumbeat of anti-chaebol criticism. DJ's policymakers, while more or less siding with the chaebol in favor of lower interest rates, carried their crusade for reform into the heart of chaebol authority, the offices that exercised day-to-day control. Meeting with executives of the 30 largest chaebol, an emergency economic committee made up of members of the incoming and outgoing administrations told them on February 9 that it wanted them to disband their coordination and planning offices as part of restructuring.[20] Legislation effective March 1 would outlaw cross-guarantees of loans. The system of cross-guarantees as well as cross-ownership of companies within each group held a typical chaebol together in the absence of any central holding company.

To Lee Hun Jai, incoming chairman of the Financial Supervisory Commission, the superagency to be formed in March to oversee reform, the attack on the coordination and planning offices was vital to breaking down family control. Chaebol chairmen should revise their control structures at open shareholders' meetings, said Lee, and chaebol executives and their banks should face "criminal punishment" if they violated the law that would "completely ban new cross-payment guarantees of conglomerates."[21] Theoretically a policy of openness would encourage the chaebol to jettison companies on the brink of bankruptcy rather than support them secretly. The committee recommended enforcing the policy by exacting penalties for "excessive" borrowing and giving minority shareholders the right to file class-action suits and study company records. "The most important element is the rights of shareholders," said You Jong Keun. "The chaebol will have difficulty defending their control."[22]

The cushion provided by the IMF bailout and foreign banks, however, meant that policymakers were under no compulsion to revolutionize a historically nationalist economic policy, much less do away with an economic structure dominated by a handful of chaebol. The ambivalence in outlook emerged in a plan suggested by a newly formed emergency committee for "hostile mergers and acquisitions" of Korean companies by foreign firms. "Basically our policy is to make mergers and acquisitions for foreign investors free and open," said Kim Min Sok, a National Assembly member who served as liaison between the committee and the incoming administration.[23] The plan permitted a foreign investor to increase its stake from 10 percent to 33 percent in any company without the approval of the management of the company. Nor would an investor have to obtain permission from the ministry of finance to invest in companies with assets of more than two trillion won, $1.3 billion at the current exchange rate.

The western phenomenon of the "hostile M&A," however simple on paper, was unimaginable in reality. The emergency committee's recommendation included a catch that made a serious attempt at hostile takeover unlikely. The committee would do away with the regulation forbidding companies within the 30 largest chaebol from investing more than 25 percent of equity in other concerns within the same chaebol. For the country's top 30 chaebol, the change in the rule against cross-ownership of more than 25 percent of equity meant they would only have to get rid of the losers while clinging to top-heavy prestige companies still wallowing in debts. Revocation of the 25 percent rule, rammed through several years earlier to curb the growing size of the chaebol, relieved chaebol chieftains of some of their worst nightmares. The fear was that foreign interests, armed with the new regulations, would take over some of the largest industries, including motor vehicles and electronics.

That kind of fear lay behind a battle among bureaucrats that led Dow Corning on February 11, 1998, to give up a plan to build a huge chemical plant in Korea and go to Malaysia instead. The problems encountered by Dow Corning in negotiating tax and tariff concessions epitomized the difficulties foreigners faced in penetrating this society despite the professed desire to lure foreign investment. Foreigners were scouring the depressed economy in search of bargains but were either coming up empty or waiting to see whether rules and attitudes would shift after DJ's inauguration.

"The bureaucrats should change the way they handle foreign investors," said You Jong Keun, in his role as governor of North Cholla Province, where Dow Corning had considered investing more than $2 billion on a facility on reclaimed land by Korea's southwest coast. "I am very disappointed." The president of Dow Corning Korea, Kim Sun Mo, cited "inconsistent policies" among bureaucrats. The finance ministry opposed significant tax concessions of the sort usually granted big investors in foreign countries. It was also reluctant to approve reductions in the cost of electrical power and huge cuts in tariffs on materials that Dow Corning would have had to import for the plant, which would have made silicon. "There was not a good deal of cooperation and coordination," said You. "We have learned how to deal in such matters."[24]

Just how much the Koreans were learning, however, was far from clear to foreign investors and consultants. "Right now we've heard a lot of talk, and there's no real action," said Peter Underwood, a business consultant who had spent most of his life in Korea. "My concern is whether they really know what is a free market economy." Underwood, the great-grandson of the missionary who founded Seoul's prestigious Yonsei University, contrasted the difficulties American investors found in Korea with those of Koreans setting up facilities in the United States. "When a Korean company goes to Georgia, they are offered tax breaks along with a place to build," he said. "Here you don't get anything, and you're restricted in how much you can invest. This is not the place to begin if you want to go overseas. A lot of countries are easier than this one."[25]

A colleague who specialized in investment opportunities told of negotiations with a major chaebol over the purchase of one of its smaller companies. The asking price: $100 million. "We looked at their books and discovered the company was worth no more than $23 million," he said. "They thought they could make a killing off us. The chaebol say they are downsizing, and government officials say they are relaxing the rules, but finding your niche is going to be tough."

Defending themselves against the specter of hostile M&As, the chaebol began buying more shares in some of their largest companies. "There will be no hostile takeover at Hyundai," vowed a Hyundai source. The chaebol managed to water down the new rules on hostile M&As after fighting for weeks against a proposal by some of DJ's advisers for what they called "Big Deal," under which the chaebol would "swap" companies. "Big Deal is simply the exchange of items between chaebol," said a spokesman for Kim Dae Jung's transition team. Under one fanciful scenario, Samsung Electronics, the biggest electronics firm, would take over Hyundai Electronics, smaller and shakier financially; Hyundai Motor would get Samsung Motor, doomed even before beginning production.

Whoever bought out whom, foreign name signs were not likely to be shining above many of the big factories. Since the Korean War, foreigners had invested only $26 billion in facilities in Korea, a minuscule figure for an economy with a gross national product of $500 billion a year. You Jong Keun was confident, however, that the incoming administration would satisfy investors. "We need to create an entity that will be exclusively responsible for assisting them," he said. "Formerly we had such a thing as a one-stop service center, but it turned out to be just one more stop. It really didn't work."[26]

Kim Dae Jung at his inaugural on February 25 placed himself on a collision course with the chaebol by saying that reforms, including transparency, downsizing, and shareholders' rights, would be "carried out by all means." He also suggested that foreign investment might be just as important as chaebol restructuring. "Inducement of foreign capital is the most effective way to pay back our foreign debts, strengthen the competitiveness of business and raise the transparency of the economy."[27] As visible proof of the point, pop singer Michael Jackson, on the dais during the speech, hugged the new president afterward, then traveled to the nearby port of Inchon to inspect reclaimed land owned by the Dong Ah group.

Two days later, on February 27, Ssangbangwool Development said the singer had agreed to build a theme park for children called Neverland Asia in the new president's Cholla region. The theme park would emulate the Neverland Park that Jackson was establishing in his Neverland Valley Ranch in Los Olivos, California. Thus Jackson came to be the first potential investor from the names invited to the inaugural. One year on, however, the gloved one had invested nothing in Korea. The case symbolized the difficulties of bringing foreign and Korean interests together.

Financier George Soros was also on hand at DJ's invitation, considering a large-scale investment that members of DJ's team hoped would lead other foreigners to bet on Korea's future. Soros told Korea's leading business newspaper, *Maeil Kyungje,* that low stock prices and devaluation of the won would lure reluctant foreigners.[28] Soros was a better prospect than Jackson. Later in the year, he actually invested in a securities firm in Seoul.

Within hours after his inauguration, Kim faced a litmus test of his ability to govern effectively. At issue: would his followers in the National Assembly be able to persuade the majority—but opposition—party to approve Kim's choice of Kim Jong Pil, one of South Korea's most feared, longest-lasting political figures, as prime minister? DJ, in his inaugural speech, made clear his concern about approval of JP as well as economic reform legislation that he wanted to push through the Assembly within the first month of his presidency. No sooner had the dais cleared than the Grand National Party threatened to derail DJ's program by opposing JP's appointment.

Assembly members, after watching DJ's inauguration in the morning, took up the battle in the afternoon. Members of both DJ's National Congress for New Politics and JP's United Liberal Democrats hoped for a split in the GNP, which held 162 of the Assembly's 299 seats. With a combined total of 121 seats, the minority parties looked for support from splinter parties as well as the GNP to provide the 150 votes needed to ensure approval for JP. The fight would be tough if the leader of the GNP, Lee Han Dong, persuaded his members to observe party discipline and vote en bloc. Lee demanded "party unity," suggesting that party members blockade the entrances to the Assembly floor to prevent their opponents from voting at all.

Why, however, was DJ so dead set on JP? The simple answer, released in a pre-inauguration statement, was that "Kim Jong Pil had the experience to overcome all difficulties." DJ needed JP's United Liberal Democrats to push his legislation. Lee Hoi Chang, as honorary chairman of the Grand National Party, still the majority in the Assembly, signaled the showdown ahead by calling on DJ, the day after his inauguration, to name a substitute. The Assembly under the constitution had to approve whomever the president nominated by a majority vote. Failure to name a prime minister could delay creation of a new cabinet— and government.

GNP politicians planned to contest as unconstitutional any attempt to appoint Kim Jong Pil even as acting prime minister. The GNP at first boycotted the Assembly session on March 3 but returned to the floor after GNP president Cho Soon pledged, "We will fulfill our party stance with dignity through a secret ballot," and assured DJ the GNP would not block a vote. Pandemonium broke out, however, when members of DJ's National Congress for New Politics and JP's United Liberal Democrats charged that GNP members were casting blank ballots so JP would still not receive majority approval. "It's blank, it's blank," shouted one of the assemblymen, claiming GNP members were filing in and out of the voting booths on the edges of the Assembly floor too quickly to mark "yes" or "no."[29]

With that, members of JP's party dashed to the booths to prevent further balloting, and the fighting started. The floor quickly filled with black-suited men flailing at each other, shoving one another across desks, or, in some cases, pulling combatants apart, all broadcast live on all four television networks. "People are watching," shouted the Assembly speaker, Kim Soo Han. "Order, order. Please be dignified. Please be quiet." Dismissing the Assembly after midnight, Kim Soo Han told members milling on the floor, "In the history of the National Assembly, I have never seen such chaos." With 201 of the 299 members having voted, he ordered the ballot boxes sealed and guarded. "This is a prelude to instability," a commentator for the independent Munwha Broadcasting Corporation predicted as furious members left the floor and the Assembly adjourned.[30]

Still without a prime minister, the president named a new cabinet. DJ's choice for the all-important post of deputy prime minister and finance minister was almost as controversial as that of Kim Jong Pil. Reflecting JP's influence, he appointed Lee Kyu Sung, an economist who had served as finance minister under Roh Tae Woo. Lee said that under Roh he had recommended the measures now demanded by the IMF and the government—"to improve the transparency of the chaebol, to lower leverage, to eliminate cross-guarantees."[31] Lee was remembered, however, for having forced the Bank of Korea to loan the equivalent in won of more than $3 billion to Korean trust companies, which in turn had funded major listed companies facing bankruptcy. The rationale was that the companies had to have the money to bolster the stock exchange.

Lee advocated chaebol reform, but would he move as swiftly and as strongly as DJ had promised? "Lee's personality is quite contrary to Kim Dae Jung's reform program," said Suh Jin Young, a professor at Korea University. "I don't know what to make of that kind of inconsistency." Chung Kap Yung at Yonsei described Lee as "one member of that circle going back to the middle 1960s, the people who were claiming to build up Korea."[32]

The ruckus in the Assembly, however, was a sideshow quite apart from the problems of the economy. The market broke the 570 barrier on March 2 for the first time in four months, closing at 574.35. Investors were blasé about the furor over DJ's stubborn support of JP, and DJ's team denied any notion of telling the chaebol how to conduct their business in a manner reminiscent of the autocratic style of Park Chung Hee. "The government is not going to involve itself in restructuring the chaebol," said Assemblyman Kim Min Sok, acknowledging "confusion on this issue." Rather, he said, "we will make overfinancing more difficult." By enforcing banking regulations, the chaebol "will have no choice but to sell less necessary parts."[33]

The most talked-about method was the requirement that all the chaebol produce "consolidated statements" showing their companies' profits and losses. "Accounting systems will be much improved to meet international standards," said Bae Ie Dong at the Federation of Korean Industries. Wanting to sound cooperative, the FKI said the chaebol needed group planning offices "so they can continue to work out and implement group-wide restructuring plans as required by the next government."[34] The rhetoric was a prelude to another round of "restructuring" news.

SK Telecom, South Korea's telecommunications giant, yielded to intense pressure from minority shareholders on March 20 and promised to reserve two board seats for directors representing shareholders outside the inner circle of members of the owning family. The company, flagship of the SK group, surrendered to demands that were threatening to turn a shareholders' meeting on March 27 into a battle for shareholders' rights. Although SK Telecom was far from bankrupt, the company was struggling to pay its bills. The SK group was burdened by a debt-equity ratio of nearly 5 to 1, considerably above the national average of 3.8 to 1. SK's decision to appoint directors supported by minority shareholders was incorporated in a formal agreement with four funds united for their campaign under Tiger Management in New York. The group, including the Korea Fund, the Oppenheimer Global Fund, the Oppenheimer Valuable Account Fund, and the TEI Fund, held nearly 10 percent of SK Telecom's stock. The agreement also included the right of minority shareholders to name auditors. The purpose was to achieve the "transparency" needed for understanding profits, sales, losses, and expenses.

The trend was catching. Samsung Electronics, global leader in semiconductors, planned to name a foreign director in anticipation of outside pressure. Shareholders' groups were going to court in pursuit of their rights at other companies. Outside directors now would tell company chieftains "to start rewarding shareholders rather than friends and relatives," said Daniel Harwood at ABN AMRO.[35] Their presence might inhibit such habits as diverting funds from company to company within a group, exploiting the facilities of one company for the benefit of another, and favoring companies within the same group with contracts.

How to deal with companies or entire chaebol on the edge of bankruptcy was one of the most troublesome issues before the chaebol and the government. Lee Jae Woo, an economist with the FKI's Korea Economic Research Institute, predicted that established founding-family shareholders would lose more and more power as they sold assets to investors in order to get the cash to pay off their debts. "People are worried a lot about change," he said. "In the past the controlling shareholders had all the power. Now the characteristic of Korean management is changing, voluntarily or involuntarily. They have to adjust."[36] Most of the companies considered on the brink, unable to pay their bills, at the mercy of the courts, were still humming away, often quite confident that real failure was only a bad dream. One great risk was that companies forced out of business would never repay their loans. The corollary risk was that the owners would lose their entire investment, which they persisted in thinking they could recoup if the crisis would just go away.

The fear of chaebol failure translated into a cumbersome system, based on the Japanese model, in which a company or group, when unable to pay its bills, applied for court receivership. The court might appoint a new chairman but then take its time deciding whether to accept the application for receivership, under which a company renegotiated its loans with creditor banks. Officials might state that troubled banks and corporations must be liquidated, but they could do little to hurry up the system.

One precedent was the court's decision to grant receivership to Halla Heavy Industries. The court rejected an application from a sister company, Halla Merchant Marine, but the Halla group appeared likely to survive despite its 20-to-1 debt-equity ratio. Even where the court ordered liquidation, dissolution was far from inevitable. Hanbo Steel, part of the Hanbo group, whose top executives were jailed in the 1997 bribery scandal that ensnared Kim Young Sam's son, still produced specialty steel. The danger was that companies might view the courts as offering insurance policies against failure. After Midopa, the department store chain owned by the Dainong group, applied for receivership, Midopa stores remained open as usual. Nobody expected them to close soon. The availability of low-interest "cooperative" loans harked back to the days when central planners controlled the credit from banks. The government in the 1990s had pulled back from direct intervention but held final power through the banks.

Bureaucrats now were more concerned than ever by the growth of the top five in contrast to groups down the list and the failure or mediocre performance of small and medium-size enterprises. Figures provided by the Fair Trade Commission in 1997 were cause for alarm. While the top five had all made net incomes in the billions of dollars the year before, four of those ranked fifth through tenth in total assets suffered income losses. Sixth-ranked Ssangyong lost more than $120 million in 1997, most of it from motor vehicles, while seventh-ranked Kia lost $150 million. Among the "top 30" chaebol, 13 were in the red for the year. "The top five are getting bigger," said Kang Hee Bo, who had served as a special assistant on such matters under Kim Young Sam. "The lower ones are not doing so well."[37]

Repercussions of the bankruptcy of Hanbo, the fourteenth-largest chaebol, were still reverberating. The greatest fear was that some of the other chaebol were about to go under. "There is a showdown in business," said Park Nei Hei, economist at Sogang University. "The biggest chaebol can extend their business enterprises like an octopus."[38] For all the talk about slashing costs, downsizing, and merging, the chaebol now were digging themselves deeper into debt. The ratio of debt to equity for the 30 largest chaebol by March 1998 was soaring above five to one as the chaebol defied official efforts to persuade them to stop tapping their banks for low-interest loans.[39] The chaebol had to replenish accounts amid declining domestic sales, reduced income, and the need for more loans to stay afloat.

The mounting debt problems of the chaebol emerged as a 14-member IMF team negotiated with the finance ministry on fine-tuning the bailout package. Government officials on April 17, 1998, pressed for reducing the high interest rates that the IMF had set as a prime requirement. The government position was that interest rates, now about 18 percent, discouraged the borrowing needed to keep companies from going bankrupt. Since December, an average of 2,000 firms had gone bankrupt each month. Pleading that "a reduction in the rate is urgent," Finance Minister Lee blamed high rates for having "threatened to push scores of economically viable companies into bankruptcy" while driving up unemployment. Finally, he managed to wrest the IMF's reluctant approval. "Interest rates will continue to be lowered in line with market conditions," said a memorandum that the finance ministry said was worked out with the IMF.[40]

Lee embarrassed the government, however, by seeing the World Bank as a source of ever more billions for the chaebol. In vain, he begged the World Bank to plunge several billion dollars into funds for reducing the mounting debts. Arguing that Korea's "current corporate debt is simply not sustainable," he let it be known in mid-April that the government wanted to set up two such funds worth ten trillion won, about $7 billion, with the help of "multilateral and bilateral donors." One fund would provide "equity participation" while the other converted short-term into long-term debts. The funds, he said sanctimoniously, would be "subject to strict accounting reviews" and would not be "available to insolvent companies."[41] Sri-Ram Aiyer, the World Bank's Seoul representative, noted that the World Bank president, James Wolfensohn, had decreed in Washington that he did not want "public funds used for bailing out industries."[42]

The chaebol blamed the rising ratios of debt to equity in large part on the devaluation of the won, now worth about 1,400 to 1 dollar. "Mostly the debts result from a rapid increase in foreign exchange losses," explained Bae Ie Dong of the FKI. He acknowledged, however, that the chaebol relied on "cooperative loans" from friendly banks. Typically, the interest on a "cooperative loan" was 6 percent below the benchmark three-year bond rate. Lee Jae Woo, at the Korea Economic Research Institute, said the government's plea for reducing the debt-equity ratio to two to one was "ridiculous."[43] The government tried to appear flexible while remaining firm. Kim Tae Dong, a senior secretary at the Blue House, said the chaebol should "maintain the internationally recognized standard for debt ratio," not any specific requirement.

The Hyundai group typified the pattern of the top-30 chaebol with a ratio of 578.7 percent, nearly six to one, up from about four to one a year earlier. One reason for Hyundai's rising debt-equity ratio was Hyundai Motor, its largest entity. Car sales on the domestic market had fallen by 50 percent, and the company was now attempting to slash its 40,000-person workforce through an early-retirement scheme that had met with strong opposition from its union. Hyundai's debt-equity ratio was the worst of the big five but considerably better than that of other groups on down the list. By spring 1998, the debt-equity ratio of the Hanjin group, Korea's sixth-largest chaebol, was 909.8 percent, more than nine to one. Losses incurred by Korean Air, its largest entity, were primarily responsible for Hanjin's precarious condition. Hanwha, eighth-largest, suffered a debt-equity ratio of 1,214.7 percent, more than 12 to 1, largely as a result of the problems of its oil refinery.[44]

Again, the chaebol issued great pronouncements to convince the government and the public that they were doing the right things. The Samsung group on May 6 announced a plan to reduce its "core businesses" from ten to four or five while looking for $5 billion in foreign investment. The group also said it would reduce its debt-equity ratio from the current level of 350 percent to 124 percent in five years while paying off about 29 trillion won, more than $20 billion, in debts. The Samsung plan on one level appeared as a serious attempt at coping with the worst problems. By the vagueness of its wording, however, the plan betrayed difficulties in grappling with the need to shed prestige companies. It avoided the issue of the future of its newest, most troubled entity, Samsung Motor, while listing "core businesses" as electronics, finance, and "services," unspecified. At group headquarters, now called a "restructuring team," and at Samsung Motor, officials were not certain if Samsung Motor fell under the category of "services."[45]

Critics questioned the willingness of the chaebol to engage in reforms despite intense government pressure. "The chaebol haven't shown anything yet," said Jang Ha Sung, professor of finance at Korea University. "The government has not found the means to implement its policy." Jang, as chairman of a committee of minority shareholders challenging the chaebol at board meetings, was suing Samsung for giving shares to the son of Samsung Chairman Lee Kun Hee. He charged Samsung with falsifying records of a director's meeting, claiming that Samsung Motor had amassed 4.7 trillion won, about $3.3 billion, in debts and had only 800 billion won, approximately $500 million, in capital. At one meeting of Samsung shareholders, he alleged that a dummy company in Ireland had loaned Samsung Motor $300 million in funds that had come from Samsung Electronics.[46]

Not to be outdone by Samsung, three more of the top five chaebol announced massive restructuring plans on May 7, 1998. The announcements, however, elicited more jeers than cheers. Hyundai, LG, and SK said they were seeking a total of $20 billion in foreign investment, $8.5 billion for Hyundai, $6.5 billion for LG, and $5 billion for SK, while cutting down the size of their groups. Basically, they were talking about merging affiliates with affiliates, forming bigger affiliates with perhaps new names. A number of other chaebol were expected to announce similar plans soon in answer to demands from the government for point-by-point proposals for cutting back on hundreds of billions of dollars in debts.

So far, however, the restructuring plans of the chaebol confirmed their reluctance to engage in IMF-style overhauling. As the central point of restructuring, Hyundai cited a plan to separate nine subsidiary companies from "core" companies including construction, motor vehicles, shipbuilding, finance, and petrochemicals. Hyundai had announced a similar plan five years earlier but had never carried it out after an initial flurry of publicity. Sons and relatives of the Hyundai founder and honorary chairman, Chung Ju Yung, would remain in charge of companies that were no longer technically part of the group.[47]

The vagaries of business and politics inspired widespread cynicism. Candidates for local and provincial elections on June 4, 1998, were unnoticed by most people. Choi Jong Young, chairman of the National Election Commission, cautioned the nation's 32 million eligible voters against giving up "your rights by hiking, fishing, or going on a trip" on the day off set aside for voting.[48] For many, however, constant debate among politicians, bureaucrats, and business leaders about what to do to resolve a possibly worsening economic crisis had inured them to almost anyone running for any office.

Policymakers and politicians saw the election as a test of the popularity of Kim Dae Jung. Although the election had no direct bearing on the makeup of the central government or the National Assembly, DJ wanted an endorsement of his policies. He had reason to be concerned after sternly opposing sporadic strikes by militant workers fearful of losing their jobs. Aides admitted he would be "a little embarrassed" if his National Congress for New Politics fared badly as he was preparing to leave on June 6 for a nine-day visit to the United States, highlighted by his first summit with President Clinton. He wanted his hosts to believe he was in control.

Everywhere, however, apathy remained the worst enemy of the 10,222 candidates, running for governor of nine provinces, mayor of seven metropolitan cities and 232 small cities, and more than 4,000 seats on provincial councils. Much to the government's relief, candidates of both the National Congress for New Politics and Kim Jong Pil's United Liberal Democrats fared fairly well, winning a majority of the races with a voter turnout of 50 percent. The fragile coalition might yet come apart as JP, once approved as prime minister, angled for more than ceremonial power, but the government now could focus its struggle against the chaebol and the vague forces of other hard-hit Asian economies.

Emboldened by the success of his ruling party in the local elections, Kim Dae Jung the next day, June 5, 1998, escalated his campaign against money-losing companies sheltered by the bloated chaebol. "Business is for profit," said the president, savoring especially the triumphs by his party's candidates in mayoralty and gubernatorial contests in metropolitan Seoul and Inchon, a megapolis that was home of nearly half of South Korea's population. "If they are not making profits, it is meaningless." Those words constituted a blunt warning that the chaebol had to stop their delaying tactics and dump scores of companies blamed for weighing down the economy. Talking to journalists on the one hundredth day after his inauguration, DJ insisted "the government will not intervene in business" but called on banks to discipline the chaebol through their power over credit. The government could force compliance by managing infusions of funds from the central bank, buoyed by IMF loans.[49]

DJ endorsed a demand by the Financial Supervisory Commission for the banks to include money losers from the top five chaebol on a blacklist of companies no longer eligible for loans. The president's tone bore out reports that he was angered when he learned that the list included companies from none of the leading chaebol even though a number faced serious difficulties. Hyundai Electronics and Hyundai Engineering and Construction, both core entities of the Hyundai group, and Samsung Motor, the biggest loser from the Samsung group, were examples of companies dragging down their groups. Despite Kim's remarks, the banks' hit list would never touch companies of such size and prestige.

South Korean companies, always behind on research and development, now jeopardized all they had achieved by economizing. Nowhere was the struggle more critical than in semiconductors. The transition from 16-megabit to 64-megabit chips assumed special importance for Korean manufacturers, who had to advance to compete with the Japanese but were cutting back on R&D. A revolution was exploding beneath the bland exterior of the desktop computer. The era of the 16-megabit DRAM (dynamic random access memory) semiconductor was fading. Manufacturers of memory chips were anxiously awaiting "bit-cross," the point at which it would be cheaper to install a single 64-megabit DRAM into a computer than to load it with four 16MB chips.

For South Korea's three major semiconductor manufacturers, Samsung Electronics, LG Semicon, and Hyundai Electronics, bit-cross could not come soon enough. "All Korean manufacturers are now losing money on memory products," said Choi Hye Bum, an analyst with the Korea Semiconductor Industries Association. "They've experienced a 70 percent decline in two years. They have to move to the 64MB to survive."[50] Samsung Electronics, amid mounting losses from the 16MB, was waging a tough battle with NEC of Japan, the second-largest memory-chip maker, in hopes of profiting again from semiconductors by 1999. "We're ahead of our competitors by about one year," said Chung Eui Yong, marketing director for Samsung Electronics. "We are already producing eight million units of 64MB chips a month. We hope high-density demand will increase beginning in April."[51]

Economic turmoil in Korea, however, forced Samsung Electronics to slash its research budget from $2 billion in 1997 to somewhat more than $1 billion in 1998. Korean semiconductor manufacturers, which produced 38 percent of the world's semiconductors, would need all the edge they could muster against Japan's NEC, Toshiba, Fujitsu, Hitachi, and Mitsubishi Electric, which accounted for 40 percent of the market. Although Samsung ranked first in production of memory chips in 1997, it was seventh in total semiconductor manufacturing, with sales of $6 billion for the year. Intel, focusing on nonmemory chips, led semiconductor manufacturers with $21.1 billion in sales, more than twice that of seventh-ranked NEC, whose sales totaled $10.7 billion. LG Semicon, the only other Korean semiconductor manufacturer in the world's top 20, ranked nineteenth with sales of $2.1 billion. The 64MB DRAM provided the best hope. Both LG Semicon and Hyundai, making 3 million and 5 million 64MB DRAMs in March 1998, planned to increase output to 10 million 64MB DRAMs a month by 1999.[52]

Cost cutting was dangerous. LG was three months behind in construction of a new semiconductor plant in Wales and would not have it ready for production until 1999. Hyundai, having suspended plans for a $2 billion plant in Scotland, was far behind in opening a $1.3 billion plant in Eugene, Oregon. How soon before the 64MB DRAM, like the 16MB

DRAM and the 4MB DRAM that preceded it, were history? Already Samsung and NEC were producing the next generation of memory chips, the 128MB DRAMs, in limited quantities. Would Korean companies have time and funds to develop the technology for another transition?

As if Korea's own economic turmoil were not enough, the depreciation of the Japanese yen around the same time added urgency to the drive to trim the losers. There was an instinctive sense that the Japanese were conspiring to destroy Korea's major industries while Korea was searching for foreign investment. "The Japanese government is now seeking to boost the Japanese economy through increasing exports," said Ohn Un Ki, senior research fellow at the Korea Institute for Industrial Economics and Trade, a government research organization. "A sharp decrease in the Japanese yen will kill Korean industry." Ohn's remarks, on June 10, 1998, reflected a widespread Korean view that Japanese government and business wanted to drive down the yen against the dollar so Japanese companies could reinvigorate their own stagnant economy by defeating foreign competition on world markets.[53]

Now Koreans saw the hard-won gains of the past generation of competition with the Japanese evaporating with every move downward of the yen. Exporters hoped that the won would fall in tandem with the yen, counterbalancing the impact of a declining yen. Hardest hit would be shipbuilding. Hyundai Heavy Industries, already diverting profits to other companies in the group, might suffer declining orders as buyers turned to Japan. Lesser Korean shipbuilders, including Samsung Heavy Industries and Daewoo Heavy Industries, faced more severe problems. Halla Heavy Industries hung on through funds from HHI and a resurgence in orders before the yen had begun its decline.

Korea's motor vehicle industry could not survive in its present form even without yen depreciation. The Daewoo founder and chairman, Kim Woo Choong, who had invested several billion dollars in expanding Daewoo Motor at home and abroad, conceded on June 9 that the industry was "overinvested" but said manufacturers should "focus on exports and reduce domestic production." As prices of Japanese motor vehicles gradually decreased worldwide, however, would large foreign companies still be interested in investing in Korea, and could Korean vehicles still compete overseas, where they were far overshadowed by the Japanese? "The export of motor vehicles falls 1.6 percent when the yen depreciates by 1 percent," said Kim Chang Uh at the ministry of commerce.[54]

Yen depreciation was immediately visible in June in the widely varying response of Korean versus Japanese semiconductor manufacturers to the worldwide glut of memory chips. On June 3, Hyundai Electronics suspended production at its plant at Ichon, south of Seoul, sending all 7,000 workers home for a week. "We're reducing, not stopping, production," said Park Young Man, a manager in the memory division.[55] "Hyundai Electronics is a real high-risk company," said Del Ricks, head of research for ABN AMRO in Seoul. "It's got a debt-to-equity ratio of nine to one. It's been borrowing heavily."[56] Samsung Electronics was not in such trouble but suspended its line for a week. LG Semicon, like Hyundai suffering heavy losses, also planned cuts. Samsung, Hyundai, and LG produced one third of the world's memory chips in 1997, but prices for 64-megabit and 16-megabit chips had dropped 50 percent since January.

The prospect of angering trading partners by action that might speed yen depreciation was a consideration in the Bank of Japan's reluctance to lower interest rates still further. At the same time, the Japanese did not view a rise in interest rates as an option even though they admitted it would arrest the decline of the yen. Yoshio Nakamura, managing director of the economic bureau of Keidanren, the Japan Federation of Economic Organizations, did not believe that low-priced Japanese products would force the collapse of Korean companies. "Trade friction results from problems in the other countries," he said. "If you have a very good economy, there is no political reason for collapse. What is important is the economic welfare of the country, not the trade imbalance. If people can buy reasonably priced goods, people enjoy themselves."[57]

Spurred on by the threat from Japan, authorities on June 18 decided on a list of 55 firms deemed "nonviable," ineligible for credit, and ready for liquidation.[58] Analysts denounced the list, however, for failing to include the names of any of scores of major companies also in severe difficulty. "In the end even the president had to get involved," said David Kim, senior researcher at Indosuez W. I. Carr, "but the actual list did not contain one major company." Indeed, said Kim, "I don't recognize two thirds of the names."[59] Lee Hun Jai, chairman of the Financial Supervisory Commission, was apologetic after he and the president of the Commercial Bank of Korea released the names. Although 20 affiliates of the country's top five chaebol were included, Lee admitted "the results of evaluation of the top five were a little unsatisfactory."[60]

Lee said that the Financial Supervisory Commission and creditor banks would "prepare follow-up measures for restructuring," notably "exchange of business operations" among chaebol. That observation confirmed rumors of negotiations under which large entities would merge. According to one report, the Hyundai group might yield Hyundai Petrochemical to LG Petrochemical. At the same time, LG would yield LG Semicon to Hyundai Electronics or Samsung Electronics. Samsung in turn would yield the fledgling Samsung Motor, which had sold only 7,090 vehicles by the end of May, to Hyundai Motor or Daewoo Motor.

The question raised by the choice of only minor companies, many of which would have gone out of business under any circumstances, was how quickly the government would move on the promised restructuring. Noting that the 55 companies accounted for 25 percent of corporate loans as of the end of March, Edward Campbell-Harris at Jardine Fleming Securities asked, "What about the other 75 percent of loans?" Campbell-Harris wondered whether the companies on the list would really be denied credit, as demanded by the Financial Supervisory Commission with the concurrence of the banks. "We've all been waiting for this hit list," he said, "What does it mean? Do these companies close their doors?" Only ten, he noted, were listed on the stock exchange.[61]

The chaebol, meanwhile, were relieved that the list included none they wanted to keep. Shares of Samsung Aerospace, rumored on June 17, 1998, as a candidate for extinction, rose by 265 won, closing at 2,475 won, when word leaked on the morning of the 18th that it was not on the list. "Everybody wanted to buy Samsung Aerospace shares after the rumor went around that no Samsung company on the stock exchange was included," said a Samsung official.[62] Only one of the four Samsung companies on the list, Samsung Watch, even bore the Samsung name.

The largest Hyundai company on the list, Hyundai Livart, which manufactured furniture, would become a division of Hyundai Engineering and Construction. Hyundai Aluminum, also on the list, was already becoming a unit of Korea Industrial Development, another Hyundai company, while Hyundai's Suneel Shipping was being absorbed by Hyundai Merchant Marine. Park Dae Shik, manager of the international business team of the FKI, indicated strong resistance to further intervention. "I don't think it is reasonable for the government to directly intervene," said Park. "I don't think this kind of announcement can solve any problem. It is just symbolic."[63]

The recalcitrance of the chaebol was disturbing to American officials, who wanted Korea to abide by the agreement with the IMF, an institution that often served as an instrument of American policy. Treasury Secretary Robert Rubin, visiting Seoul on July 1, 1998, on the last stop of a four-country swing through Asia, met with some of the wealthiest tycoons, and asked them if they were considering a "Big Deal" for getting rid of money losers. He received an ambivalent reply from Daewoo Chairman Kim

Woo Choong, newly named chairman of the FKI. "It will take a long time," replied Kim. "We have many other technical problems to discuss in detail." When Rubin told Kim that many foreign business people were waiting to see how the chaebol were doing on reform, Kim replied, "We're moving as quickly as we can," but gave no details.[64]

The treasury secretary's session with chaebol leaders marked a low point in an otherwise upbeat day. Kim spent most of the meeting complaining about the difficulties of obtaining credit and justifying the debt-equity ratios of the chaebol. Predicting that the chaebol would reduce the ratios to 200 percent by 1999, Kim said the average ratio was 360 percent before the crisis—far too low a figure. After intervention by the IMF, Kim argued, "Our foreign exchange appreciated and the debt ratio increased to 500 percent." Otherwise, he maintained, the debt ratio would be "about 300 percent"—another distortion. Kim asked Rubin if he might help in a scheme for foreign banks to coordinate with Korean companies in streamlining the banking industry—a ruse to funnel more money into the chaebol.

Under such prodding from the west, the government wanted to move quickly to show its seriousness about reform. DJ's advisers were already developing a plan to completely privatize the showcase Pohang Iron and Steel, Posco, respected for sound management, and four other large government-invested entities, Korea Telecom, Korea Heavy Industries and Construction, Korea Gas, and Korea Tobacco and Ginseng. Park Tae Jun, the Posco founder who now served as a leading economic adviser as well as chairman of the United Liberal Democrats, said the government wanted to give up its stake in Posco. The government owned 19.57 percent of the company, and the government's Korea Development Bank owned 14.6 percent. Officials were talking about a special offering overseas of shares in Posco after deciding they were not likely to find enough buyers on the domestic market while Korean banks and companies struggled to meet debt obligations.

On July 3, 1998, two days after Rubin's visit, the Planning and Budget Commission said the government would privatize 11 state-owned companies. The state's shares in Posco and the four others, plus their subsidiaries, would go on sale "immediately."[65] Another six would be privatized over a four-year period, with the government hoping to earn as much as $8 billion by 2000. The announcement set off a wave of speculation about how and when the government would carry out the program and who would be the most interested buyers.

News that the government would sell its entire stake in Posco shocked workers at the company headquarters in Seoul. "It's a small revolution," said Lee Jong Sung, an assistant manager. "We hardly expected that."[66] Ultimately the possibility loomed that Posco, founded in 1968 as the country's single most prestigious industrial project at the height of Park Chung Hee's rule, could fall under foreign control. The plan called for the government to withdraw restrictions on individual as well as overall foreign ownership of Posco by the year 2001. The first stage called for raising the limit on individual ownership of Posco shares from 1 to 3 percent while foreign ownership would be held to 30 percent.[67]

The decision to sell the government's stake in Posco and other government-invested companies, some of them criticized as extremely inefficient, reflected not only the government's need for cash but also an evolving free-market philosophy. Privatization of national companies, like the shutdown of money-losing private firms and the closure of heavily indebted banks, fit in with the IMF prescription. Privatization, however, drew quick criticism from labor unions, fearful that investors, seeking to turn marginal operations into profitable ones, would lay off many of the 150,000 people working in the state-invested companies and their subsidiaries. The Korean Confederation of Trade Unions said the plan would be the death knell for an agreement reached in January among government, business, and labor leaders.[68] "Unite and fight," said the headbands and signboards held by several thousand fist-waving workers in the vast square in front of Seoul Station.

Most worrisome to labor was the plan for issuing 10 percent of the shares of Korea Telecom, 71.2 percent owned by the government, to a single "strategic investor" that would exercise effective control over the company. The plan also called for offering up to 30.6 percent of Korea Telecom shares in a public offering with 18 percent for foreign investors and the rest for Koreans. Finally, the government would offer the remaining 33.4 percent in its hands for general sale after 2001. "The only way for investors to make money is through manpower cuts," said Kim Ho Sun, president of the union at Korea Telecom, with 58,556 people on its payroll—the largest of any of the companies affected.[69]

The first question, however, was exactly when the government would begin selling its shares in Posco and the four other companies in which it said it would sell its stake "immediately." A spokesman for Posco, capitalized at $2.6 billion, with 19,294 employees, said the company had received no directives. The four other companies in which the government would sell its shares "immediately" were Korea Heavy Industries and Construction, Korea General Chemistry, Korea Technology Banking, and

National Textbook. The commission outlined formulas for unloading the government's equity in the six other companies, Korea Telecom, Korea Electric Power, Korea Gas, Korea District Heating, Daehan Oil Pipeline, and Korea Tobacco and Ginseng, which controlled local tobacco as well as ginseng, a root said to have aphrodisiacal qualities. The plan for Korea Tobacco and Ginseng sought to answer complaints of foreign tobacco companies by removing the company's "monopoly of cigarette production rights in the year 2000."[70]

Next on the list of priorities was the need to stimulate the economy in a hurry as jobless rolls increased daily. The finance ministry on July 5, 1998, called for raising the budget deficit in order to carry out public works projects and spread money through an economy mired in a slump that was now worse than anticipated. The ministry said the deficit under its new program would have to be 4 percent of South Korea's gross domestic product rather than 1.7 percent, the figure on which officials had settled with the IMF in April. The difference in percentages meant that the deficit would rise from 7.8 trillion won to 17.5 trillion won, nearly 10 trillion won or $7.2 billion.[70]

Ministry officials acknowledged that the plan to increase the deficit represented an abrupt revision of the earlier deficit ceiling worked out with the IMF. The ministry disclosed its plan amid reports that the gross domestic product for the second quarter was down 4 percent from the second quarter of 1997. The GDP had decreased 3.8 percent in the first quarter of 1998 from the first quarter of 1997. The revised budgetary needs were revealed after DJ and chairmen of the chaebol agreed on a pact obligating the government to pump ever more funds into their beleaguered organizations. In return, the chaebol leaders agreed, as always, to move faster toward "Big Deal."

"The government will do its best to ensure the stable supply of funds to finance exports and imports," said point one of the nine-point agreement. The agreement pledged an extra 500 billion won, about $350 million, in credit guarantees and another one trillion won, about $700 million, for credit for expanding exports and imports by small and medium-size enterprises. Government and industry leaders disagreed, however, on a basic question: where to get all the money? Daewoo's Kim Woo Choong was only too glad to expand on the concept he had pushed with Rubin for a megabank with foreign partners. Financial Supervisory Commission Chairman Lee Hun Jai warned against building such a bank on the basis of ever more loans and demanded full disclosure of the sources of funds.[72]

Signs of the stabilization of the won provided little comfort. Once the victim of runaway depreciation, the won had lately gained so much against the dollar that economists feared its strength was dragging down recovery by discouraging import of capital equipment and raising the price of exports. The won closed on July 9 at 1,307 to the dollar en route to 1,200. "The corporates have a lot of dollars because of the decrease in imports," said research analyst Leland Timblick. "The only thing they have been importing is a small amount of raw materials they absolutely have to have."[73] The government was doing all it could to stimulate exports and slash imports. "Domestic consumption is shrinking so much, companies think the best strategy is not to spend money on plant investment," said Philip Uhm, analyst at ABN AMRO. "Exporters are accumulating foreign currency cash. They don't need foreign currency as long as the foreign banks roll over their debts."[74] The reluctance of companies to import anything they did not need right away was one reason foreign exchange reserves had shot up from $7 billion in December to $47 billion while imports in June were $7.78 billion, down 36.6 percent from June 1997.[75]

With corporate debt estimated at more than $600 billion and hundreds of companies on the brink of bankruptcy, however, the government again called on banks to lower interest rates. Four of the biggest, Cho Hung Bank, Commercial Bank of Korea, Hanil Bank, and Korea Exchange Bank, cut their prime rates by 1 percent on July 9, and the Kookmin Bank cut its rate by 0.5 percent. The rate of the benchmark three-year corporate bond was now 13.7 percent after having reached 30 percent in December 1997. The government also sought to stimulate the economy by announcing a supplementary budget of $4.5 billion. With the government forecast to operate at a deficit in 1998 of about 4 percent of the gross national product, planning officials hoped to make up part of the difference by issuing state bonds valued at nearly $6 billion.

Economists realized the crippled economy was likely to decline by 6 percent for the year rather than the target of 4.6 percent set by the IMF and the Bank of Korea. On August 10, 1998, researchers at the bank said they expected the economy to dip more than expected under the impact of one of the worst storms of the century as well as delays in restructuring major industries.[76] As if all that were not enough, the stock market on August 10 closed at 311.95, its lowest level since July 14. Analysts attributed the fall largely to fears that China would devalue its currency, making Chinese products far more competitive abroad. Figures for the second quarter, released by the Bank of Korea on August 27, confirmed the estimates. The period from April 1 through June 30 showed a 6.6 percent drop in gross domestic product from the same quarter of 1997. It was the worst quarter since the last three months of 1980, the year of a power shift climaxed by the Kwangju revolt. The gross domestic product dropped 5.3 percent in the first six months of 1998 for the biggest half-year decline since the bank began reporting such statistics after the Korean War.[77]

Among the hardest-hit areas was manufacturing, on which the country relied for the exports that accounted for steady economic gains since the war. With factories running at only 63.7 percent of capacity, manufacturing declined by 10 percent in the second quarter after falling by 6.4 percent in the first quarter. Construction, a bellwether of economic health, dropped 12.1 percent in the second quarter after falling 7.5 percent in the first quarter. Yet another sign of the times was an 8.5 percent drop in wholesale and retail earnings, on top of a 5 percent decline in the first quarter.[78] The negative statistics reversed more than 30 years of almost unbroken progress in which the gross domestic product had shot up by an average of 8 percent a year, evidence of the "Korean miracle." Even in the last quarter of 1997, as the economy slipped into crisis, the GDP had risen by 3.9 percent. The country was in recession if recession was defined as two negative quarters in a row.

The government on September 2, 1998, revealed a broad package for pumping new life into a dangerously sluggish economy. Finance Minister Lee, presenting the plan, warned that "increasing economic uncertainty and the continuing credit crunch are feared to be leading to a vicious spiral of recession."[79] The plan called for investing heavily in government projects, easing credit restrictions, lowering interest rates (again), and increasing the money supply in a bid to stimulate everything from industry to consumer spending to real estate and banking. The plan also sought to reverse a downward trend in corporate investment by extending $4.7 billion in new loans, including $2.7 billion from the Korea Development Bank and $2 billion from the U.S. Export-Import Bank. One of the plan's top priorities was to encourage consumer spending, down 12.9 percent in the second quarter. The plan also sought to bolster small and medium-size enterprises by raising the total ceiling on loans by nearly $2 billion.

"We should not be too cautious in this time of a rapid deflationary spiral," said economic adviser You Jong Keun, drawing analogies from the Vietnam and Gulf wars. "We should not replicate the U.S. policy of a piecemeal reaction. We should take the Desert Storm approach. If you commit enough resources, you win the war."[80] The top five responded by agreeing on mergers that fell short of the "Big Deal" demanded by government. Hyundai Oil would take over Hanwha Energy. Hyundai Heavy Industries was prepared to join Samsung Heavy Industries and Daewoo Heavy Industries in producing ship engines. Hyundai Precision & Industry would join Daewoo Heavy Industries in producing railroad cars. Hyundai Electronics' semiconductor division would merge with LG Semicon.[81]

The merged semiconductor operation had the potential of surpassing Samsung as global leader in semiconductor production, capable of producing more than 300,000 DRAM chips a month. The companies, with debt-equity ratios of about eight to one, employed about 15,000 people in their semiconductor plants, 8,000 at Hyundai and 7,000 at LG, even though LG had a larger capacity. Executives wrangled, however, over exchange of equity. Hyundai wanted to assume basic control, which LG was reluctant to relinquish. "They agreed in principle but not in detail," said Jwa Sung Hee of the Korea Economic Research Institute. Jwa was "pessimistic" about trade-offs in general. "The private sector has come up with this plan only to comply with the government," he said.[82] Media were also skeptical. There was, Munwha Broadcasting observed, "no big trade among companies."[83] The result, said KBS, "was not the expected Big Deal."[84]

While calling on the chaebol to downsize, however, the government persistently demanded a decrease in interest rates. DJ coupled lower rates with a wide-ranging program to pump billions of dollars into the stagnant economy. With a current account surplus expected to reach $37 billion by the end of the year, DJ on September 28 said the government had enough funds to inject $8.7 billion into banks and spend billions more on public works. The result would be to stimulate demand and provide work for some of the unemployed, expected to crest at nearly two million. No one believed the economy could plummet to the levels of December 1997, when reserves were $3.8 billion and the current account deficit was $2.2 billion. "We have stable consumer prices, and our exchange rate is stabilized," said DJ. Korea had enough to pay back $36 billion in loans due in 1999 and still increase its reserves and current accounts surplus.[85]

The top five chaebol sopped up more than 80 percent of available credit, while lower-ranking chaebol got the rest and small and medium-sized enterprises were lucky to be able to roll over loans. "The laws and systems have changed, but they haven't brought results," said Jang Ha Sung at Korea University. "The top five are getting bigger and getting higher debts."[86] Factories were now operating at 62.9 percent of capacity, the lowest level since the National Statistics Office first revealed such figures in 1985. The sense of panic ranged from industries such as motor vehicles, where production in August fell by 60 percent from the year before, to finance, where five banks had shut their doors and six more had agreed to merge.

The World Bank added to the pressure. The government agreed on October 2, 1998, to a $2 billion loan package from the bank but had to accept conditions to force the chaebol to move. The loan required prompt action on such topics as shareholders' rights and the right of independent directors to sit on boards. "Korea's problems are across the board," said the World Bank representative, Sri-Ram Aiyer. "Our rule is to influence them, to tell them what's happening in the best countries, to make them do things a little faster."[87]

The ministry of finance agreed on some of the conditions set by the World Bank after tough negotiations. One of the hardest-fought debates focused on efforts by the bank to obtain the government's agreement to sell all bad loans, worth several hundred billion dollars, at huge losses over the next three years. Instead the finance ministry committed itself to getting rid of more than half of such loans. In a letter to World Bank President Wolfensohn, Finance Minister Lee said the government would "limit 'emergency' loans," a primary device for extending credit to nearly bankrupt companies. Lee also said the government would "reduce cross-guarantees" and would not offer "special support," demanded by chaebol chairman, as a reward for selling off entities.[88]

The access of the major chaebol to credit gave them the confidence not to cooperate with government demands. Losing patience, the government attacked the chaebol on October 7 for a new "restructuring" regime that it said had "failed to ensure the satisfactory execution of previously announced goals." The FSC said the chaebol had fallen short in "improving debt structure, eliminating duplicate and excessive investments, and the ultimate aim of significantly enhancing industrial competitiveness." The commission denounced the plan after the chaebol said they had not agreed on joint operations in three major fields.[89] The government viewed the chaebol announcement as an effort at putting off mergers of heavily leveraged companies floating on emergency loans. "The government will not just sit at the sidelines and wait for the chaebol to come up with specific plans," said the commission, threatening "liquidation or sell-off of marginal companies," "suspension of loans," and "claims on guarantee liabilities."[90]

The largest chaebol, the Hyundai group, was involved in the worst failures cited by government officials. Hyundai and LG could not come to terms on control of a merged semiconductor company after agreeing that one should acquire 70 percent, the other 30 percent. Then Hyundai Precision backed out of the plan to form a consortium with Daewoo Heavy Industries and Hanjin Heavy Industries for producing rolling stock. Hyundai Heavy Industries failed to merge its power generator unit with the state-owned Korea Heavy Industries and Construction.

To show it was serious, the government escalated its struggle against the largest chaebol on October 11, refusing to back down on $53 million in penalties for antitrust violations. The Fair Trade Commission charged that each of the top five had sought to keep money-losing entities alive by favoring them with special deals. The Fair Trade Commission threatened to broaden the investigation to cover more companies among the five as well as some of the lesser chaebol, which also owned enormous companies. The FTC claimed that the SK group had channeled funds from its profitable mobile phone and chemical companies into SK Securities at only 5 percent interest, far below the prevailing local market rates. Samsung Life Insurance, Korea's largest life insurance company, was said to have purchased bonds issued by Samsung Motor.[91]

The Korea Development Institute, a government think tank, forecast a deteriorating economy for the final quarter of 1998 with both consumption and investment "declining at record paces." The institute drew a picture of frustration in combating recession while calling on policymakers to speed up restructuring of both chaebol and financial institutions. "Economic uncertainties remain high," said the report, "because domestic structural reforms are not yet fully implemented."[92]

# Chapter 6

# Piggy-Banking

The future of efforts at getting rid of money-losing companies depended in part on whether the banks could resign themselves to losing some of the trillions of won they had lent in the days when Korea's rise as an economic powerhouse seemed undeniable. Amazingly, the chaebol believed that they were best qualified to run the commercial banks even though they had created the crisis by forcing the banks to extend bad credit. A government commission in 1997 called for reforms that would tighten the easy credit but permit the chaebol to own up to 10 percent of a bank. That was far, however, from ownership and control. Inside the government, antichaebol sentiment was fierce even before the crisis. "The big companies have very bad financial structures," said Won Bong Hee, director-general for financial policy of the ministry of finance. "The chaebol already are borrowing a lot of money from banks. If they have a lot of ownership in the banks, they can abuse them."[1]

The chaebol were only waiting for the government to yield. Kwak Manh Soon at the Korea Economic Research Institute was oblivious in mid-1997 to the signs of disaster. Predicting that the chaebol would be in commercial banking "within five years," he denied they would turn the banks into "piggy banks" from which they could shake out cash as they wished. "Even though the chaebol may participate in management, there will be restrictions on lending and borrowing," he said. He believed that debt-equity ratios would improve. "The weak chaebol will die," he said. "That's a good sign. That's competition."[2]

Over the years, South Korea's banking system had evolved into a behemoth like none other in a country with aspirations of ranking as an equal partner among major industrial powers. The chaebol chieftains ruled over industry and commerce like dukes and barons in a latter-day version of a medieval system. At least until the government had to appeal to the IMF for rescue, they could and did tell their favorite banks how much they wanted for which projects, and they got it.

The bottom line was that among the top 30 chaebol, the average debt-equity ratio was more than five to one. Bankruptcies in early 1997 of a couple of chaebol were an early warning. Dire reform was needed. The banks could not go on extending easy credit. The "system" had to change. The warning assumed greater urgency when considering other realities. These ratios did not take account of the credit extended the chaebol by banks overseas. Nor were they all that realistic as measurements of chaebol indebtedness, since a company might own shares worth up to 25 percent of its value elsewhere in its group.

A sign of the seriousness of the credit problem was that a presidential commission in mid-1997 came up with a "report" on "financial reform." The most important element called for "a more independent central bank" along with "an effective supervisory system." The Bank of Korea, under this scenario, would not be subject to inspection by the ministry of finance—a powerful entity tainted with politics and longstanding relationships with the chaebol. For just that reason, this recommendation would be extremely difficult to include in legislation. The message it carried was one of discipline that the chaebol—and the banks—had long ignored. It implied, as the commission stated, "an increasing importance of prudential regulation over banking," and it required "the independence and autonomy of the financial supervisory body from the policy-setting functions" of bureaucrats in the ministry of finance.[3]

Such reform would represent a revolutionary divergence from the system on which modern Korea was built. Park Chung Hee, during his 18-year rule, told the chaebol what industries to build and ordered the banks to supply the money. The purpose of agencies like the finance ministry was to facilitate the deal. The doyens of the ministry did not like to see their authority diluted—or lost. The old system did have its merits. Without strong central leadership, without the tight top-down authority once imposed on all sectors, without empire-building incentives given chaebol founders and their heirs, would Korea have reached such productivity and prosperity? Could any other system have worked so well for so many? For alternatives, one had only to look north, above the demilitarized zone, to see how a misguided system, in the name of *juche* (self-reliance), the principle, basically, that had guided Park, could destroy an economy.

By 1997, however, Korea had reached a new stage economically. Comparisons with the rest of the capitalist world were needed. Debt-equity ratios in the United States and European Union countries were about 135 percent. In Japan, where debts were once much higher, the ratio was about 130 percent. Taiwan, sometimes compared with Korea, did better than any of them—about 50 percent.[4] The chaebol, in a period of economic slump led by the drop in price of semiconductors and a glut, on the domestic market, of cars, were not generating the equity to cover payments. The problems surfaced in numerous ways. "Where else in the world," a foreign executive in Seoul asked me in July 1997, does a major company "pay you with a check that's 120 days postdated?" Where else were companies so often known to elevate their assets just by raising what they said was the cash value of holdings?[5]

Clearly the chaebol dream of controlling their own banks was not realistic. As it was, a chaebol could now own only 4 percent of a commercial bank. Under the commission's recommendations, ownership in some cases could go to 10 percent. The argument for the chaebol owning commercial banks outright was that the chaebol, many of which owned securities firms, investment trusts, merchant banks, and financial firms overseas, were more experienced in high finance than the banks themselves. The government might no longer name bank presidents, as it had before most banks were privatized, but top posts were often filled by de facto political appointees. The counterargument was that the chaebol chieftains when possessed of their own banks would dispense funds more recklessly than ever, and if payoffs were needed to grease the way, they could also be arranged. As a top foreign executive put it to me, "The guys walk into the bank, they walk out with the deal in their pockets."[6]

The most important recommendation of the presidential commission—after strong central bank control—was to demand that every chaebol with total bank credits of more than 500 billion won, $550 million at the time, report consolidated financial statements as opposed to the "combined statements" the chaebol now issued. Consolidated statements, including "balance sheets, profit/loss statements of cash flows for all the companies belonging to the business group," would "discourage the corporations' high dependence" upon ever more credit.[7]

Foreign bankers felt very strongly the need for this much transparency—and much more. "Transparency would show how each subsidiary does," said a senior European banker. "It would show how much each owes, including the outlook for each subsidiary. Right now, you don't know each company, and you don't know how much they owe over-seas"—a point not in the commission's recommendations. How backward was Korea in this area? Responded the banker: "In the States, France, Japan, of course you know."[8] Transparency was against all the instincts of the chaebol owners. The chaebol did not list a majority of their companies on the stock exchange. Disclosure might reveal weakness rather than strength.

The failure of banks to impose stringent controls before the IMF crisis was widely seen as a contributing factor. Suspension of nine merchant banks on December 2, 1997, was needed to finalize negotiations with the IMF. The IMF, in the agreement signed the next day, ranked at the top of its demands the liquidation of insolvent banks, including both merchant and commercial banks. Finance Minister Lim Chang Yuel, not one to admit trouble ahead, denied on December 7 that two of the top commer-cial banks were about to merge on the basis of a secret understanding with the IMF.[9] The two, the Korea First Bank and SeoulBank, had four months to show they could get over their problems. Korea First Bank, Kia's sec-ond-largest creditor after the government's Korea Development Bank, had borrowed one trillion won from the Bank of Korea in September after Kia defaulted on loans of more than $700 million.

Korea First Bank and SeoulBank were saddled with bad loans totaling more than $4 billion each. Korea First Bank's bad debtors, besides Kia, included the bankrupt Halla, Hanbo, Jinro, and Sammi groups; SeoulBank had extended smaller loans to hundreds of enterprises now in bankruptcy. While Lim pledged "all-out efforts" to keep them from failing, they needed life support. On December 8, as the won sank to a low of 1,330 to the dollar and stocks fell 4.61 percent to 414.83, the government rushed to rescue SeoulBank, pumping in the equivalent in won of about $901 million.

The government, on December 10, 1997, fought to bolster the currency by suspending another five merchant banks that bureaucrats blamed for touching off the panic. Finance Minister Lim charged the merchant banks with "nearly paralyzing the entire financial system" by having "exces-sively depended" on high-interest short-term loans. The five were driving companies into bankruptcy by demanding they pay off loans issued for periods ranging from a day to 15 days at about 25 percent. Thus the banks hoped to pick up the money they needed to pay off their own immediate debts, totaling about $700 million. The Korea Deposit Insurance Corpo-ration provided a degree of protection from bad loans, but the government had to issue $16 billion in bonds to cover them.[10] The government by now had suspended 14 of the 30 merchant banks, giving them until the end of December to develop viable plans—or to merge with other banks.

Analysts questioned whether either an infusion of funds or a merger would help either SeoulBank or Korea First Bank, but Lim on December 15 was still promising significant foreign investment in the two banks. He believed the government could sell at least one of them to a foreign institution. Why, however, would a foreign bank want to rescue a bank steeped in bad debts and worse habits? From the outset, it was clear that the plan to sell one or both of these banks to foreigners would be easier said than done.

Initially, the idea of a sale encountered a resounding lack of interest abroad. The worst problem, said a Citibank executive, was that "nobody really knows what is the quality of their portfolios." Citibank, he said, would want "a complete review of every single loan" that either of the banks had made—an impossible task in a system still shrouded in secrecy. Yet another reason for foreign lack of interest in the banks was their reputation for high labor costs and technological inefficiency. "With modern technology, you don't need all these people," said the Citibank executive. He estimated that SeoulBank and Korea First Bank employed between 18,000 and 25,000 people nationwide, all protected by the law that banned layoffs. Citibank, with 11 branches in Korea, found it possible to operate a typical branch with 9 or 10 people—far fewer than equivalent Korean branch banks.[11]

Rather than buy into a troubled Korean bank, Citibank envisioned expansion on its own—either by opening new branches or by taking over individual entities of troubled operations such as a credit card portfolio. Citibank for years had been the only American institution that did retail banking in Seoul, and it was interested in expanding. "Everything here is going to open up big-time," said the Citibank executive, but much depended on whether the National Assembly passed a banking reform bill that would give the Bank of Korea a badly needed supervisory role. "These banks are not heavily supervised," he said. "They've been talking about this legislation for a year. Now they have to do it." In fact, the IMF, in the bailout agreement, had called for passage of banking legislation as one of its requirements.[12]

The finance minister's suggestion that a foreign bank take over at least one crippled Korean bank was partly responsible for the stock market's rising again on December 16 for the second straight day—this time 4.78 percent to 404.26. "Banks led the upsurge because they said foreign banks could come in and buy one or two basket cases," observed Victor Kang, head of sales at ABN AMRO. "It was a financial big bang."[13] The rise in stocks paralleled another day of recovery of the won—the first day after the government decided the won could float freely rather than remain within a 10 percent band. The value of the dollar at the end of the day was 1,425 won, down from 1,563.9 the day before. Long lines formed at banks as customers, some of them carrying thousands of dollars in $100 bills, waited to change dollars to won.[14]

Government officials, anxious to attract as much hard currency as possible, refrained from asking the identity of those changing money even when they were carrying bundles of bills totaling as much as $100,000. Increasing numbers of Koreans were expected to begin changing their foreign currency hoards if the won held steady for a few days. Such optimism, however, did not seem likely to persuade foreign bankers of the wisdom of helping the system by buying out banks that many believed should close.

Michael Brown, general manager of the Seoul branch of the First National Bank of Chicago, reflected the sentiment of many foreign bankers. Korean banks "should have known how dependent they were on volatile money market funding to support longer-term loans and growing trade finance," said Brown. "Either profitability needs to improve dramatically or debt levels need to come down," Brown told the FKI. In Korean banks as well as companies, "profitability and capital levels are too low to support current levels of bad debts, let alone the higher levels of bad debts which are to be expected in a restructuring and slowing economy."[15]

Brown spoke for many of his colleagues in his sweeping critique. "Acts of collusion between the government, banks, and corporates fuel the foreign perception that business practices are not transparent," said Brown. The "path to market confidence is Korea's commitment to change and economic transparency." Brown suggested the need for tight scrutiny, remarking that "consolidated financial statements are generally not readily available to foreign creditors and investors"—and that "special allowances are periodically made to make financial statements difficult to compare with the performance of previous years or peers outside of Korea."[16]

Brown wound up with "recommendations" that included "passage of a significant financial reform bill" and deregulation. Attempts by the government to maintain "absolute control over borrowing limits, licenses, or other business activities," he warned, "can foster collusion"—a reminder of the cozy relationships by which banks lent far too much to the chaebol. He also agreed with other foreign bankers that too many banks were vying for too small a market. "Try not to fight domestic consolidation," he said, suggesting that closure of some banks would lead to stronger Korean institutions.[17]

The government, however, was not listening. On December 22, 1977, the ministry of finance ordered Korea First Bank and SeoulBank to improve their management—a prelude to taking them over. The government two days later, on Christmas Eve, announced in conjunction with the IMF that it would reform the banking industry and promised to sell Korea First Bank and SeoulBank before November 15, 1998. On January 30, 1998, the government took over both banks, buying nearly 95 percent of the shares of each of them for about 1.5 trillion won apiece. Korea First Bank and SeoulBank, judged too big to die, were now government banks while the government sought foreign buyers.

Government officials opposed a proposal for selling bonds to cover more than $90 billion in short-term debt owed by banks and financial institutions. Instead, they counted on foreign banks to stave off disaster. Deputy Treasury Secretary Lawrence Summers, in Seoul, suggested on January 16, 1998, that Korean negotiators come to terms in New York as a condition for receiving $8 billion in emergency funds from the "group of seven" industrial nations. Korea would receive "bilateral support" only "in the context of an extension" of short-term loans to Korean financial institutions, Summers warned.[18]

A team of economic advisers left for New York on January 18, 1998, to discuss financing with creditor institutions. The 12-man delegation hoped to persuade banks led by J. P. Morgan to roll over $24 billion in short-term debts into long-term ones and compromise on demands for converting the debts into government bonds at interest rates from 11 to 13 percent. Kim Yong Hwan, on the delegation, argued that Korea's economy was now "showing signs of recovery" and "there was no need to stick to high interest rates."[19] The delegation was against converting bank debts to government bonds but might yield if the American banks agreed to the "call option" under which interest rates would go down if Korea's bond ratings improved from "junk" status. Delegation members said they would rather the government "guarantee" private debts than issue bonds to cover them and strenuously opposed a counterproposal for the government to issue bonds covering 60 percent of the debt, $15 billion. Senior aides of both the outgoing and incoming presidents widely criticized the proposal as placing too heavy a burden on the government.

The bargaining over the terms and method by which Korean institutions aimed to win another reprieve on short-term debts estimated at $92 billion was reflected in official reports on Summers's meetings with the president and president-elect in January. The Blue House quoted Kim Young Sam as having asked for U.S. support for decreased interest rates. YS acknowledged the inevitability of high rates on extensions of short-term debts in view of Korea's low ratings by international investors' services but called for readjustment "when our credit ratings are restored."[20] Summers did not respond positively to the plea, according to the Blue House. Instead, he observed it was "important for Korea to raise its international credibility by continuing reform efforts."[21]

While the negotiators were talking in New York, Kim Dae Jung on January 21 promised executives of multinational companies in Seoul to cut through the maze of regulations that discouraged foreign investors and promote a "user-friendly" atmosphere in which to do business. He avoided any mention of the New York talks but pledged his unstinting support of a "free market system" in which foreign companies could compete equally with Koreans. It was the first time he had offered such encouragement to foreign businessmen. He was, he said, reflecting "the concerns of the people who doubt the sincerity of the chaebol's will to reform themselves." He vowed, after his inauguration, to do away with "the collusion between government and business" and guaranteed that "only those companies that have competitiveness will win in the free market."[22]

A week later, South Korean negotiators got almost exactly the terms they wanted. On January 29, Seoul time, creditor banks agreed to turn $24 billion in short-term debt into loans maturing in one, two, and three years. Korean officials were overjoyed by their success, in the midst of a three-day lunar New Year holiday, in persuading 13 international creditor banks to abandon their earlier demand for the Korean government to cover the bonds at double-digit interest. Instead the creditor banks settled for a deal in which the government agreed to "guarantee" the bonds at interest rates of 8 percent, 8.25 percent, and 8.5 percent for loans maturing at one, two, and three years respectively.[23]

Chung Duck Koo, vice finance minister and leader of the Korean team, hailed the outcome as "a critical step" in efforts "to surmount its current liquidity crisis and to restore stability to the Korean banking sector." The plan, he said, "will provide a stable source of funding to the Korean banks on commercial, market terms and on a voluntary basis." William Rhodes, vice chairman of Citibank and senior coordinator of the talks, called it "a key step toward Korea's goal of returning shortly to the international capital markets." The deal had the support of the 13 banks involved in the talks, including Bank of America, Bank of Nova Scotia, Bank of Tokyo-

Mitsubishi, Chase Manhattan Bank, Citibank, Commerzbank, Deutsche Bank, HSBC Holdings, J. P. Morgan, Sanwa Bank, Société Générale, West Deutsche Landesbank, and Warburg, Dillon Reed.[24]

The Korea Stock Exchange was closed for the final day of the holiday but was expected to rebound on January 30. The stock market, recovering for most of the month from levels of below 400 in December 1997, closed on January 26, the last day of trading, at 518.64, up nine points. "This agreement will contribute greatly to the stability of market psychology," said Lee Ung Bak, manager of foreign markets at the Bank of Korea. DJ, visiting a textile factory south of Seoul, told workers the deal was "struck in a harmonious fashion that accommodated most of the terms and conditions demanded by the government." The talks could be "successful only if they are completed with single-digit interest rates," he claimed his experts had told him.[25]

No amount of rollovers, however, could salvage all the banks that were crippled with non-performing loans. The Bank of Korea on February 26, 1998, placed a dozen commercial banks on a watch list on the basis of statistics exposing the dire straits of the entire banking system. The bottom line, said the central bank, was that the 27 commercial banks, as of January 1, had extended 22.6 trillion won, $14 billion, in loans they did not expect to recover. That figure represented 6 percent of the credit extended by all the banks. "The non-performing credit ratio has sharply increased to 6 percent as of the end of December 1997 from 3.9 percent as of the end of 1996," said the Bank of Korea. The bank named four groups, Hanbo, Kia, Halla, and Sammi, all of which had ratios of debt to equity that far exceeded the national average of 4.8 to 1.[26]

The Bank of Korea, notoriously slow to discipline banks for overextending credit, gave the 12 banks until April 30 for providing plans for restructuring and reducing debts within six months to two years. As evidence of its desire to hew to the letter of the IMF agreement, the bank adopted the standard of judgment on the banks prescribed by the IMF. Banks placed on watch had all fallen below the BIS capital ratio of 8 percent set by the Bank of International Settlement in Basel, Switzerland. At the same time, the finance ministry shut down two more merchant banks after giving them failing marks on their plans for restructuring. The ministry gave three other suspended banks another month to come up with realistic plans.[27]

The Bank of Korea's effort at disciplining the banks showed the new administration's determination to abide by the IMF. Other statistics, however, were disturbing. The total foreign debt of Korea's corporations and financial institutions amounted to $186 billion, according to the finance ministry, up from the $154 billion that the ministry had been using for weeks.[28] Banks were beginning to require signed agreements by chaebol leaders on restructuring. Hyundai was negotiating with its largest creditor, Korea Exchange Bank, on a plan for reducing its overall debt-equity ratio from four to one to two to one. Hanil Bank, one of those disciplined by the central bank, said it had reached agreements with six of the chaebol and hoped for similar deals with five more. The finance ministry estimated, however, that Korean companies, after having borrowed $53.2 billion overseas by the end of 1997, would have to repay $9.5 billion in debts by the end of 1998.

A positive sign was that the financier George Soros might open a consulting firm in Seoul. In an interview with the financial daily *Maeil Kyungje,* he said that he had promised to invest a "surprising" figure. Soros Fund Management's QE International in February formed a consortium to take over 27 percent of Seoul Securities, giving it control of the firm. The optimism generated by that report was credited with buoying the stock market. The market closed at 524.98, up 8.6 points, 1.6 percent, on February 26 after having fallen for three days.[29]

By April 1998, however, Korea faced a second financial showdown. Debts were mounting rapidly while the government lagged painfully behind in implementing reform. "The possibility of a credit crunch can lead to a default risk in Korea," said Kim Jun Kyung, one of the authors of a foreboding 200-page analysis put out on April 23 by researchers at the Korea Development Institute. The country's total debts were now estimated at a stupendous 100 trillion won, approximately $800 billion, including more than $150 billion owed to foreign banks, while nonperforming loans were at least 48 trillion won or $34 billion.[30] "The debt situation has been deteriorating, and if it continues there is no way to get out of this crisis," warned Jwa Sung Hee at the Korea Economic Research Institute.[31]

The sounds of alarm overshadowed claims by government officials that the country, with the aid of the IMF, international creditor banks, and, most recently, a $4 billion government bond issue, had stabilized the economy. Officials had hoped to shed comparisons with Thailand and Indonesia, the two far less advanced Southeast Asian countries hardest hit by the regional economic downturn, but the Asian Development Bank on April 23 listed all three as likely to show negative growth in 1998.

Industry leaders and economic analysts wrestled with how to jumpstart an economic reform program that Kim Dae Jung had been struggling to implement since February. Daewoo Chairman Kim Woo Choong talked at a forum on April 23 sponsored by the *Financial Times* in the Daewoo group's Seoul Hilton. "The possibility for another financial crisis still remains due to the weakness of financial institutions and ever-decreasing profitability of companies," he said. "The current situation could be worsened in the wake of the consecutive collapse of insolvent businesses and mass unemployment creating social unrest."[32]

The need to get rid of bankrupt firms was universally recognized, but the question was how to do it. The government, the chaebol, the banks, and opposition politicians were at odds. Critics charged that the government was threatening the chaebol by telling chaebol leaders how to run their business, but Kim Dae Jung, at the *Financial Times* forum, reiterated his pledge "to transform the Korean economy into an open economy based on free market principles." A backstage debate over liberalizing Korea's foreign exchange control law illustrated the problems of easing restrictions. While the country's leaders promised to open the doors wide to foreign trade and investment, officials confirmed on April 24 that there was disagreement among them over the wording of a new foreign exchange control law that the government hoped to present to the National Assembly in May.

Finance Minister Lee promised that the new law would "fully liberalize foreign exchange transactions," but Lee Jung Yung of the Korea Institute of Finance, working on the draft, said proposals ranged from "extreme liberalization to more conservative."[33] One question was whether Koreans would be able to move money in and out of the country as easily as foreigners in a period when the country was attempting to earn and save foreign exchange. The finance minister said he was giving top priority to the act in the drive to attract investment. The current law would be repealed and replaced by one based on what he called "the negative list system" that would only restrict items such as "national security and money laundering."[34]

Lee discussed ways to lure foreign investment in response to skepticism among foreigners as to the wisdom of investing in Korean companies. Foreign and Korean analysts criticized the government, the banking system, and the chaebol for failing to move. "Korea has been moving rapidly toward a complete liberalization of the capital market," the finance minister told the *Financial Times* forum. As of June 30, he noted, "mergers and acquisitions, both friendly and hostile, will be fully liberalized." Laws pertaining to foreign investment, he promised, "will be streamlined and incorporated into a single legal framework represented by the Foreign Investment Promotion Act."[35]

Officials expected foreign investors to shore up the market, at its worst in 11 years, after lifting the ceiling on foreign ownership of private companies from 55 percent to 100 percent beginning May 25, but investors were more impressed by the negative signs. The market closed on May 26 at 311.99, down 5.8 percent from May 25, on a slide toward the 300 barrier. "There was a misguided belief there would be a strong interest in the blue chips on the expectation of a buying binge from foreign investors," said Edward Campbell-Harris at Jardine Fleming Securities. As word spread that the foreigners were not buying, he said, the market went into "a bit of a free fall."[36]

One problem was the government's ongoing policy of issuing loans to companies on the brink of bankruptcy. The government in the first five months of the year had doled out $2 billion in emergency loans, extending several hundred million dollars in recent weeks to save the Dong Ah group, a chaebol that had grown on construction contracts. "If you don't discontinue this," said Richard Samuelson at Warburg, Dillon Read, "you're putting the banking system at risk." Investors also worried about the long-term worth of bonds totaling 40 trillion won that the government was issuing to help cover domestic debts approaching 800 trillion won and foreign debts of $150 billion. "They opened up rather hurriedly to foreign investment thinking that would be a panacea," said Samuelson. "The market is the most severe disciplinarian, and the market is telling them that they are not behaving responsibly."[37]

While Korean banks refused to extend more loans to relatively healthy companies, under government pressure they were providing credit for companies already in receivership just to keep them alive. By mid-1998, it was clear that neither the banks nor the chaebol were moving quickly to get rid of the losers. Voices both in and out of the government criticized the banks for failing to single out companies in the leading chaebol for its widely publicized "hit list" of endangered companies no longer eligible for loans. The Financial Supervisory Commission on June 3 moved closer to a "survival-of-the-fittest" policy, rejecting the list proposed for extinction by the banks. The commission complained that the list included not a single company from the top five chaebol, forcing banks to postpone from June 8 to June 20 the news of which companies could not get loans.[38]

"The banks' death list of companies is a joke," said Del Ricks at ABN AMRO. "Why did it take them so long to come up with such a list?" He suggested that banks were reluctant "to pull the plug" on clients who would otherwise never repay enormous loans.[39] Dozens of major companies in fields ranging from electronics to construction to petrochemicals to iron and steel to motor vehicles were equally distressed, despite efforts by the chaebol to defer either restructuring or closing them. Shin Bok Young, chairman of SeoulBank, acknowledged that "bad companies are being helped in a vicious cycle." The result was the banks did not have funds to lend to healthy companies, which then lacked cash to import raw materials and components for their products. Shin blamed the major chaebol for making matters worse by supporting weak companies. "The obstacle is cross-guarantees," he said. "That's the biggest stumbling block to mergers and acquisitions."[40]

Sohn Byung Doo, deputy FKI chairman, blamed the banks for impeding imports of supplies. "Many companies have received orders from abroad, but they are having a very difficult time importing raw materials," he said. He warned, however, against attempts to force the chaebol to abandon companies that seemed to face too much competition. "It doesn't make sense if we say one company must concentrate on this sector, another on that sector. There should be many different teams in competition. Only then will a company succeed."[41]

Foreign financial experts were pessimistic about the banking system as they witnessed halfhearted efforts at curing pervasive ills created by the insatiable appetite of the chaebol. Corporate Korea was "rapidly crumbling under the crushing weight of a half-trillion dollars of debt, making a banking crisis imminent," said Stephen Marvin at Jardine Fleming. "We expect that the crisis will hit by autumn."[42] The depreciation of the Japanese yen increased fears among local banks that Japanese banks might recall short-term loans while denying new credit in order to decrease their Korean exposure. Korean institutions owed $22 billion to Japanese banks, more than their debt to the banks of any other country.[43]

In an effort at elevating confidence among its clients, Japan's Sumitomo Bank promised on June 10 to guarantee letters of credit up to $100 million for five Korean banks in coordination with the International Finance Corporation, an arm of the World Bank. Although the amount guaranteed by Sumitomo was small in terms of the size of Korea's debt, Sumitomo said its basic purpose was to enable Korean companies to obtain the credit they needed to conduct foreign trade. "It's difficult for Korean companies to get trade credits," observed James Rooney, president of Templeton Investment Trust Management. "That's affected both import and export." The problem would add to the crisis "when there's no more raw material with which to make products for exports."[44] The Sumitomo plan called for Sumitomo to guarantee 60 percent and the International

Finance Corporation 40 percent of letters of credit from Korean banks for local manufacturers.[45]

The show of support by one major Japanese bank, however, was not expected to create confidence in the viability of the banking system or the companies that relied heavily upon the banks for credit. Nor did promises of support given Kim Dae Jung during his visit to the United States the same week have much impact. As evidence of the worries in Korea about the declining yen, the stock market fell 4.3 percent, closing on June 10 at 324.54. Pressure mounted on the chaebol to accept the need for overhaul.[46]

As one solution to the banking crisis, Daewoo Chairman Kim Woo Choong reiterated his megabank proposal for joining several enormous banks with funds from Hyundai, Samsung, Daewoo, and LG. The megabank could take over troubled local banks and set up joint ventures with foreign banks. That idea, however, represented another effort to avoid serious reform while collaborating against government pressure. "Whenever a chaebol has owned a financial institution, it turns into a piggy bank for the chaebol," said Rooney. "Where do they get the money from? They have to borrow it. Then they'll lend it back to themselves. It's a shell game."[47] Chaebol leaders, still barred from owning more than 4 percent of a commercial bank, did not relent in their demands for relaxing the restriction.

Such a campaign endangered a banking system already in jeopardy. "The leaders of the top 50 chaebol are responsible for this mess by overinvesting and overborrowing," said Marvin. "Now basically the government is holding this together with bubble gum." Marvin called for "quick, forceful action" to liquidate many chaebol and raise substantial funds off-shore for recapitalizing the banks in order to avoid "a full-blown banking crisis."[48]

For the smaller commercial banks, the death knell was not long in sounding. In another government-enforced test of survival of the fittest, five ailing banks with capital adequacy ratios below 8 percent faced liquidation on June 29, 1998, for extending dangerously excessive credit. The assets and liabilities of each liquidated bank were to be assumed by a strong bank, but the debts of the weak ones might be written off. The banks, with about 7 percent of the total assets of the banking system, employed about 10,000 people. Although the assets were small, liquidation created major headaches for acquiring banks as well as for customers who discovered their banks were closed.

Financial Supervisory Commission Chairman Lee Hun Jai, calling their liquidation "the first step toward restoring the health of the financial system," said their closure was needed "to speed up financial restructuring." Liquidation of the weakest banks, picked from a watch list of a dozen troubled banks, marked another halfway attempt to get rid of debt-ridden institutions. These banks were obvious candidates for oblivion; the government still was reluctant to attack major institutions. Bigger banks all "avoided the worst in consideration of their size," Yonhap reported, but they remained on the watch list. Paradoxically, the FSC, finance ministry, and Bank of Korea agreed on June 28 on a program for rolling over about 84 trillion won, $60 billion, in loans for small and medium-sized firms.[49]

The closures drew protests from thousands of bank employees in danger of losing their jobs. More than 1,000 workers gathered on June 28 in the lobby of the Daedong Bank in Taegu, shouting that the government had no right to interfere. "No shutdown," read the words on red headbands and signs carried by the demonstrators.[50] Workers at the four other banks set for liquidation also threatened protests even as authorities entered their banks that night to avert any efforts at thwarting liquidation. Officials from the relatively healthy banks named to absorb the weak ones worked quickly to make certain they had control of the victims' computer systems and vaults.

Thousands of newly unemployed bank workers refused to cooperate on June 29 in the takeover of their liquidated banks, creating technical problems that might take months to overcome. "We didn't have to break up any computer files," said Moon Kyung Hee, a clerk at Daedong Bank, after getting the news that Daedong Bank was one of five ordered to merge with a stronger bank. "We just changed the code to enter the program." Moon recounted her role in obstructing the closure of her bank as she joined more than 1,000 other bank workers in a demonstration on the slope leading to Myongdong Cathedral. All of them had journeyed to Seoul, by bus and train, early in the day from the bank's headquarters in Taegu.[51]

The demonstration in front of Myongdong Cathedral was one of scores of protests around the country; all the liquidated banks faced problems. At the headquarters of Donghwa Bank in Seoul, 1,500 employees wearing headbands saying "Stop the Forced Restructuring" blocked entry to representatives of Shinhan Bank, designated to take it over. Shinhan officials got into some of Dongwha's branches but were refused keys to the vaults. In the port city of Inchon, employees at the headquarters of Kyungki Bank, after blocking the entrance, changed the passwords for access to the bank's most important computer systems. Officials at Dongnam Bank left the premises rather than cooperate with representatives of the Housing and Commercial Bank when they arrived to take over the records. Officials at Chungchong Bank had a ready reason for why they did not cooperate with Hana Bank: they said they had already resigned.

The order to close the banks "will face resistance and will have dissat-isfied people," said DJ's press secretary, Park Jie Won, but "this is the first time in our history we have shown banks can go bankrupt." The decision got an enthusiastic endorsement from John Dodsworth, IMF representa-tive in Seoul. "The first step is very good," he said. "The policy direction is correct"—even though "Koreans are having a hard time."[52] Financial analysts and union leaders were not so sanguine. "We don't know if the shutdown of five weak banks will be good or bad," said Kang Hun Goo of ING Baring Securities. "Foreign investors are already worried about their stakes in blue-chip banks like Kookmin Bank and Housing and Commer-cial Bank."[53] At the Korean Confederation of Trade Unions, spokesman Yoon Young Mo said the government did not have a blueprint that antici-pated the consequences of its actions. "It does not target the big banks but creates a worse credit squeeze," he said. "This will cause bankruptcy for many small and medium-size companies."[54]

The figures bore out the dismal state of the economy. The Bank of Korea reported on August 27 that the gross domestic product had fallen 6.6 percent in the second quarter of 1998 from the second quarter of 1997. The decline marked the second straight quarter of negative growth, a definition of recession. With domestic debt believed to exceed the equivalent in won of more than $700 billion, the government on October 16 threatened to significantly increase the pressure for chaebol reform by restricting corporate bond issues, on which they now relied for fresh capital. The Financial Supervisory Commission planned to cut chaebol bond issues by ordering banks not to guarantee them. "That may be the only way to persuade the chaebol to go through with the restructuring they have to do," said an FSC official. "Otherwise, they do not want to listen."[55]

The threat to use the government's power over the banking system escalated the struggle in which the chaebol had firmly dug in against change of the sort demanded by DJ and his advisers. Daewoo Chairman Kim Woo Choong, as chairman of the FKI, called on chaebol chairmen to coordinate on a "voluntary" approach—meaning another way to postpone what the chaebol hoped was not the inevitable. The government responded that Kim as always was buying time. The FKI countered "there was "no legal basis" for the threat to cut off access to credit and doubted "if such a policy can be implemented."[56]

Economic policymaker Park Tae Jun, whose hostility toward the chae-bol dated from his days as chairman of the state-invested Pohang Iron and Steel, called them "a major stumbling block to the country's reform struggle." On October 17, he urged DJ to show that the government, not the chaebol, controlled economic reform. "Corporate sector restructuring is the most basic task," said Park. "If voluntary efforts for readjustment are not sufficient, the government will have no other choice but to do something." Park, primary architect of "the Big Deal" for merging insolvent chaebol companies, complained that the chaebol, failing to go through with "Big Deals," were propping up money-losing entities on credit that was not available to smaller firms.[57] The top five chaebol were issuing almost 80 percent of the corporate bonds on the domestic market.[58] The top five had hit the ceiling on credit available from banks, now estimated to hold more than $110 billion in bad debt, but the banks routinely rolled over their loans.

The fear of a second banking crisis increased despite a $50 billion government program to rescue the banks from the crushing weight of nonperforming loans that banking experts by late 1998 estimated at more than $200 billion. Foreign analysts in November said that the amount earmarked for salvaging the country's commercial banks would not be enough and predicted that some would fail after running out of funds as a result of bad loans. With five commercial banks already forced out of business, the 22 remaining banks were scrambling for survival. "Some more of the troubled ones have to go under," said Jason Yu, banking analyst with Indosuez W. I. Carr Securities, estimating that nonperforming loans now totaled 270 trillion won, approximately $210 billion. That figure, he said, amounted to 36 percent of the domestic debt owed by Korean companies but did not include more than $150 billion owed foreign banks.[59]

Stephen Marvin at Jardine Fleming likened the crunch confronting the banks to the crisis that had forced the government a year earlier to go to the IMF. Marvin based his prediction on both the shortage of funds for saving the banks and lack of discipline in extending credit. The government's plan "is conceptually correct but fatally flawed," he said. "A large portion of the money will be lent to borrowers who can't repay the money." He criticized the government for lacking "the political will or fortitude" to block credit for money-losing companies even though officials claimed to have imposed tight constraints on chaebol in "workout programs" to recover viability.[60]

Marvin cited the effort at saving SeoulBank and Korea First Bank as indicative of the government's inability to cope effectively with the problem. "They pumped money into them to save them from bankruptcy," he said, "and now they're almost bankrupt again." The government still hoped to auction them off by the end of the year while acquiring equity in other banks, including several that were merging. The banking system, however, was not prepared for the mountain of nonperforming loans. "The policy of the banks has been not to permit writing off debts," said Peter Bartholomew, longtime business consultant in Seoul. "That's the problem of all attempts at mergers and acquisitions right now. A lot of M&A buyers have come, but not many are buying. The banks have no experience in writing off debts."[61]

The government was dividing a total of 64 trillion won—$50 billion—between issuance of bonds to provide fresh funds for ailing banks and purchase of nonperforming assets from commercial banks at between 30 percent and 45 percent of value. The government promised to complete the program in 1999 after a series of mergers and the sale of SeoulBank and Korea First Bank to foreign interests. The program meant that the government was taking over more banks. Besides owning most of Seoul-Bank and Korea First Bank, the government had 94 percent of Hanil Bank and the Commercial Bank of Korea, whose merger on January 4, 1999, would form Hanvit Bank, Korea's second-largest. "The government is holding equity ownership in return for capital injection," said Lee Sang Mook at the finance ministry's task force on restructuring. "It's to cover bad loans. They are concerned about newly emerging nonperforming assets. Most of their clients are in trouble." As Richard Samuelson put it: "In recapitalizing the banking sector, a lot of private debt is replaced by public debt."[62]

The program reflected the urgent need to prop up a banking system that might otherwise collapse under at least $700 billion in loans, most of them routinely extended rather than paid back. The figure for domestic commercial loans, nearly double the gross domestic product, accounted for 50 percent of Korea's total debt. While buying time for "voluntary" restructuring, the chaebol were daring the government to make good on threats to force the banks to cut off credit. Park Yung Chul, overseeing the merger of Hanil and CBK into Hanvit Bank, remarked on December 4, the day after the anniversary of the IMF agreement, that such threats would not last long. "I'm sure they will put pressure on those conglomerates to pay back if they don't do what the government wants," he said. Government officials "will exercise their leverage." In the end, "Interference will disappear, especially in loan management. Once restructuring is completed, they don't have reason to interfere with bank management."[63]

In the quest for funds, Kookmin Bank, one of the biggest, was also considering the issuance of $500 million in new shares and was soliciting foreign banking institutions for another $100 million. At the same time, the government put public funds into Kookmin and the Korea Long-Term Credit Bank to facilitate their merger on January 1, 1999, into Korea's largest commercial bank. "Many banks will be recapitalized through such a scheme," said Lee Sung Gun, general manager of the international finance department of the government's Korea Development Bank. "They will be officially government-owned. The banks are under restructuring. Their business now is very slow." The basic purpose of the program was to replenish accounts depleted by nonperforming loans. Typical were those issued to Kia, whose creditor banks had agreed to write off nearly half its debts. The government was also pumping another five trillion won, $3.5 billion, into the KDB, the largest creditor to Kia and many others.[64]

While the commercial banks floundered, the government still had the means, through the KDB, of rescuing companies that it wanted to keep in business. The KDB in turn had tremendous leverage over the commercial banks, to which it could extend loans separately from those issued by the Korea Deposit Insurance Corporation. The government's takeover of banks reflected a certain confidence that the banks would recover as the economy improved, companies increased exports, and the government pumped more money into the economy through public works projects. Recognizing that international investors believed "corporate reform should be expedited," Kim Young Mo, deputy director of international finance at the finance ministry, said the banks were issuing "emergency loans" only to chaebol that had agreed to "workout programs" under which they were to drop money-losing entities.[65]

Choi Buhm Soo, counselor at the Financial Supervisory Commission, denied the result would be direct government supervision of commercial banks. "Even though a bank is owned by the government, it will be managed commercially," he said. "The government won't be involved in the daily business of bank management." If the government got more closely involved, "nobody in the world would want to buy their shares."[66] The future of banking in Korea might depend on the extent to which foreign banks were interested. It was one thing for a foreign bank to roll over loans that would not have been paid under any circumstances but another to plunge into the Korean system without guarantees of reform in habits and attitudes as well as rules and regulations.

Thus preliminary agreement on December 31, 1998, for the sale of Korea First Bank to an American investment group, Newbridge Capital, appeared at the time like a major triumph. Fashioned by Richard C. Blum & Associates and David Bonderman's Texas Pacific Group, Newbridge had a reputation for picking up failed enterprises and making them profitable. Under the deal, Newbridge would acquire 51 percent of the bank, reportedly for $600 million. The agreement symbolized Korea's efforts at drawing massive foreign investment into failing entities.

Anxious to regain credibility lost in the Asian economic crisis in 1997, the major international credit-rating agencies were now ready to risk reputations for independence and accuracy in betting on Korea's economic rehabilitation. Fitch-IBCA was the first, elevating Korea's sovereign rating to investment grade on January 19, 1999. The agency, which had downgraded Korea's ratings in November 1997, now said Korea's program for economic stabilization was "well advanced" while the economy was "beginning to recover." The ability of local industry "to export world-class products at attractive prices remains undimmed," said a statement from the agency's London headquarters.[67]

One week later, on January 26, Standard & Poor's, the New York rating agency, also raised Korea's sovereign rating to investment grade, predicting Korea's credit standing "could continue to improve in a one- to three-year horizon if private sector restructuring continues."[68] Only Moody's Investor Service, also based in New York, held back, awaiting final checks in February before reaching a judgment that Korean bureaucrats and foreign investors alike saw as crucial to the drive to recover from near bankruptcy 14 months earlier. "We don't want to overstate or understate the risk," said Tom Byrne, Moody's lead analyst on Korea. "Our default studies are our record. What we don't want is for countries or companies that we rate single A or triple A to go into default. Then we aren't doing our job right."[69]

Whatever they said, whenever they said it, the word of the rating agencies was key to investors with millions of dollars to pump into the economy. "Whenever they change their ratings on Korea, there's a market reaction," said Byun Yang Ho, director of the finance ministry's international finance division. "Whenever they announce good news, the market price gets better for our government bonds."[70] In the face of crushing debts and rising unemployment, Korea was slowly emerging from more than a year of economic crisis as a strong prospect for investment, according to a growing consensus among foreign investors and securities analysts. Ultimately, however, investors wanted to see the views of all the rating agencies before deciding whether or how to risk their clients' fortunes. That rule applied especially to managers of vast retirement funds and institutions such as insurance companies, constrained from risking assets on anything below investment grade.

Some critics believed that Moody's, Standard & Poor's, and Fitch-IBCA all had overreacted in downgrading Korea's sovereign rating in late 1997 and early 1998. "They placed Korea way below investment grade when there was no objective basis," said Philippe Delhaise, president of Thomson Bank Watch Asia. "We downgraded Korea before the others when we saw the crisis coming in banking, but we never placed Korea below investment grade. When we saw the extent of the damage, we reasoned that Korea's background strength was still there and we decided the market had priced Korean debt at a level that was totally unrealistic."[71]

Analysts confessed to difficulties discerning the truth about finances in Korea. "The world missed the vulnerability of Korea because there was not a lot of transparency and the data was not complete," said Stephen Hess, a senior analyst at Moody's. "The crisis has changed them on the issue of transparency." He was less confident, however, of the willingness of the Korean banks and chaebol to divulge information. "They're still not up to the standards one might see in other countries. It will take a while before the level is up to what one might like to see."[72]

As if to confirm all the analysts were saying, Moody's on February 12, 1999, upgraded its sovereign ratings from junk bond to investment grade. The rating of Moody's was viewed among Koreans as the final international seal of approval. With the Moody's endorsement, officials expected to attract both equity and foreign direct investment viewed as essential to fulfilling the forecasts of both the finance ministry and the IMF of a 2 percent increase in the gross national product in 1999 from 1998. Moody's cited Korea's "vastly improved external liquidity position" and steps to liberalize foreign investment and "put into place a comprehensive framework for reform and restructuring of the financial and corporate sectors." Tom Byrne at Moody's saw the official optimism as helpful. "They've gone from a state of pain to feeling, 'Maybe we'll pull out of this,'" he said. "The psychology is changing here. If people feel their jobs and income are more stable, they'll start spending again."[73]

The improved ratings came with qualifications, however, reflecting a widespread view among experts that Korea still had far to go—and could suffer a relapse if it failed to follow through. "This is an economy in a recovery phase," said Byrne. The Moody's report balanced praise for the buildup of foreign exchange reserves, stabilization of the exchange rate, and the decline in interest rates with worries about rising unemployment and the failure of the chaebol to restructure.

Nonetheless, the upgrade capped off a series of upbeat assessments by influential visitors more than 14 months after the Korean economy had sunk to its lowest depths since the period immediately following the Korean War. "Korea has accelerated reform progress and has moved rapidly to redress serious structural issues," said Donald Johnston, secretary-general of the Organization for Economic Cooperation and Development. Johnston cited strengthening of the banking sector, industrial restructuring, and improved social welfare as major needs. He argued, however, against the view that Korea was not sound enough to qualify for membership in the OECD when it was admitted in 1996. Rather, he said, "Korea should have joined sooner." Then, "these reforms would have moved more quickly, they would have begun a capital liberation process." The result would have been "strong foreign reserves, which might have solved the problem" of inability to pay short-term debts.[74]

Elevation to investment grade was like an award that confirmed the impression of Korea's improving viability as a place to invest. Foreign investors were clearly the major force behind a steady upward push from the worst of the crisis in December 1997. The danger now was that excess liquidity would weaken the market. "There's a lot of money in the Korean market," observed Park Nei Hei at Sogang University. "People don't find a good place to invest so the money just flows into the market. I worry about the bubble effect."[75]

With financial assets 6.5 times its gross domestic product, "Korea's got too much liquidity," said James Rooney at Templeton Investment Trust. One harsh reality was that the chaebol had done little to reduce debt-equity ratios averaging four or five to one and might be undermining the market by offering overpriced shares in a bid to pay off debts. "You'd have to burn the paper for debts to shrink," said Rooney.[76] Still, the finance ministry set a target of $15 billion for foreign direct investment in 1999, nearly double the $8.85 billion invested in 1998, which in turn was 27 percent above the 1997 level of $6.97 billion. "Industrial investment has proved to be a good place to invest," said Jonathan Dutton, securities analyst at Warburg, Dillon Read. "Why? The market overreacted initially to the crisis. Investments were cheap."[77]

Adrian Cowell, in charge of Kleinwort Benson in Seoul, saw rising portfolio and direct investment as paralleling one another. "A lot of money has come in over the past three or four months," he said. "Progress in terms of Big Deals for restructuring the chaebol will lead to further spin-offs of companies."[78] The test, however, would be the extent to which the banks could operate independently from the chaebol, confident of their ability to assess credit risks on merit, to refuse credit at any sign of risk. The danger was that the chaebol, once the initial crisis had subsided, would never restructure while the banks and the chaebol resumed traditionally cozy relationships.

It seemed anticlimactic when, less than two months later, on February 22, 1999, the government reached preliminary agreement on the sale of SeoulBank to HSBC Holdings, the Hong Kong and Shanghai Banking Corporation, which had originally wanted Korea First Bank. HSBC would get "an initial 70 percent stake for consideration of approximately $700 million" in addition to "an up-front facilitation payment to the government of $200 million." The bank's nonperforming assets and liabilities were to go to the Korea Asset Management Corporation. The government in turn retained a 30 percent stake and additional warrants equivalent to 19 percent of the shares, which it could sell to HSBC as well. The government agreed, however, not to "take any part in the management of the bank."[79]

HSBC already had three branches in Korea with assets of $3 billion. SeoulBank as of the end of 1998 had assets of $24.8 billion in 292 branches with 4,809 employees. "One of HSBC's objectives is to acquire a significant local presence," said Gokul Laroia of Morgan Stanley Dean Witter, adviser on the deal to both the government and to SeoulBank. The understanding was the government would cover the bank's nonperforming assets of about $3 billion while HSBC invested more than $1 billion to turn the bank around.[80]

Government officials were euphoric. "This is one of the most important achievements that Korea would like to boast to the rest of the world," said Kim Joon Hyok, in the economic information bureau at the ministry of finance. "This is the fulfillment of the pledge that Korea made at the outset of the crisis. These two banks were in deep trouble. They had to be closed or to have surgery." The bidding for SeoulBank was open to Korean investors, but foreigners were welcomed for their expertise as well as funding. "They can bring in advance high-tech management skills," said Kim. "Korea First Bank was auctioned off to Newbridge. We prefer to sell SeoulBank to another foreign investor. The rest of the world has been watching how the Korean government will pull this deal off. For a good part of the past year, the international community remained rather skeptical. The sale of these two banks was a litmus test."[81]

The symbolism inherent in the sale of such large entities was a factor in the bargaining. Other huge foreign interests bought major portions of Korean banks, but none of them had majority stakes. Germany's Commerzbank became Korea Exchange Bank's second largest shareholder in 1998 after the Korean government when it paid $270 million for a 30.4

percent stake. Goldman Sachs agreed in March 1999 to become the major shareholder of Kookmin Bank, investing $500 million for 16.8 percent. ING Groep NV in July picked up 10 percent of the Housing & Commercial Bank for $280 million.

Completing the deals for Korea First Bank and SeoulBank, however, was a far more difficult proposition. No foreign interest had actually taken control of a Korean bank. Newbridge Capital and government negotiators quickly bogged down in talks on Korea First Bank's debt load, deadlocking on selling back nonperforming loans to the government. Weijian Shan, managing director in Hong Kong for Newbridge Capital, spearheading the deal, negotiated with a toughness shared by few if any of his adversaries. Born in China, Shan had survived Mao Zedong's Great Cultural Revolution as a teenager dispensing medicines in Inner Mongolia. He liked to reminisce about riding horseback across the Great Gobi Desert, warding off hordes of mosquitoes in the summer and digging trenches in the ground for protection against sub-zero temperatures in the winter. In his spare time, Shan learned English, devouring foreign books that replaced the formal education denied him by the rampaging Red Guards. It was not until he got to San Francisco, then studied at Berkeley, eventually earning a doctorate, that he became a western capitalist. He first taught finance at the University of Pennsylvania's Wharton School, then joined J.P. Morgan, and ultimately moved to Hong Kong with Newbridge Capital.

Shan also learned how to send messages through the press, making himself available when Korean officials snubbed him. At a press conference on April 27, 1999, countering rumors that the deal was dying, he predicted "a perfect marriage between the Korean government and Newbridge."[82] Shan denied that Newbridge had demanded that the government cover nonperforming loans of seven trillion won, double the total listed by the bank. The deal would be "good for everyone, for the Korean taxpayers, for the customers, and above all for the bank," he said. Newbridge would "value our customers, including the chaebol, and would like to develop a long-term and strong business relationship."[83]

As the government pumped in another three trillion won, $2.5 billion, on May 13 to prop up Korea First Bank, banking experts contemplated the potential impact of collapse of the agreement. The government had spent about $41 billion to buy off nonperforming loans and recapitalize banks since December 1997. "If we fail in the Newbridge deal, the confidence of foreign investors could fall down," said Koh Sung Soo, a researcher at the Korea Institute of Finance.[84] "These negotiations are of life-and-death importance to us," said Choi Won Ku, manager of the privatization team at the Korea First Bank. "We would like the deal done."[85] The Financial Supervisory Commission said the bank could no longer extend fresh loans since it now had an overall deficit in assets of about 1.5 trillion won—$1.2 billion—as a result of unpaid loans.

The suspicion remained, however, that somewhere in the bureaucracy lurked a fundamental distaste for selling the bank to foreigners, at least at a reasonable price. "The question is whether they have the political will to follow through with banking reform at all," said Shan, in a mood of exasperation. "The fundamental problem with this economy has to do with the banking sector. If you don't have real banks, you can't have a market economy. There was a need to send a signal to the world that Korea was going to reform the banking sector." Failure of the deal, he believed, would cast doubt on whether the banks were reforming properly.[86] Finance Minister Lee Kyu Sung was not sympathetic. "Business is business," he said, admitting differences in the agonizing process of "due diligence" of the books. "Korea is looking for the best deal."[87]

The complex over foreign interests in general, and foreign banks in particular, quickly surfaced on a much larger scale in a double standard set by the government in coping with the loans of the Daewoo group. The fast-mounting troubles of Daewoo were inextricably linked to the fates of Korea First Bank and SeoulBank, neither of which would have been in such trouble themselves if they had exercised restraint in extending credit to Daewoo, whose debts now totaled more than $57 billion. The FSC on July 19, 1999, gave Daewoo a three-month reprieve, compelling domestic creditors to roll over 7 trillion won, $5.9 billion. Then, pouring good money after bad, the creditors had to provide 4 trillion won, $3.4 billion, *more* credit, all to cover nearly $10 billion in domestic debts falling due almost immediately.[88] Foreign banks, including HSBC, ABN AMRO, UBS AG, Chase Manhattan, Citibank, Bank of Tokyo-Mitsubishi, Daiichi Kangyo, National Australia Bank, and Arab Bank, holding nearly $10 billion of Daewoo's loans, howled that they were left out. They did not agree to rescheduling, were not interested in exchanging debt for equity, and threatened to pick up collateral in the form of Daewoo holdings overseas.

There was sympathy abroad for the position of the foreign banks. "Ominously, there are indications that Daewoo intends to treat its foreign creditors worse than its local ones," grumbled the *New York Times* on August 7.[89] Worried about a rising crescendo of foreign criticism, the FSC sought to allay such concerns by promising that the foreign banks would be treated as "equals to local creditors" under the terms of an agreement reached on August 16 for spinning off all but six Daewoo companies. It was disturbingly unclear, however, whether a fund of 10 trillion won, $8.3 billion, set up under the agreement to cover loans that could not be rolled over, would also protect foreign banks. Only domestic creditors, led by Korea First Bank, participated in the negotiations. Nonetheless, the FSC asked foreign creditors to "support the agreement."[90]

Foreign bankers, however, demanded a firm guarantee. Under the circumstances, the government could not stand to see SeoulBank, a domestic institution, go under the control of one of Daewoo's major foreign creditors, which then could exploit that position to ensure that foreigners did indeed receive equal treatment. Thus Korea's quest for "the best deal" resulted in the collapse of the agreement with HSBC—but, for the same reason, not of that with Newbridge. FSC Chairman Lee Hun Jai, over lunch with foreign correspondents on August 30, revealed that talks with HSBC on SeoulBank had ceased even though the government still hoped to sell the bank. The differences, he said, were "huge" and "difficult to narrow."[91] The news stunned the banking community. Foreign bankers believed the government might have become overconfident amid signs that the economy had recovered. The problems of the sale of SeoulBank came to symbolize not the success but the frustrations of arranging deals between Korean and foreign interests. There was speculation that some officials were reluctant to sell either SeoulBank or Korea First Bank to foreign interests, but the cases were not quite the same.

Differences with HSBC revolved around the criteria by which loans were classified as nonperforming. An FSC official said HSBC wanted to apply "more stringent standards" than those set by the government. The FSC planned to keep SeoulBank alive with a transfusion of more than $4 billion. Nahm Sang Duck, in charge of bank restructuring for the FSC, said the government hoped to "increase the value" of SeoulBank before attempting to attract another bidder. Confronted by widespread criticism over the failure of the deal with HSBC, Nahm said the government also would seek a foreign bank or at least a foreign banker to manage Seoul-Bank.[92]

At the heart of the impasse was a problem whose enormity should have been apparent when the FSC and HSBC signed their memorandum six months earlier. That was how to classify $600 million worth of loans extended by SeoulBank alone to companies in the Daewoo group, on the brink of long-overdue collapse. "The main issue is the Daewoo restructuring," said Todd Martin, chief of research on Asian banking for Warburg, Dillon Read. "HSBC is not going to compromise." Martin believed

Korea would suffer a loss of confidence even though the value of Seoul-Bank shares rose 15 percent on the news the government would bolster it with more funds. "HSBC could have added a lot of credibility by taking over SeoulBank," he said. "It was more a question of financial policy-making. This is coming down to whether a bank is willing to classify loans that are currently nonperforming."[93]

Standard & Poor's, on August 31, sided with the foreign bankers when it downgraded the credit rating of the Daewoo Corporation, the central company in the group, from "CC" to "D." The "D" rating meant that Daewoo Corporation was in default for nonpayment of loans even though Korean officials did not place it in that category. The agency coupled the downgrade with a strong warning of the repercussions of the Daewoo dilemma not only for Korea but for the entire region. "Drawn out or ineffectual restructuring of borrowers in default is a considerable risk for Asia's recovery from the economic crisis," said the agency. "The restructuring of Daewoo will be an important gauge of the prospects for a solid and sustainable recovery both in Korea and the region at large."[94] The government's plan to bolster SeoulBank with more money was an effort to maintain credibility. "They can't let a public bank fail," said Bernhard Echweiler, vice president in charge of Asian economic research for J.P. Morgan. "They have to cover up the liabilities. With the Daewoo case, it became apparent the bank's assets all looked worse."[95]

Daewoo and foreign creditors haggled bitterly over whether the foreigners should go along with the program for putting off repayments of debts, $2.95 billion of it due by January 1, 2000. In one meeting, on September 16, the leading foreign creditor banks, representing nearly 200 creditors abroad, insisted the government, or an agency backed by the government, guarantee on-time payment of their loans. They did not appreciate a proposal by Daewoo for rolling over $5.48 billion of their debt for seven months, until April 2000. Nor were they consoled by assurances by Daewoo and a "corporate restructuring coordination committee," closely linked to the FSC, made up of Korean creditors, professors and consultants, that they would share equally with domestic creditors in the 10 trillion won collateral pool. Some foreign creditors, including Natexis des Banques Populaires of France and Bank Brussels Lambert, were determined to follow through on action to collect collateral from Daewoo subsidiaries in Hong Kong and Europe. They counted on orders by foreign courts to freeze Daewoo bank accounts and seize shipments of Daewoo cars on the docks.[96]

The demands of foreign bankers triggered completion overnight of the sale of Korea First Bank, Daewoo's lead creditor with loans totaling $2.5 billion. Nahm Sang Duck at the FSC announced on September 17 that Newbridge would acquire 51 percent of Korea First Bank for the bargain price of 500 billion won, slightly more than $417 million. Why did Newbridge get a deal while HSBC failed? Until the deal was announced, the conventional wisdom had been that HSBC would succeed because it was a bank. Newbridge, since it was not a bank, it was often said, did not have the expertise the Koreans wanted to rescue Korea First Bank. The exact opposite of this reasoning, however, prevailed. Newbridge won out because it was not a banker, was not among Daewoo's querulous foreign creditors, and would not align with them. Difficult though Newbridge might be, Weijian Shan could promise not to make so much trouble for the government, the banking system, and underlying relationships with the chaebol. The FSC would find no better option. HSBC, in contrast, would demand stringent standards—and would surely have exploited SeoulBank as a mechanism for enforcing them.

The deal extracted by Shan, tempted more than once as the talks wore on to pull out, was so good as to displease Korean critics. For one thing, the amount that Newbridge agreed to pay up front was considerably less than the $600 million figure that Korean officials had previously told local journalists was the sale price. For another, the government by the time of the sale had injected $4.2 billion into the bank to keep it on life support. As if that were not enough, the government, retaining almost all the remaining 49 percent stake in the bank, agreed to cover bad loans falling due in the next two years. True, Newbridge could invest another 200 billion won, nearly $170 million, in two years but was under no obligation if the bank did not do well. As a kind of compensation prize, the government had the option of acquiring another five percent of the bank's shares after three years—a chance to recoup some of its losses.[97]

Almost as important as streamlining the banking system was overhaul of Korea's debt-ridden insurance companies. The FSC in early 1999 named six life insurance firms with debts approaching $2 billion as candidates for extinction in the next round of reforms. On February 19, 1999, the FSC charged that the six companies had vastly understated their debts in response to a request for statements on their viability. FSC chairman Lee Hun Jai included life insurance companies along with investment trust companies and mutual funds as targets for another wave of investigations. Lee promised, however, that the next phase of restructuring would "wait until the money markets have regained stability" to avoid causing "a major shock" to the economy.[98]

The race was on for selling off the ailing life insurance companies in a bidding war seen as vital to the financial health of the Korean conglomerates that relied on insurance companies to buy their bonds. On May 12, 1999, the process for disposing of them began in earnest with the opening of bids for Korea Life Insurance, the nation's third-largest, with 18 percent of the market. The LG group looked like an easy winner, but the FSC ruled LG's finances were too shaky and called for fresh bidding. AXA of France and Metropolitan Life Insurance, the American company, failed to submit bids after lengthy studies of the books. They clearly were disturbed by the jailing of Choi Soon Yong, the former chairman of the Shindongah group, which owned the company, on charges of secretly shipping huge amounts out of the country, bribing executives to get them to put funds into the company, and bribing officials to overlook financial irregularities. The company's debts totaled 14 trillion won, $11.6 billion, against assets of 11 trillion won, $9.2 billion. The net deficit of $2.4 billion about equaled the funds that Choi was charged with having pilfered from his firm's coffers.[99]

Korea Life's problems seemed likely, however, to get worse before they got better. After the FSC failed in two more auctions to find a buyer willing to pay an acceptable price, Choi, sentenced to five years, waged a bitter struggle to prevent the government from taking over the company. In a lawsuit filed on September 16, two days after the FSC declared Korea Life "non-viable" and a "weak financial institution" as a prelude to nationalizing it, Choi charged that he had been deprived of a lawful opportunity to revitalize the company.[100] He claimed he had the solution in the form of a massive investment by a New Jersey firm, Panacom, but the FSC doubted if Panacom had the funds and saw the Panacom proposal as a ploy by Choi to run Korea Life from jail. The FSC was determined to put Korea Life on its feet with a cash injection by the Korea Deposit Insurance Corporation of 2.7 trillion won, $2.3 billion. Unless Korea Life was nationalized, said the FSC, the firm would "suffer unrecoverable damages"—with a devastating effect on the Korean insurance industry and the entire economy. The court ruled against Choi, who lost his 28 percent stake when Korea Life was nationalized at midnight, October 1, 1999.[101]

Lee Jong Koo, director-general of the FSC, said the commission also planned to hold auctions for the six other companies and welcomed bids from foreign companies into a market that was closed to foreigners before Korea's economic crisis began in 1997.[102] Thus the government hoped to minimize the injection of public funds that would otherwise be needed to cover the companies' liabilities, estimated at $10 billion. Another objective was "to introduce world-class insurance expertise" by the influx of foreign life insurance companies with skills that insurance experts said were lacking at most Korean insurance companies. Many of Korea's 29 life insurance companies had suffered losses by risky practices not countenanced in the west. By contrast, Korea's two largest life insurers, Samsung Life, a mainstay of the Samsung group with 35 percent of the market, and Kyobo Life, with 20 percent of the market, had been healthy.

"There's a lot of potential to grow here," said Tim Ferdinand, deputy managing director of Crédit Lyonnais Securities Asia, advising the FSC and six of the ailing insurance companies. "Korea is the sixth-largest life insurance market in the world. Liberalization coupled with the size of the market makes it particularly attractive. Previously it's been difficult for foreign insurance companies to break in here. Now that the government has liberalized, it will be simpler."[103] The government had begun easing regulations as a requirement for joining the OECD on January 1, 1997, and had opened up as one of the conditions of the IMF loan package in December 1997.

The FSC had begun to mail material on five other ailing companies to prospective bidders in Korea and abroad. Separately, Allianz AG of Germany, Europe's second-largest insurer, signed a memorandum of understanding to buy Korea First Life Insurance, Korea's fourth-largest life insurer with 4 percent of the market. New York Life and the International Finance Corporation agreed to buy two thirds of Kookmin Life Insurance, the tenth-largest, with 1.5 percent of the market, while the government purchased the remaining one-third.

The selloffs were needed as a measure of security for the chaebol, which relied on insurance and investment trust companies as major sources of income. "They're a magnificent way of milking the public to support your chaebol," said a foreign analyst. Koreans, with one of the world's highest savings rates, had long been accustomed to buying endowment policies from life insurance companies in hopes of attaining interest rates higher than those of commercial banks. "The Korean domestic bond market is heavily dependent on the insurance companies for liquidity," said Ferdinand. "They're among the biggest buyers of Korean corporate bonds. If they stop buying, the Korean corporates go bankrupt."[104]

On another level, however, the banks displayed a prosperity that belied fears of bankruptcy. The words "private banking" at a major bank in Seoul opened the customer to a vista of carpeted private rooms with burnished conference tables and super-polite secretaries eager to serve tea and coffee. The signs on the way to what some banks called "the VIP floor" said, in English, "Private Banking Center" or "Private Banking Team" or perhaps just "Private Banking." Private banking in Korea was hardly what it was in the west, but bankers hoped restrictions would change.

"Private banking is focused on discretionary services in the States," said Son Won Kyung, manager of private banking for Citibank in Seoul. "They take the fee and invest." In Korea, "the private bank does not have discretionary service"—that is, the authority to invest funds on behalf of a client in return for a commission. "We sell only a retail service."[105]

In a sense, private banking in Korea was a fantasy world whipped up by the banks to make well-to-do customers seem important. Despite the restrictions, banks found it enormously profitable. Hana Bank, one of the largest in private banking, had more than 30,000 customers in 1999 with accounts of more than 100 million won, said Moon Soon Min, manager of the bank's "personal banker support team." He estimated that 4,000 of them had deposited more than 500 million won. "I advise them on convertible bonds and stocks," said Moon, who also dispensed advice on radio and TV.[106]

The reason for such success was the rise in stock market values as the country pulled out of crisis mode. Despite the crisis, individuals as well as institutions had a great deal of cash. "The government and companies both need money for financing," said Moon. "The government needs money to inject into the sick financial sector, and companies need money to lower their debt ratios." So who had the money? "Individuals like you and me," he responded. "The money goes to the companies and the government through the channel of the stock market. That's why the market should be bullish." The result: "People who have money in banks are withdrawing their money and investing in the market."

The Shinhan Bank, one of Hana Bank's major competitors, saw its version of "private banking" as a growing field even if the government did not deregulate to the point at which banks could charge commissions. "Private banking is very popular among rich people here," said Lee Kyu Won, a manager on the bank's PB team. He said the bank had about ten "PB centers," each of which had between 600 and 900 customers with deposits of at least 100 million won. Lee analyzed the impact of "the IMF crisis" on the financial world. "After the IMF came in, the rich are getting richer and the poor are getting poorer," he said. "The stock market is booming because the corporates are undervalued."[107]

For all Korea's financial troubles, a great deal of money was in motion in Seoul. "The private sector is recovering, the corporate sector is gaining momentum," said Finance Minister Lee Kyu Sung in mid-1999. "The velocity of money is rising. Money is circulating more rapidly to support these conditions. During the first quarter of this year, listed equity companies raised 6.2 trillion won ($5 billion)." Amid pressure from the chaebol to take over the banks, said Lee, "We cannot exclude the possibility that increased ownership by the chaebol can be utilized to use the banks as their private vaults or to expand their market share." The government, he said, had to "make sure this does not occur."[108]

The chaebol were eager to dominate the financial market. One highly visible example of this drive was the "Buy Korea Fund," set up by Hyundai Investment Trust Management in March 1999 with a goal of amassing 100 trillion won by 2002. Hyundai Securities, controlled by Hyundai Merchant Marine, with 17 percent of the shares, and Hyundai Investment Trust and Securities, a satellite company of Hyundai Electronics, which held more than 30 percent of its stock, were sales agents for the fund. Investing in stocks and bonds issued not only by Hyundai but also by other chaebol companies, notably high-flying Samsung Electronics, the fund helped to power a 70 percent surge of the stock market in the first eight months of 1999.[109] Investors, many of them well-to-do women, were inspired by the patriotic fervor surrounding the fund's sales campaign as well as the advice of their banks.

The activities of the Buy Korea Fund were increasingly suspect. On September 9, 1999, after three days of questioning, Lee Ik Chi, chairman of Hyundai Securities and mastermind of the Buy Korea Fund, and Park Chul Jae, senior vice president of Hyundai Securities, were charged with manipulating the price of shares in Hyundai Electronics. Hyundai Securities had bolstered share prices, said prosecutors, by using 220 billion won or $184 million from two other major Hyundai entities, Hyundai Heavy Industries and Hyundai Merchant Marine.[110] Although the specific offense was committed before the Buy Korea Fund was established, the links between the fund and Hyundai Merchant Marine and Hyundai Electronics were incriminating.

Lee, nicknamed the "compdozer" for bringing the attributes of a computer and a bulldozer to his work, was following time-honored tradition. On September 19, his boss, the cochairman of the Hyundai group, Chung Mong Hun, fifth son of the group's founder, Chung Ju Yung, was questioned for 12 hours in his capacity as chairman of Hyundai Electronics. Mong Hun was no stranger to prosecution. He had been jailed and fined heavily for shifting company funds to aid his father's ill-fated presidential campaign in 1992.[111] Cornered again by prosecutors, the affable Mong Hun knew what to say, professing sorrow for everything, about which he regretted he knew nothing. Also under interrogation was Park Se Yong, chairman of Hyundai Marine Merchant. Park, who was also chairman of the Hyundai Corporation, had gone to jail twice before—first in a bribery scandal in Saudi Arabia in 1979 and again in 1992 for helping to shift funds in the 1992 campaign. Others questioned included Kim Hyong Byuk, chairman of Hyundai Heavy Industries; Lee Kye An, president of Hyundai Motor, and Noh Jung Ik, chief of the strategic management team, among nine Hyundai executives ordered not to leave the country.[112]

On September 21, Lee Ik Chi, Park Chul Jae, and three lesser executives were indicted. Prosecutors said they had to act firmly to prevent others from following their risky example—but cleared all the other executives. Lee Ik Chi's brainchild, the "Buy Korea Fund," had raised more than 11 trillion won, nearly $10 billion, and done much to propel Hyundai stocks. Housewives, who had helped popularize the fund, pouring in money from savings accounts and their husbands' salaries, demonstrated in front of Hyundai headquarters, demanding the arrests of other members of the family of Chung Ju Yung. The fear was the banking system would again be under duress, this time by strains placed on Hyundai. Such investigations, however, were familiar stories in modern Korean history—flare-ups in newspaper headlines, soon forgotten.

The real issue was whether the Korean financial system, notably the banks, had learned much if anything from the shocks of the previous two years. Two days after disclosing the terms of the Newbridge agreement, on September 19, the FSC announced establishment of a fund of 20 trillion won, $17 billion, to allay the fear of a second Korean crisis set off by Daewoo's troubles. The Bank of Korea would back up contributions from local financial institutions, forming, in effect, a huge pool to cover Daewoo's debts when they again fell due.[113] Daewoo's bonds would always be good, as far as investors were concerned, even if the Daewoo companies that issued them had no funds. The plan meant salvation for investment trust companies, in danger of toppling under the weight of upwards of $20 billion in Daewoo bonds.

Nonetheless, the scheme raised the specter anew of unseen forces conspiring in a vast shell game in which the economy remained at risk amid recovery. The recalcitrance of foreign creditors inspired an angry response from Korean officials and creditors, who charged the foreigners, not the Koreans, were the ones demanding privileged treatment. Oh Ho Gen, executive director of the corporate restructuring coordination committee, declared in a letter to Chase Manhattan Bank, Bank of Tokyo-Mitsubishi, and HSBC Holdings, after the disastrous meeting of September 16, that

Korean creditors and government officials could not "accept that foreign creditors be given preferential access to assets of Daewoo or other preferential treatment." Rather than risk "formal court protection and all that entails," the letter urged "meaningful discussion" as "the best way to come to an accord on the terms of a forbearance agreement with Daewoo's foreign creditors—without which there can be no out-of-court restructuring." Thus foreign creditors could "participate fully in the restructuring process." In response to those "pursuing judicial remedies," the letter stated, "we are prepared to commence local insolvency proceedings"—file for bankruptcy in the courts of the countries where Daewoo holdings were threatened—"to protect the interests of all lenders in the assets of the affected company."[114] As Oh explained it, a creditor institution had little chance of recovering the full amount of a loan in any country where a company got itself declared legally bankrupt.

The letter was a blatant attempt at intimidation. Oh believed that foreign banks had already "enjoyed a great deal of profit by lending in Korea" and had "enjoyed a subtle preferential treatment."[115] Now he was determined to make them behave, to curb their greed, to bring them into line, as he saw it, with Korean institutions. This nationalist desire, however, ignored an underlying problem. Whatever equality the Koreans were promising, the final decisions would rest with Korean officials and Korean banks. The foreigners, with no power, would have to wait anxiously in the wings in hopes the Koreans would make good on their pledges. Foreign banks, two years after Korea had plunged into crisis, learned one lesson from the Daewoo affair: the more things changed, the more they stayed the same. They would always be second-class noncitizens in a culture that saw them at best as sources of funding and prestige, at worst as exploitative interlopers to be viewed with the deepest suspicion—and spurned when their goals appeared to conflict with those of big government, big chaebol, and big banks.

The foreign banks, however, were increasingly reluctant to endure such treatment. In a meeting with Daewoo officials in Tokyo on October 28, representatives of 71 foreign creditors adamantly refused to go along with the government's plan for holding up debt payments by the 12 leading Daewoo companies long enough for most of them to recover. The plan amounted to a glorified version of the type of program prevalent in Korea under which companies were relieved of having to pay their debts while in receivership. Foreign bankers realized, however, that assent to such a plan would be tantamount to writing off most of the loans they had made.[116]

The revelation five days later, on November 3, that Daewoo's debts were really 86.8 trillion won, $73 billion, and not $57 billion as estimated earlier, showed how meaningless were the promises made by Daewoo or the government. Daewoo Corporation, which had dispensed funds around the group, owed 40 percent of the debt while Daewoo Motor, Daewoo Heavy Industries and Daewoo Electronics together owed about 30 percent. Adding to the uncertainty were reports of audits showing that all four companies, as well as Daewoo Telecom, had lost more than one third of their book value in the months since Daewoo's troubles had surfaced in July. In fact, the 12 companies under workout were basically worthless. With assets of 61.2 trillion won, not 77.8 trillion won as Daewoo had claimed earlier, their negative net worth was 25.6 trillion won, equal to $21.5 billion. The biggest loser, again, was the Daewoo Corporation, whose value was 60 percent lower than Daewoo had admitted, for a negative net worth of 14.5 trillion won.[117]

The Daewoo crisis threatened the entire banking system. Commercial banks and investment trust companies alike were at risk. Fearful that investors might try to redeem billions of dollars worth of Daewoo bonds, the government on November 4 announced a "stabilization package" that included 3 trillion won, $2.5 billion, for the two largest investment trusts—two thirds for Korea Investment Trust, one third for Daehan Investment Trust. The package also included a scheme for the Korea Asset Management Corporation to buy otherwise nearly worthless Daewoo bonds. Commercial banks, however, posed a much greater risk. Net losses for all commercial banks during the third quarter, June through September, came to $2 billion, according to the FSC. SeoulBank was by far the biggest loser, accounting for three-quarters of the losses, while Korea First Bank and Hanvit Bank absorbed most of the rest. If foreign creditors did not suspend debts, the FSC warned , the government might place the once-mighty Daewoo Corporation and other Daewoo companies in receivership, in which case the banks would get little or nothing.[118]

Such threats did not restore confidence in a system in which the banks, domestic as well as foreign, remained the first victims of chaebol greed. The latest plan for bucking up a couple of investment trusts and sucking in Daewoo bonds was just another stopgap. As for the 64 trillion won bequeathed the banks in the early days of the crisis, that figure covered only about 10 percent of bad loans—not including Daewoo's debts, which had yet to be categorized as "non-performing." Banking reform, Moody's warned, had still "not gone far enough to ensure that the country and its financial sector would be able to withstand another shock."[119]

# Chapter 7

# Unrest in the Workplace*

S outh Korea clung to a rigorous Confucian heritage. While madly seeking to catch up industrially, an authoritarian system of non-questioning obedience permeated business and government. That was not just the rationale for why the owners of the chaebol had to pay homage—in the form of huge gifts—to Chun Doo Hwan and Roh Tae Woo during their terms in the Blue House. It was also why, in keeping with ancient tradition, the owners in turn expected unswerving devotion from the legions of executives, managers, and workers who served them.

The discipline of a Korean company, whether in the office or on the assembly line, was both a strength and a weakness. Obviously Korea could not have achieved the miracle without allegiance to growth and production. Park Chung Hee as president ordered the leaders of business and industry to produce results at home and abroad in a patriotic drive intended to reverse the wrongs of history. The downside was that Park was a dictator who threw thousands of foes into jail, imposed martial law, and scorned democratic forms, including workers' right to strike and businessmen's right to compete and innovate.

South Korea's unparalleled growth had a lot to do with easing restraints. Big business, as much as radical demonstrators, led the way. Chaebol leaders, as their companies grew, needed the freedom to go into new fields and expand on old ones without bureaucratic hassles. They also discovered that their minions would not always come up with new ideas while under the gun of managers demanding immediate results. Those realities, however, bumped up against the old ones—of the need to kowtow before the ruler to do business, to follow the manager's orders unquestioningly in order to keep a job and get promoted.

In this milieu, layoffs were difficult if not impossible at large companies as long as the employer remained solvent. Unions made this point at the beginning of 1997 when 400,000 workers staged a general strike that lasted four weeks. The target of their ire was the most sweeping labor reform legislation since the Korean War. The purpose was to protest the law's most feared provision, giving companies the right to dismiss workers and hire new ones to replace those on strike. The layoff provision, in the view of leaders of both the Korean Confederation of Trade Unions and the Federation of Korean Trade Unions, the two major umbrella organizations, justified a showdown.

The government said it needed the law to comply with requirements of both the International Labor Organization and the OECD, which South Korea was joining after a lengthy campaign for membership. Government and business analysts viewed the new law as an attempt at streamlining business and industry. "We cannot fire, lay off, workers even though the employer pays everything," said Park Il Kwon, a Hyundai manager, alluding to the system under which companies paid salaries and provided office space for union staff members. "I read in the papers that American companies fire 1,000 workers at a time."[1] After the month-long strike was over, however, the law was dead.

The economic crisis resurrected the issue of the right to dismiss workers. Kim Dae Jung, on December 22, 1997, four days after his election, reversed his opposition to layoffs after government officials briefed him on the severity of the economic condition. "I have suffered all my life, but after being briefed I realized our economy is completely at the bottom," said Kim, who would never dare as a candidate to go along with legislating layoffs. "Our economy is in such bad shape, we could go bankrupt even tomorrow." Reluctantly, he conceded, "When businesses cannot save themselves from bankruptcy by reducing salaries, they may inevitably resort to redundancies."[2] His revised position put him in line with at least one of the IMF requirements for digging Korea out of crisis. DJ hoped to wrest labor's consent to limited layoffs through a tripartite commission representing government, the chaebol, and the two big unions.

The National Assembly convened on January 15, 1998, for debate on a bill to permit layoffs from financial institutions. The Assembly then had to debate a bill to legalize layoffs at large companies—a measure that government and business leaders feared would ignite strikes and riots. That fear was of paramount importance to the chaebol. "Layoffs will only come as a last resort," said Hyundai Chairman Chung Mong Ku, worried about the repercussions on Hyundai Motor, which had 45,000 people on its payroll, and Hyundai Heavy Industries, with 27,000 employees.[3]

The unions believed that agreement to participate in the commission would force the Assembly to postpone debate on the new laws. "There will be no legislation," said Yoon Young Mo, international secretary of the 600,000-member Korean Confederation of Trade Unions, as the Assembly opened a four-day session on January 15, 1998. Under current law, only companies with 300 or more workers were unable to dismiss employees. Thousands of smaller companies had dismissed workers neither covered by the law nor protected by unions. In the past two months 300,000 workers had lost their jobs. Companies were closing down every day. In addition, companies had failed to pay wages while eliminating overtime and bonuses, sometimes half a year's wages, and cutting salaries.[4]

Labor officials hoped the tripartite commission, to include leaders of the two unions, two members of chaebol management, two top government officials, and four representatives of political parties, would come up with a formula by which the chaebol admitted responsibility for the crisis. "There will be very strong controversy on the law," said Ahn Dong Sul, director of the international division of the Federation of Korean Trade Unions, the nation's largest labor organization, with about 1.2 million members. "We will discuss how to amend it. We strongly demand the government make the labor law strict. Companies must not try to use the labor law to lay off workers." At the top of the FKTU's demands was that companies sell off assets and stocks, downsize managerial offices, and if forced into bankruptcy "consult with the trade unions" to see if layoffs were necessary. Finally, said Ahn, the government should extend unemployment insurance from three months to one year and raise unemployment checks from 50 to 70 percent of wages, including bonuses.[5]

The unions appeared to have hardened their stance while chaebol leaders were reaffirming the need for change. Chairmen of the top 14 chaebol, after a meeting at the Federation of Korean Industries, vowed "to terminate businesses with poor potential, sell off unnecessary assets, and engage in mergers and acquisitions as part of swift and voluntary restructuring efforts." The chaebol leaders sought to show their willingness to cooperate with unions by acknowledging that "all segments of society must share the burden of the current financial and economic difficulties," but how far they would go beyond words was unclear.[6]

The outlook of both union and chaebol officials raised the question of whether both sides still hoped to avoid more layoffs and bankruptcies as the economy recovered from the worst of the crisis. After two days of wrangling in meetings of the tripartite commission, the two national labor unions won an agreement with top business and government leaders on January 20, 1998, that avoided compromise on mass dismissals and blamed business and government for Korea's economic mess. The much-heralded "joint declaration" on "fair pain-sharing to overcome the crisis" emerged amid strike threats and protests in the form of a five-point accord. The key points were that the three parties should "create the atmosphere to induce overseas capital" and "cooperate to reach an agreement" on the labor issue in the National Assembly.[7]

The unions got the declaration to place most of the onus on government and business, beginning with an opening that declared, "Government and corporate management are responsible for the crisis." The five-point declaration began by calling on the government to "reinforce conservation measures" and to "prepare a policy for unemployment," then called on companies to come up with "substantial restructuring measures" and stop "abusing labor by illegal layoffs." At the same time, the declaration demanded labor unions "cooperate on the problem of wages and working hours" and "avoid strikes and protests."[8]

The crucible of the labor movement was Ulsan, on Korea's southeastern coast, where nine Hyundai factories whirred away in the widest range of industries of any industrial complex in Asia. The Hyundai factories that had risen there over the past generation had earned for Ulsan the nickname "Hyundai City" and made it a thriving hub of commerce and industry. Nowhere was the disillusionment so pronounced as in this one-time fishing village, now swollen to more than a million people, the majority dependent on Hyundai for a living. In Ulsan, the workers had strength in numbers. Hyundai employed about 100,000 people inside the city, which covered a geographic area larger than Seoul. The proximity of the factories contributed to the crusading spirit.

"IMF" became a catchphrase for doubt and despair about the future of the two million people in the city and surrounding villages. Asked how was business, a shopkeeper in 1998 was certain to respond that it was "slow" if not "dead." Asked why, the answer was likely to be "IMF"—even if the shopkeeper had no idea what the initials stood for. "IMF" was reason enough for a round of "sales"—the word emblazoned in English on store fronts—as Hyundai factories shut down for the week on Monday, January 26, for the three-day lunar New Year. "We are conducting an IMF sale," said the soft voice over the loudspeaker in Hyundai Department Store, across a busy street from the walls of Hyundai Heavy Industries. "We are giving extra discount on selected items." In the bars and restaurants up and down streets and alleys, "IMF discounts" and "IMF lunch" were the order of the day, and "IMF drinks" were on sale everywhere. In the Hotel Diamond, another Hyundai enterprise, next to Hyundai Department Store, foreign engineers and contractors selected dinners from an "IMF special menu"—including the requisite "IMF discount."

Managers at Hyundai Department Store explained that customers were hoarding their money—even if the crowds before the lunar New Year seemed fairly large. "People are buying presents only because of the holiday," said Cho Hyung Ryu, behind a counter laden with electronic products, mostly made in Japan. "Clothes are about all that anybody is buying. Overall sales are down by about 70 percent. Nobody wants to buy anything from abroad."[9]

Executives and managers at the Hyundai companies, most of them from Seoul, no longer flew home on weekends. The cavernous new Ulsan airport, with its long picture windows giving a view of the runway and surrounding hills, was almost empty except for a brief flurry whenever a plane was due. Air hostesses offered water during the 45-minute flights to and from Seoul but no longer served snacks or drinks. "IMF," said a hostess when asked why she had no tea or coffee. She covered her mouth with her hand to suppress a giggle. "IMF," she repeated.

On the ground, the most critical test case, for government, business, and labor, was Hyundai Motor. In the headquarters of the company's main plant, a video urged visitors to "take a look at Hyundai Motor Company in the twenty-first century."[10] The official picture of a motor vehicle company on its way to "the global top ten" among world manufacturers contrasted with the fears of Hyundai workers as they grappled with the threat of mass layoffs and long-term unemployment.

In the streets on the way to the assembly lines, workers talked in anger and fear of the chances of losing their jobs as sales dwindled and prices rose in the new era. "Although we now have jobs, houses, and cars, we live day-to-day on our pay," said Lee Sang Yong, a veteran of four months in prison for his role in a strike in which workers had seized the plant for several days in 1992.[11] To rank-and-file workers, layoff legislation revived memories of a decade-long struggle, first for the right to strike, then for higher wages and working conditions, and finally for job security.

To leaders of Hyundai unions, most of them affiliated with the KCTU, one problem was the hereditary leaders of the chaebol. At Hyundai Motor, the target was the company's 35-year-old chairman, Chung Mong Gyu.

"He is a novice," said Kim Kwan Soo, talking in a union office festooned with caricatures of chaebol owners. "He has no ability to control Hyundai Motor."[12] The attack on Chung Mong Gyu, who had replaced his father, Chung Se Yung, as HMC chairman in 1996, reflected the widespread view among union members that the owners were responsible for overexpansion that had led to economic crisis.

Chung Mong Gyu, an Oxford University graduate, groomed for the chairmanship in a series of managerial positions, still dreamed of leading the company's drive to become one of the world's top ten motor vehicle producers by 2000. The company's three plants now had the "capacity" to produce 1.8 million vehicles a year but in 1997 sold only 1.24 million vehicles, approximately 600,000 shipped overseas and 640,000 for the domestic market. "This year," said Lee Byong Gil, an official in the plant headquarters, "our production plan is flexible"—so flexible that the company had eliminated overtime and sent most of its 45,000 workers home to celebrate the lunar New Year.[13]

At Hyundai Heavy Industries, union leaders were not yet worried about layoffs. All told, 26 ships, ranging from a 300,000-ton supertanker to a 40,000-ton container vessel, were under construction in 1998 in a facility that had produced more than 730 vessels in 24 years—and had contracts through the year 2000. Even if the jobs of his own members were not at risk, however, HHI union leader Yoon Jae Kun opposed any deal that jeopardized workers at other Hyundai companies. "Rather than strike, we would cooperate to revive the economy," said Yoon inside the HHI compound on a street leading to dry docks beneath goliath cranes hovering high above the hulls of half a dozen vessels, "but if it's a one-sided thing, we are not going to accept it."[14]

HHI union leaders, like those at the motor company, placed top priority on "new management, not family management," as their first demand. Their primary targets were HHI's two leading stockholders, Chung Ju Yung and his sixth son, Chung Mong Joon. The latter, a member of the National Assembly, held the title of "adviser" to HHI while focusing on his job as chairman of the Korean Football Association, preparing to cohost World Cup 2002 with Japan.

After the New York bankers agreed in the last week of January 1998 to roll over debts, however, labor leaders no longer hoped to embarrass the government by stonewalling. Rather, their best hope lay in bargaining for the best terms for the hundreds of thousands about to lose their jobs. Finally, on February 6, 1998, the tripartite commission announced a sweeping agreement that would legalize mass layoffs. After three weeks of negotiations and an all-night bargaining session, representatives of labor, business, and government finished a draft of the legislation to present to the Assembly. Leaders of the two national unions abandoned their no-layoff stance in return for several concessions, including unemployment insurance, legalization of a national teachers union, and the right of unions to participate in politics.[15]

"It's a very fragile agreement," said Yoon Young Mo at the KCTU. "It requires the government and the president-elect to do everything possible to win the confidence of workers."[16] The group called off plans for a three-hour sit-down strike while the chaebol hailed the agreement as "a gallant act." The FKI said it showed "our strong will to overcome the current crisis" and promised "every effort to reinforce management transparency," one of the IMF demands. Ahn Dong Sul, at the FKTU, predicted, accurately, that the unemployment figure of 660,000 would soon double.[17]

The unions demanded provisions under which owners and executives would suffer along with the rank and file. That effort was reflected in the claim of the tripartite commission that all sides had agreed on "the social contract to equally and sincerely share the pain in the course of economic restructuring." Management could dismiss workers "only if it is unavoidable." Companies had to give 60 days' notice of layoffs both to the ministry of labor and to the workers and their unions and had to rehire as soon as possible. Mergers and acquisitions would count as a "management crisis" under which layoffs might be necessary.[18] The legislation would cushion the blow of dismissals with unemployment insurance, welfare benefits, and retraining.

The 300 chiefs of the company unions that made up the KCTU, in a raucous six-hour meeting that began on the evening of February 10, 1998, and lasted into the early hours of the 11th, called for "reopening negotiations" and agreed on a set of demands. First among them was that the government double the unemployment insurance fund provided under the legislation from five trillion to ten trillion won—$6.5 billion at the time—to guarantee benefits for laid-off workers. Second, the KCTU demanded the resignations of the chaebol chairmen and owners and said they should invest their own privately held funds in their cash-strapped companies. Finally, delegates voted two to one to reject a bill proposed by the commission, including the KCTU's president, to legalize layoffs under "emergency" circumstances. Dan Byung Ho, the rabble-rousing head of an "emergency task force," announced a "general strike" by the union's 600,000 members beginning February 13. "We will fight to protect the security of the workers," Dan told the meeting, the original purpose of which had been to "ratify" the union leadership's decision to endorse lay-off legislation.[19]

Representatives of more than 60,000 workers at Hyundai company unions, all affiliated with the KCTU, announced separately that they would strike on February 13. Workers at Mando Machinery, the country's largest manufacturer of automotive components, said they too would strike if the company failed to provide a "special agreement" on job security. A strike at Mando, one of the few profitable entities of the Halla group, could have a ripple effect. "It's inevitable we will stop production if Mando goes on strike," said a spokesman for Kia. Mando, which employed about 8,000 workers, hoped to maintain 70 percent of production by putting white-collar workers on assembly lines. The move threatened a standoff between management and labor that could set a precedent for other industries.

Members of the commission were to have signed the agreement on February 11 as the Assembly began considering it in hopes of passing it by February 14. The defiance of the Korean Confederation of Trade Unions came as a blow to DJ's advisers, who believed they had overcome the main objections of workers through the commission. The president-elect, however, still counted on the larger Federation of Korean Trade Unions, which had supported him in his campaign. The FKTU, historically allied with the government and less combative than the KCTU, had not objected formally to the agreement.

The demands of labor were expected to gain momentum as companies were forced to lay off tens of thousands of workers. Although the National Assembly had legalized layoffs in January, the KCTU threatened strikes and walkouts if companies abused or overlooked requirements for discussing layoffs with union leaders and dismissing workers only in an "emergency." Workers at Hyundai Motor's main plant in Ulsan on February 22 said they would protest suspension of five of its lines amid sagging domestic sales. A Hyundai spokesman said workers would receive holiday pay for much of the time they were off the job. Workers at Mando canceled one strike in February only after the company promised not to lay off workers after failing to meet debt payments in December 1997.

Government and business officials believed, however, that the KCTU would relax its demands. "We still have some time to reach some breakthrough," said Bae Ie Dong at the FKI. "Eventually they will sign it. We can avoid the worst scenario." Bae scoffed at demands of unionists that chaebol owners resign and give seats on the boards of their companies to union members. "Nobody will pay attention except their own unions," he said.[20]

Preaching the inevitability of layoffs, DJ, after his inauguration in February 1998, was transformed in the eyes of workers from hero to villain. On the May Day holiday, more than 20,000 union members and student sympathizers confronted lines of police armed with truncheons and tear gas. "Kim Dae Jung is the enemy of labor," said a headline in a newspaper published by radical workers. "The Kim Dae Jung government is deceiving labor," said a pamphlet distributed by the KCTU on May Day.[21] The police said they were seeking Lee Kab Yong, newly elected KCTU president, and seven other organizers as well as radical student leaders for their roles in the riots. After the police had dispersed his followers with volley after volley of tear gas, Lee led several hundred of them in a slogan-shouting rally on the steps of Myongdong Cathedral several blocks away. "Never weaken, let's keep going," was the theme of his remarks, echoed by the men as they shouted back his words.

"We will prove ourselves by action, not words," said Lee Kab Yong, interviewed at the cathedral. "If we demonstrate in cities around the country as we did on May Day, it will show our power." Jailed in the early 1990s for leading labor disturbances at Hyundai Heavy Industries, Lee cited Hyundai Motor's plan to lay off 8,000 workers as an example. "They should cut down hours, not lay off workers," he said. "Only workers are asked to suffer," said one of the demonstrators, Chang Hwa Shik. "The company people and the government people stay where they are." The workers expected DJ to be as tough as previous presidents in fighting labor unrest. "There's no difference between him and the others," said Kim Chong Rae, a subway train driver. "He is not governing the country for labor but for the chaebol and the rich."[22]

The government's underlying concern was that the May Day demonstration, the first violent protest against Kim Dae Jung's policies, might be the precursor of a prolonged labor revolt. Approximately 1.4 million workers, more than 6.5 percent of the workforce, were unemployed, and the number was rising as small and medium-sized companies went under and larger firms slowly laid off workers. The consequence of labor strife could be both to undermine the tripartite agreement and to scare away hesitant foreign investors. The fears were reflected in the numbers. With foreigners selling off stocks, the Korean Stock Exchange descended to its lowest level since the depths of the crisis in December 1997. The market closed on May 2, 1998, at 406.56, down 14.69 points.

The pressure for political calm before the storm was such that Hyundai Motor on May 6 disavowed reports that it planned to lay off 20 percent of its 45,000 employees. As a portent of the labor strife feared by the industry, about 5,000 workers went on strike at Mando Machinery. The KCTU on May 26 called for a general strike the next day, claiming that 120,000 of its 600,000 members at 90 heavy industrial concerns would join the walkout after 32,000 workers at Hyundai Motor's Ulsan plant voted to strike for two days to protest layoff plans. Shin Hyun Kyu, Hyundai Motor's public affairs director, denied the company had decided to lay off workers but said that 8,000 of its 45,000 employees were no longer needed, since motor vehicle production had fallen more than 50 percent since the end of 1997. "The Korean economy is going down," he said. "We don't know when it will hit bottom."[23]

The company asked authorities to stop any protest, and reinforced police units were poised outside the plant in Ulsan. Workers at two other Hyundai companies in Ulsan, Hyundai Precision, which manufactured cargo containers, vans, and sport-utility vehicles, and Hyundai Motor Service, which operated the sales network for Hyundai vehicles, were also expected to join the strike.

At the epicenter of the workers' revolt in Ulsan, the issue was survival. "We will fight for our families, our fathers, our mothers, our wives, our children," Kim Kwang Shik, leader of the Hyundai Motor union, orated before about 15,000 of his followers gathered on the afternoon of May 27 inside the plant. "We are fighting to avoid death." While the workers waved fists and shouted, a panoply of tear gas tanks, buses filled with policemen, and rows of police on foot guarded the Hyundai compound. On a five-mile march from the company compound to a riverside park, workers threatened worse to come. "When the company needs people, they make us work all the time," said Kim Yong Jin, an assembly line worker. "When they don't need us, they abandon us. We cannot let our close friends lose their jobs." The wife of a worker forced to "retire" spoke of the humiliation. "Since my husband is not working, it's embarrassing to my children," she said. "Financially we are in trouble. Therefore all Hyundai workers greatly support this strike."[24]

Mindful of the strikes of previous years, the company said it would talk to the union before laying anyone off. "More than 1,400 have already applied for voluntary retirement," said a Hyundai official, noting that assembly lines had been operating at only 45 percent of capacity since January.[25]

How strongly the workers would challenge the company was uncertain, however, after the strike wound down just two days later. More than 20,000 workers at Hyundai Motor and Hyundai Precision promised to return to their jobs the next day after having served what their leaders said was a "warning." Although the government had declared the two-day walkout "illegal," the police had made no effort to stop the workers from rallying inside the Hyundai Motor compound or marching to the park. Officials appeared to be counting on a lukewarm response from ordinary citizens to a walkout that many believed would further undermine the economy. The walkout demonstrated the ambivalence of Koreans, including many workers, about a prolonged strike. "Traditionally Ulsan has the most radical unions," said Park Sam Yuel, a Hyundai Precision manager, observing the confrontation. "Today's strike is very important because the labor movement here is the first and the strongest in the country. Everybody keeps their eyes on the result."[26]

While the walkout was a tame affair compared to the violent strikes that had rocked Ulsan in the late 1980s and early 1990s, no one could ignore the warning of Chun Chang Soo, vice president of the Ulsan council of the Korean Metal Workers' Federation. "There were no problems with the police this time," he said, "but there will be many in June." Chun admitted, however, the strike's unpopularity. "People say workers who walk off their jobs don't buy products," he said. "Some people say workers make the Korean economy look bad."[27]

To Chung Dal Ok, vice president in charge of the Hyundai Motor plant, the problem was how to convince workers of the impossibility of winning even if they closed down the plant for weeks on end. "Eventually they will know, without layoffs, many more people will suffer," he said. "To prevent our situation from getting worse, we have to take action." He said the company would produce about 900,000 vehicles in 1998, down from 1.1 million in 1998, losing money for the first time in more than 20 years. The numbers were "terrible," said Chung, a reflection of the country's economic condition as seen in figures showing that industrial output had declined by 10.8 percent in April 1998 from April 1997 while wholesale and retail sales had dropped 15 percent.[28]

The hardest hit were not the workers at the major companies but those at small and medium-sized enterprises, downsizing or closing. With nearly 2,000 people losing jobs every day, South Korea's unemployment rate soared to 1.49 million by the end of May 1998, a record 7 percent of the workforce, compared with 658,000 out of work the previous December. The number losing jobs was certain to rise as the government pressured insolvent companies to downsize or close. Nonunionized small and medium-size firms bore most of the heat; the government could avoid unsettling political pressure from newly jobless workers if most of them did not belong to unions. An army of unemployed was forming similar to that in Britain, where 4.5 million people were "on the dole."

The finance ministry on June 23 admitted that the economy would probably contract by 4 percent during the year after having previously insisted a 1 percent decrease was possible. The government had no choice but to provide extra funds for trade, housing, and unemployment even if the result would be to weaken the won. That policy was a response to both companies and workers. Companies complained that they were unable to obtain credit to import raw materials and machinery needed to produce goods for export; unions said the government was not providing nearly enough for welfare and unemployment benefits.

Union pressure rapidly increased in the face of government and business efforts at laying off approximately 500,000 more workers by the end of the year. Leaders of both the national unions, despite their frequent criticism of each other, got together on July 12 to lead thousands of their followers in a fist-waving rally in defense of job security. The rally, in a park beside the Han River in Seoul, displayed both the strength and the weaknesses of the unions as they entered a week of strikes and slowdowns. "Protect our jobs" was the slogan repeated again and again by KCTU President Lee Kab Yong. "We must fight to the end against layoffs." For all the words, however, the rally drew only about 20,000 people, far fewer than the 70,000 who the organizers claimed were there or the 100,000 they predicted beforehand would come.[29]

Most of the workers at the rally were from financial institutions, including the five banks that the Financial Supervisory Commission had liquidated. Bank employees appeared less enthusiastic than workers in heavy industries about a general strike. Choo Won Suh, president of the Korean Confederation of Banking Unions, said he preferred to pursue talks through the tripartite commission. With 30,000 of the country's 150,000 bank workers out of jobs, he favored a plan under which banks would lay off workers gradually.[30]

Among the workers' greatest concerns was takeover by foreigners. "We don't know what they will be like," said Choi Jong Kuen, a union worker. "Their style is different."[31] A pamphlet handed out by workers from the metal industry called plans to sell Posco, the Pohang Iron and Steel Company, "a betrayal of the nation"—even though Posco itself was not unionized. Another union representing the 58,000 workers at Korea Telecom, the national telephone company, also going up for sale, asked, "Do foreign investors want our economy to survive?" The answer, said the union, was "Once they get profits, they will leave."

Down in Ulsan, the tension was growing by the hour. On July 16, union leaders at Hyundai Motor resolved to revive their strike after the company announced plans to lay off 2,678 workers, a sharp reduction from the 8,000 originally marked for dismissal. Denouncing the plan as a "betrayal," a spokesman for the union said workers would walk out again the next day when the company was to announce the names of the workers losing their jobs. Another 900 workers, said the company, would be furloughed without pay for up to two years. The decision to go through with dismissals galvanized the union to action two hours after union leaders had called off a strike in the belief the company would accept demands for lower pay and shorter hours rather than layoffs.

The confrontation at Hyundai Motor epitomized the national labor crisis. As police searched for more than 80 labor leaders responsible for a series of "illegal strikes," top labor leaders again retreated to the drive up to Myongdong Cathedral. "If they arrest me, the workers will be outraged," said Lee Kab Yong. "They have had many opportunities, but they are afraid." In the sanctuary of the cathedral grounds, Lee and other union leaders planned another "general strike" beginning July 22, 1998. "Stop the layoffs, reform the chaebol, and reform the social welfare system," Lee repeated as a litany when asked what the government and the chaebol had to do to halt the strikes. Those demands, he said, were "preconditions" for negotiations.[32]

Kim Ho Sun, president of the union at Korea Telecom, one of the country's largest employers with 58,000 workers, said he had led 20,000 of them off their jobs for one day in the first strike in the company's 115-year history. In the face of government claims that strikes were illegal, Kim, also on the government's "wanted" list, directing his followers from a tent in front of the cathedral, said his union would join workers in the heavy industries as well as a number of banks in staying off the job. "The most important result is that workers gain self-esteem," said Kim, surrounded by several other labor leaders, also "wanted" for "illegal" walkouts. "Our strikes will get bigger and bigger. We will begin with 50,000, then we will have 100,000 and 150,000." He promised to lead a general strike unless the government abandoned its plan to privatize the company. "Once foreign companies come into the market, they will conflict with other companies, and many more people will lose their jobs," he said. "Then they will try to exclude what we have done for the public interest."[33]

With pledges by leaders of Korea Telecom and the Seoul subway system to walk out on July 23, union leaders predicted 100,000 workers would stay off their jobs, twice as many as the week before. While the walkouts did not have popular support, officials were reluctant to attempt to stop them with riot police for fear of more violence. In negotiations with Korea Telecom, they sought to explain the union's objections to privatization as a misunderstanding. "The plan isn't as bad as they think it is," said Kong Sung Do, director of the privatization division of the government's planning and budget commission. "They need to be careful not to equate privatization with massive layoffs."[34]

The KCTU remained adamant, however, against layoffs. A showdown was likely on Friday, July 24, the last day of work for 2,678 workers whom Hyundai Motor was laying off. Already about 2,000 workers and family members were camping out inside the factory grounds in Ulsan. The head of the Hyundai Motor union, Kim Kwang Shik, also sought by police, had retreated inside the compound. The fear was the workers would cripple the industry. The company had not produced a single vehicle since July 14. By now several thousand policemen were guarding the compound. The company, after persuading about 8,000 workers to retire or quit voluntarily, sent out dismissal notices on the 24th to 1,569 more. About 36,000 people remained on the payroll after voluntary resignations and early retirements.

As talks dragged on through the evening of August 10, there was no indication when production would resume. "We have lots of back orders from abroad," said Lee Byung Ho, a Hyundai spokesman in Ulsan. "We are facing the collapse of our suppliers. Major vendors face bankruptcy."[35] Lee said that police action to remove demonstrators from the compound might be "one of the solutions." The confrontation was a test of the ability of companies to carry out restructuring entailing heavy payroll reductions. "Corporate restructuring is happening very slowly," said Shim Sang Dahl, chief economist at the Korea Development Institute. "The economy is now going through its worst moment."[36]

Inside the Hyundai Motor compound, an air of quiet desperation hung over the plastic and vinyl tents lining the streets. Housewives led workers in song, and loudspeakers announced the news of the next rally. Children scampered in the shadows of idle assembly plants, and fight slogans were painted on walls and pavements. In furtive groups outside their tents, men explained why they had virtually taken over the compound and would never let the company make another vehicle until it agreed to their demands. "We are fighting for life," said Kim Jong Myung on August 16, more than a month after the assembly lines had produced their last vehicles. "We think responsibility for suffering lies with the company, but the company doesn't suffer. They blame the union members."[37]

The standoff intensified over the weekend with the refusal of both union members and management to change their positions. For union members, the issue was simple: it was all right to cut work weeks and possibly to lower wages but never to lay anyone off. The management position was just as plain: with production now down 45 percent from a year before, it was necessary to lay off 1,500 workers. The company would accept "voluntary retirements" from more than 6,000 others and bequeath two-year furloughs to another 2,000. "Most of the workers are for us," said spokesman Lee Byung Ho, in company headquarters across a square from the nearest workers' tents, decked out with flags and banners. "More than 20,000 of us are rallying on Monday to show the strikers are in a minority."[38]

In a company roiled by strikes almost annually in the early 1990s, the showdown appeared more intense than the strikes of years gone by. The difference was that hundreds of women and children now lived in the tents, ready to join in lines of defiance, breathing tear gas and taking body blows from thousands of policemen poised to drive them from their tents. "Yes, we are afraid of the police coming in," said Lee Young Ja, mother of three small children, sharing a tent with 100 others between a cafeteria and a supply building, "but if our husbands lose their jobs, that's also a way to die." Given the alternatives, she said, "we may as well fight to the end until the police come in."[39]

To the government, the prospect of such a denouement was alarming. On the evening of August 16, a vice minister of labor, Ahn Young Su, arrived in an eleventh-hour bid for compromise. "The labor union says there is no truce, even one person cannot be laid off," said Ahn. "We asked the union to collect opinions through the night. We will wait until Monday morning for anything to change." He denied he was offering an ultimatum. Down the street, union leader Kim Kwang Shik watched his followers from a platform atop a 60-meter-high scaffolding made of iron pipes erected on one of the factory buildings. Inside, a union spokesman said Kim would hold out through any police attack. "If there is a police raid," he said, "the union's attitude is we will fight to the death."[40]

All day Tuesday, August 18, the lines were forming in the most severe test so far of the policies of the government in confronting the conflicting demands of big business and big labor. Thousands of policemen, armed with tear gas, shielded by padded uniforms, masks, and helmets, moved in tight formations along the street running by the high walls of the compound. Inside the gate, rows of workers, armed with steel pipes and fire extinguishers for combating tear gas, moved in equally tight formation.

"Fight to Secure Jobs" was the translation of the white Hangul lettering on the red headbands worn by the 5,000 workers and several hundred wives and children still with them. "We choose to die with our husbands," women with babies in their arms shouted across barricades set up by the workers to impede the policemen whenever they got their orders to swarm into the compound. The mood intensified into the evening as police commanders barked orders to their men and union members banged steel pipes onto the pavement of the compound 100 yards away. Helicopters swept overhead, dropping leaflets headlined "Final Warning" to the workers, and union fight songs blared over loudspeakers along the streets of the compound. By nightfall, as labor activists from Seoul whipped up the strikers with slogans and speeches, workers parked a truck bearing a large steel drum in front of the barricades. The drum, and several containers of natural gas, would explode, they claimed, when the police attacked.[41]

The confrontation represented the failure of government efforts at bringing management and union to terms. "Both sides have to think seriously," said Labor Minister Lee Ki Ho, refereeing final talks between Hyundai Motor Chairman Chung Mong Gyu, also down from Seoul, and union leader Kim Kwang Shik, down from his scaffold. "Both have not changed their position. The problem of Hyundai Motor is not just a problem for Hyundai. If it continues, it creates a big ripple effect for the Korean economy."[42] Equally daunting was the fear that a police attack on the strikers, shielded by rows of women and infants, would lead to a rash of sympathy strikes elsewhere.

The statistics on the wall beside the main gate at the Hyundai Motor headquarters showed the price that Hyundai was paying for dismissing about 1,500 workers without offering the voluntary retirement packages or long-term furloughs bequeathed about 8,000 others. As a result of the strike, the company had failed to build 82,810 vehicles and lost about $600 million, according to its tabulation, while its supply network had lost another $500 million.[43] "It's very depressing," said Chung Mong Gyu, opposite union leaders at the table.[44] "We have decided other Hyundai companies will go on strike," said Huh Young Koo, KCTU vice president. "There will be a general strike." The pressure was on. "If we leave the situation any longer, it will do a great deal of damage to the economy," said Munwha Broadcasting. "The government should learn from experience and take swift action."[45]

To the disappointment of Hyundai executives, however, violence was an option that the government was determined to avoid. On August 21, 1998, as thousands of policemen remained on guard outside the compound walls, the leader of the government team, Nho Moo Hyun, an Assembly member dispatched by DJ to negotiate a settlement after all else had failed, said there was "no cause to send in the police." The reason, he said, was that the strikers had agreed on a compromise. Kim Kwang Shik, union president, won the government to his side by saying he would accept layoffs of between 250 and 300 workers and one-year furloughs for another 1,238 workers to whom the government, the union, and the company would pay benefits matching salary for six months.

Minutes after Nho and his team said they had an agreement, however, the company said the deal was not acceptable. "We cannot agree to the politicians' recommendation," said Kim Pan Kun, a managing director. The company's "last proposal," he said, was for layoffs of 400 people and no pay for those given furloughs. Chung Mong Gyu charged negotiators were "favoring the labor union." Hyundai officials said the government had abrogated responsibility. "It's the government's role to maintain the law," said spokesman Shin Hyun Kyu.[46]

In a case watched minutely by hundreds of Korean companies facing the need for layoffs, Nho indicated the government was prepared to pressure Hyundai into agreement. "There are no longer three-way negotiations," he said. "Government-labor negotiations do not exist. Now there are just government and company negotiations." The government was saying "layoffs are inevitable," said Yoo Tae Ho, executive managing director of the Daewoo Research Economic Institute, "but the problem is the size of layoffs will have a great impact on the economy."[47]

Kim, the union leader, said the government decision to send a team of politicians as negotiators represented "a shift in the government attitude." The reason he had fought layoffs, he said in the union headquarters, was that "I do not want my union members to end up with people taking their lives or end up as part of social degradation." He said that three million people were unemployed in Korea, about twice the number reported by the government, and that only 22 percent got benefits. Every day, he said, "47 jobless people take their own lives." He denied, however, that he had ever said he opposed "a single layoff." Rather, he wanted the company to consider "alternate proposals, like reducing pay across the board for everyone."[48]

Pressured by the government, Hyundai grimly agreed on August 24 to the compromise, promising to open the gates for work at its main plant in Ulsan the next day for the first time in six weeks. The final agreement, hammered out at an all-night bargaining session in Ulsan supervised by the minister of labor, was a victory for union members, who did not decamp from the compound until the settlement was announced. Under the agreement, Chung Mong Gyu had to assent to dismissing only 277 of those sent dismissal notices. The workers who lost their jobs would receive severance pay of seven to nine months' salary while the rest would be furloughed for eight months without salary and then given six months of training for new jobs. More than half of those finally let go were women who had worked in kitchens on the compound. An outside contractor would replace them.[49]

Government negotiators claimed victory by having established the precedent of layoffs, if only a few. "With this joint effort, we will do our best to normalize our factory," said Chung, shaking hands with Kim Kwang Shik, still wearing his red headband. Labor Minister Lee, standing between them, observed, "Without involvement of the police, we have talked so the union and the company can solve the problem peacefully." Hyundai Motor was the first Korean manufacturer to dismiss workers in keeping with the law enacted in February 1998 setting up the tripartite commission. "The labor side has accepted the principle of layoffs," said Kim Dae Jung, hoping to mollify business leaders.[50]

No sooner had Chung Mong Gyu returned to Hyundai headquarters in Seoul than spokesman Shin said the agreement was "not satisfactory." The company had acted properly by laying off workers as prescribed by the new labor law, he said, but the union had "occupied the factory illegally and achieved their own goal even though they broke the law." Hyundai had fallen behind in production by 105,465 units, most of them the 800-cc ATOZ (the name meant A to Z, "the car that can do anything"). Losses for the company totaled 905 billion won, about $700 million; 2,000 parts suppliers lost another 700 billion won, $540 million. Hyundai Motor sold just 10,857 vehicles in August, a drop of 90.3 percent from August 1997.[51]

Both management and labor viewed the outcome of the strike as having charted the course for intense negotiations and walkouts. Business leaders denounced it, while labor representatives were pleased. "We are very, very much disappointed," said Bae Ie Dong at the FKI. "It will set some bad precedent in many aspects. It will influence efforts in restructuring not only for private business but for many state-funded organizations. It will certainly hurt foreign investors."[52] David Young, at the Seoul office of Boston Consulting, agreed. "I'm afraid it doesn't bode well," he said. "It is going to send a lot of wrong signals to international markets. It makes me wonder how much freedom there will be. Everybody's very much confused."[53]

There was no doubt the compromise favored labor. "This solution will guide Korean labor policy," said KCTU spokesman Yoon Young Mo. Previously, he said, chaebol leaders had seen the labor law as providing "a legal pretext for layoffs under which they should have all the freedom and power to go ahead with it." He praised DJ for having held back the police. "If the police had come in, it would have sent Korea back to the dark ages," he said. "There was very strong pressure on him to do that."[54] Nor were the troubles necessarily over. The Hyundai Motor union on September 1 voted overwhelmingly against the agreement. The count showed 17,123 workers against and 9,360 for it. The workers, however, could do no more. The agreement stood as written.

If a strong union could battle the chaebol to a standstill at a major company, workers at smaller companies, those with fewer than 300 employees, had no such protection. The number officially categorized as unemployed shot up to 1,651,000, 7.6 percent of the workforce, by the end of August 1998 while exports dropped precipitously. The parallel between rising unemployment and declining exports was alarming. "Korean enterprises are in deep recession, so the bankrupt enterprises have generated unemployment while exports are going down," said Choi Jang Jip, chairman of the Presidential Commission on Policy Planning. "Most of the Korean economy cannot escape from this kind of shock." He predicted that total unemployment would rise to two million, 9 percent of the workforce, by 1999.[55]

The latest unemployment statistics, revealed by Labor Minister Lee, showed that the number of Koreans out of work by the end of July 1998 had increased by 122,000 from the end of June, when 7 percent of the workforce was jobless. It was the largest jump in a single month since March, when unemployment had risen from 1,235,000 to 1,378,000, 6.5 percent of the workforce. Lee said the figures did not include nearly 1.5 million others who were either underemployed or so discouraged they were no longer looking for work. He agreed with estimates of labor leaders that overall about three million workers were either jobless or employed less than 18 hours a week. "The social safety net is not sufficient," he said, observing that only about 50 percent of regular wage workers were eligible for benefits that lasted only two to six months. The social welfare program at this stage did not include day laborers, a major segment in need.[56]

The 7.6 percent unemployment rate was the highest in Korea since 1966, when the country was still recovering from the Korean War and entering an era of industrialization marked by massive increases in productivity and exports. The per capita gross national product, after having risen to an all-time high of about $10,000 in 1997, was now about $7,000. "The Korean economic crisis will hit bottom early next year," said Cho

Jae Hee, professor at Korea University. Almost half the country's small and medium-size enterprises had gone bankrupt in the past eight months. Industry was squeezed in a vise of low wages in China and high tech in Japan, while enterprises slashed costs by cutting imports of capital goods and raw materials needed to make the finished products for export.

As the gap between rich and poor widened in the IMF era, the spirit of revolt in the workplace was in danger of fading into the sunset of an economic miracle under which the labor movement, like corporate Korea, had come to flourish. On September 3, 1998, as Hyundai Motor assembly lines were making up for lost production, the police cracked down hard when the union at Mando Machinery tried to take over the main plant. Militants, protesting the layoff of 1,090 workers, were driven out and arrested in nearby buildings. The KCTU threatened more action, but the protest fizzled. Gradually the unemployed and underemployed were forming an underclass reliant on welfare, odd jobs, and handouts.

Amid signs of a rebounding economy, the government at the start of 1999 still predicted nearly two million people would be jobless by spring and several hundred thousand more in the category of "hidden jobless." The tripartite commission chairman, Kim One Ki, forecast unemployment at "a record high in the first half of this year," with the average time that people were out of work exceeding predictions.[57] Figures released on January 22, 1999, showed the average unemployment rate for 1998 was 6.8 percent, highest since the Korean War. The number jumped to 7.9 percent in December, a record, and might go up as a fresh crop of 300,000 high school and college graduates began looking for jobs.[58] At the same time, the Korea Stock Exchange fell by more than 10 percent on January 21 and 22, tumbling to 550.58. The market until then had been on a steady march upward, more than doubling in value over the previous year.

Lee Chung Yul, an expert adviser at the tripartite commission, cited plans by the top five chaebol to dismiss between 70,000 and 110,000 workers in the first half of 1999 as the major factor contributing to unemployment. The government and public corporations would exacerbate the crisis, he said, if they followed through on plans to lay off some of the 800,000 workers on their combined payrolls. The labor problem was likely to worsen in the spring, when 40 percent of those listed as unemployed would have been out of work for more than six months, the point at which most unemployment benefits ran out. Statistics showing the economy was improving were not always comforting. "Many people think the crisis is past," said Lee. "They don't think they have to bear the burden anymore. That is adding fuel to the fire."[59]

On February 24, 1999, Lee Kab Yong, the KCTU president, said his union was withdrawing from the tripartite commission set up a year earlier. He charged the government was "forcing only workers to make sacrifices" while building up more than $55 billion in foreign exchange reserves and forecasting a 2 percent increase in the gross national product for 1999. The withdrawal of the confederation from the commission marked a severe setback in efforts at uniting Korean workers behind the recovery campaign.[60]

Unemployment insurance and public works projects still were inadequate to protect the unemployed. Unemployment insurance lasted for only a few months, and public works programs employed about 300,000 people, often on make-work jobs such as street cleaning. "We don't have an adequate safety net," said Park Nei Hei at Sogang University. "Many of those people who are out of work are permanently unemployed."[61] Unemployment figures were deceptive, since they often did not include women who returned to their families after losing jobs and also did not cover hundreds of thousands of part-time workers.

At the headquarters of a union of construction workers, the union vice president, You Ki Soo, derided official claims. "The economy is not getting better," said You, whose union was affiliated with the KCTU. "Even when people say production is going up, it's increased not by employing people but by increasing the hours of those who are working, so it's as bad for us as before." He accused the chaebol along with the government of "taking advantage of this period by reducing employment." Construction was among the hardest-hit industries. Only 400,000 construction workers held permanent jobs, he said, while another 1.6 million moved from site to site, doing well in good times and suffering anonymously, not covered by unemployment insurance, when times were bad.[62]

At a job center supported by the union, the outlook was bleak. "Since last December (1998), 400 people have applied for jobs," said Lee Kong Suk, president of the center. "We have 350 people still waiting. The sale of apartment buildings and houses is so low, nobody wants to build them any more." A carpenter, Hwang U Young, scraping by on one or two days of work a week, explained the dilemma. "Our pay is cut in half, and the hours are longer," he said. "I have never got benefits from the government. Unemployment insurance doesn't apply to people like me. That's only for people who worked regularly for a company." Park Sang Kyu, a welder, said that he had been out of regular work for eight months. "It has been a nightmare," he said. "For day workers like us, unemployment benefits mean nothing. The law says, even if I work one day a month, I get nothing."[63]

The labor movement, however, was divided. Korea's militant unions faced division within their own ranks on April 26, 1999, at the height of a fight against corporate restructuring that might lead to dismissal of thousands more. While union leaders talked tough from makeshift tents in the sanctuary of the grounds of Myongdong Cathedral, the union of Korea Telecom called off a walkout of its 42,000 members. Union leaders cited poor attendance at rallies as proof of lack of support. Next, members of the Seoul subway union indicated they would rather return to work than risk losing jobs by remaining on strike. The strike had disintegrated when more than half the system's 10,000 employees reported for work before the strike was officially over.[64]

Leaders of the KCTU plotted quick strikes at factories marked for merger, sale, or restructuring. "Tomorrow we strike against motor vehicle factories," said Lee Kab Yong. "There is no other way to fight this mass layoff of labor." He said that 30,000 workers would participate. The union marked as a special target Daewoo Heavy Industries' shipbuilding unit, which Daewoo Chairman Kim Woo Choong had said he hoped to sell. Lee blamed Kim Dae Jung for having forced the chaebol to come up with restructuring plans that ignored the interests of workers and denounced the government for sending the police into the campus of Seoul National University the night before to dislodge hard-line unionists at the university's medical center. "If the government reacts the way it has been doing, it is no different from the previous governments," he said. "Therefore we cannot accept it as the people's government." Never, he added, would the KCTU rejoin the tripartite commission.[65]

The contrasting responses of union leaders and workers indicated the ambivalence of a movement divided between the desire of hard-liners to challenge authority and the fear that many more workers would suffer if they walked out for more than brief periods. Despite the split, government leaders believed the defiance could cripple industries just as they were recovering. Militants insisted the strikes of 1999 were different from those in previous years, in which the issues revolved around wages and working conditions. "This time it's against the neoliberalism of restructuring," said Park Seok Woon, director of the Institute for Workers' Rights.[66] "Restructuring means nothing but mass dismissals," said Kim Chul Won, an executive of the Korean Federation of Social Services. "The government demands restructuring of the chaebol and labor market flexibility but only focuses on downsizing."[67]

Anthony Michell, managing director of the Euro-Asian Business Consultancy in Seoul, predicted in early 1999 that the number of unemployed would grow, possibly reaching 2.5 million, despite an improving economy. "If unemployment peaks at 2.5 million, it could take ten years to overcome the problem," said Michell, coauthor of a study on unemployment and the economy in Korea. "That's a long prospect for an economy with no long-term social safety net."[68]

Forecasts, however, were risky. While labor strife was far from over, the drama of the struggle for workers' rights in Korea mingled triumphs with failures for all sides, business, government, and labor. The 1999 strike season never resumed after the collapse of the Seoul subway strike and the failure of the zealots in their confrontation at Seoul National University in mid-May. The statistics helped explain why the labor movement was losing intensity. Unemployment at the end of August 1999 stood at 5.9 percent, down 0.45 percent from July, 1.7 percent below the figure of 7.6 percent recorded for the previous August and a full 2 percent below the highest level of 7.9 percent for December 1998. In hard numbers, 1.241 million workers were out of jobs in August, compared with 1.349 million in July and 1.574 million in August 1998, while the economy was growing by eight percent in 1999 from the year before.[69]

The numbers, however, did not tell the whole story. A person was adjudged employed if he or she worked only one hour a week. Recent graduates were coming into the workforce in the fall while the construction industry entered a seasonal winter slump. Unemployment for the year was officially forecast at 6.5 percent, 1.4 million of 21.9 million members of the workforce. That figure represented a decrease of only 0.3 percent from 1998, the first full year of the crisis, and was more than three times as high as the last year before the crisis began. A major reason for the decline in official numbers, moreover, was that many people, especially young women and teenagers, both male and female, had stopped looking for jobs. The numbers would slide sharply downward again if Daewoo factories shut down, taking with them a network of suppliers—altogether 200,000 workers.[70]

Increasingly, companies hired workers on a "temporary" or daily basis. By the time of the second anniversary of the signing of the IMF agreement, on December 3, 1999, the ministry of labor was able to claim that unemployment had dipped to 4.8 percent—with slightly more than 1 million workers still out of jobs. The figure was deceptive, however, since at least 48 percent of the workforce fell into the "temporary" or "daily" category.

Although often in disagreement, both the KCTU and the FKTU had the same opinion of the tripartite commission. The FKTU in November followed the KCTU's lead, withdrawing from the commission in protest against the government's stated desire to privatize KEPCO, the Korea Electric Power Company, breaking it up and selling portions to foreign interests. The FKTU, representing 24,000 KEPCO workers, promised a protracted struggle mingling nationalism with social and economic protest, and was not inclined to rejoin the commission after the Assembly in early December put off action on KEPCO. Instead, rioting workers clashed with police, demanding shorter work weeks and full-time company pay for employees working full-time for the union.

Kim Dae Jung needed the votes of workers to solidify his strength in elections for the entire National Assembly in April 2000. The KCTU, undaunted by non-support of its strike calls in 1999, considered forming its own party—a minority grouping that might draw votes from DJ's machine. Radical workers counted on the timing of the elections, in the middle of the spring strike season, to galvanize their followers as they had done annually for more a decade. In one form or another, labor unrest was sure to keep exploding, forcing compromises that would undermine reforms but also defend the rights of workers whom employers would otherwise exploit, abuse, underpay, and dismiss with no notice, as they had done so often in the past.

# Chapter 8

# On/Off the Fast Track

South Korea's motor vehicle manufacturers in the late 1990s were careening down a crowded track in a frenzied race. The adversaries included the three biggest chaebol plus two second-tier chaebol that might not make it to the finish. The strongest contestants, Hyundai and Daewoo, talked incessantly of climbing into the "global top ten" of motor vehicle manufacturers by 2000, but the real question was whether the others would be making vehicles at all by then. The insistence of the second-largest chaebol, Samsung, on entering the race showed the cutthroat drive of Korea's largest groups in a time of economic slump and saturation of the local market.

A young Samsung analyst, fresh out of graduate school with a doctoral thesis on automotive problems, outraged Samsung's rivals in May 1997 by calling for "restructuring" the automotive industry—meaning wipeout for rivals Kia and Ssangyong. What angered and frightened Samsung's rivals was that Samsung had decided only a few years earlier to jump into a field that was already packed. "Samsung is not qualified to discuss restructuring of the industry," complained Hyundai Motor Chairman Chung Mong Gyu. "On the contrary, Samsung is subject to restructuring if necessary."[1] The battle, whether for number one in the industry or for survival of the fittest, illustrated the larger issue of whether it was healthy for the biggest chaebol to keep getting bigger while the underdogs zoomed toward bankruptcy.

Could Samsung, on the way to producing its first cars with technology from Nissan by March 1998, take over either Kia or Ssangyong or force one of them out of motor vehicles? The Samsung report, distributed internally at Samsung but leaked to Kia, cited not only "the necessity of restructuring the domestic auto industry" but also "probable supportive methods of the government." The allusion to government support harked back to the days when central planners allocated whole industries to different groups, controlled bank credit, and forced mergers in order to eliminate "wasteful competition."[2]

Nowhere was the competition harder or more wasteful than in motor vehicles. "I have trouble believing in the long term you could have this many auto companies," said David Young at Boston Consulting in June 1997.[3] Shakeout appeared inevitable. Production capacity stood at 3.5 million vehicles a year. Sales in 1996, domestic and export, came to 2.84 million vehicles plus another 200,000 units assembled overseas from CKD—complete knockdown—sets shipped from Korea, bringing worldwide Korean motor vehicle sales to slightly more than three million. Excess capacity was a major problem on a jam-packed highway.

No dream seemed impossible for a Korean motor vehicle manufacturer in 1997. Hyundai Motor, 9.3 percent owned by Mitsubishi, sold 1.3 million vehicles in 1996 for net earnings of $102.8 million, while Samsung hoped to sell 80,000 cars in 1998—and well over a million a year in a decade. Hyundai and Daewoo both bragged of tremendous expansion plans—"2 million cars a year by 2000" was a Hyundai slogan, and Daewoo Chairman Kim Woo Choong set his sights on 2.5 million by that date. Daewoo, since buying out its General Motors partner at the end of 1992, had bought up state-owned factories in Eastern Europe and planned to export cars to the United States in 1998. In 1996, Daewoo, including both Daewoo Motor and Daewoo Heavy Industries, which produced an 800cc minicar, produced 850,000 vehicles for a net profit of $64.3 million. Kia, 9.4 percent owned by Ford and 7.5 percent by Mazda, which in turn was one-third owned by Ford, claimed profits of a minuscule $8 million in 1996 from sales of 781,000 vehicles but counted on new models to put it back in the race. Ssangyong Motor, 3.5 percent owned by Daimler-Benz, turned out only 76,940 vehicles, suffering a large loss, but dreamed of salvation in a transfusion from another "partner."[4]

A prime candidate for bankruptcy, the sixth-largest chaebol in 1997, Ssangyong Motor had followed a typical chaebol technique, borrowing from other companies around the group. Ssangyong had produced buses, trucks, and vans for 11 years with Benz technology. The first car, a top-of-the-line model named Chairman, with Benz engine and Ssangyong nameplate, rolled out in November 1997. The fear now was that the motor company, several billion dollars in debt, would drag down the group.

Rumors of Daewoo's plan to purchase Ssangyong Motor swirled on December 6, three days after the signing of the IMF agreement, when the stock of Ssangyong's listed companies shot up nearly 8 percent, the maximum permissible gain in one day. Talks with Daimler-Benz on the sale of the company had foundered on the question of corporate debt, but Daimler-Benz still had to approve any deal.

On December 7, 1997, *Chosun Ilbo* reported that Daewoo Chairman Kim Woo Choong had virtually completed negotiations with Ssangyong Motor Chairman Kim Suk Joon, who had also become chairman of the Ssangyong group after his elder brother, Kim Suk Won, was elected to the National Assembly in 1995.[5] "Yes, we've been negotiating, and we'd like to sell the company," said a man identified as a senior Ssangyong official, talking in a disguised voice that night on Korea Broadcasting System. The reason, he admitted, was the enormous debt incurred by Ssangyong's effort to get into the business. A senior Daewoo official, also talking in a disguised voice, said Daewoo was less heavily in debt than other Korean chaebol. The inference was that Daewoo could pick up a major portion of Ssangyong's debt.[6]

Ssangyong's failure in motor vehicles was a terrible humiliation for Kim Suk Joon, forced to return the group chairmanship to his brother, Kim Suk Won, in early 1998 after the latter quit the Assembly to try to salvage what was left. The Kim brothers, inheriting the group from their father, Kim Sung Kon, successful in construction, cement, and oil refining, had for years had a hankering for motor vehicles—and prided themselves on forays into sports cars before allying with Daimler-Benz. Analysts wondered if Ssangyong had persuaded Daewoo to pick up all its debts—a move that would save the rest of the group from collapse. "The debt question has to be settled," said Hank Morris, whose own firm, Coryo Securities, had gone bankrupt on December 5. "If there is a huge transfer of debt to Daewoo, it's hard to see how Daewoo could cope."[7]

On December 8, 1997, Daewoo, suffering from a five-to-one debt-equity ratio, agreed to buy 53.5 percent of the shares of Ssangyong Motor for the price of $1.6 billion of Ssangyong Motor's $2.8 billion debt. The purchase included most of Ssangyong's shares, leaving the balance in the hands of minority shareholders, notably Daimler-Benz. "Daewoo's takeover of Ssangyong Motor is the first friendly merger and acquisition after the IMF rescue package," exulted Daewoo Motor Chairman Kim Tae Gou. "This will contribute to South Korea's restructuring." Daewoo would operate Daewoo Motor and Ssangyong separately—and use Ssangyong's four-wheel-drive vehicles, Korando and Musso, and its Istana van as part of its drive to enter the American market.

The deal, however, was risky. Adding to the burden of the banking system, Cho Hung Bank gave a ten-year moratorium for repayment of the debts picked up by Daewoo from Ssangyong Motor and gave Ssangyong Motor a five-year moratorium on payment of the rest. The government had pressured Cho Hung into agreeing to the deal, which saved the Ssangyong group from collapse. With the purchase, Daewoo technically became Korea's second-largest chaebol in assets, surpassing Samsung and LG, the only one of the top four that did not have a motor vehicle company. The impact on the motor vehicle industry, however, was minimal. Daewoo in 1997 was producing more than ten times as many vehicles as Ssangyong. Samsung Motor, meanwhile, still planned to produce its first car to hit the showrooms well before the March launch date.

By early 1998, Daewoo Motor, with a debt-equity ratio of six to one, had to look to its former partner, General Motors, for a bailout. GM opened talks with Daewoo from a vantage markedly different from that of most other would-be investors in Korean companies. Daewoo Motor was 50 percent owned by GM from 1978 until 1992, when Daewoo bought out GM's share for a pittance of $170 million. Daewoo Chairman Kim Woo Choong had chafed at the time under GM's conservative outlook toward his global expansion plans, but GM and Daewoo remained on speaking terms with GM, continuing to sell engines and components to Daewoo for approximately $1 billion a year.

Kim Tae Gou sought to win GM back as a full partner. Under one proposal, GM would buy back its former 50 percent stake in Daewoo Motor for about $330 million. "The two companies have decided they need each other," said *Chosun Ilbo,* reporting one name for the merged company was "Global Strategic Cooperation."[8] A re-merger between GM and Daewoo Motor would fit in with recommendations of the IMF for opening Korea wide to foreign investment, including mergers and acquisitions, even though Daewoo Motor, under the initial re-merger plan with GM, would probably retain operational control.

On February 2, 1998, GM and Daewoo said publicly that they had agreed to hold "intensive discussions" on a deal. "Merger and acquisition is certainly a consideration," said Alan Perriton, president of GM Korea, after signing a memorandum of understanding with Kim Tae Gou. GM was open to options ranging from marketing to financing to distribution and coproduction. Asked whether merger was a priority, he replied, "We haven't decided what is a solution or an outcome, we haven't set a number, we haven't set a target—we haven't decided there will be a major investment." Perriton said he had been talking "off and on" with Daewoo for a year and a half but indicated that Korea's economic problems had sped up the pace. "The Korean economy has precipitated an increased interest level" among a number of foreign companies, he noted. GM was the first to sign a memorandum of understanding "to go forward and have some fairly serious discussion."[9]

The overwhelming factor behind what Daewoo called an agreement "for strategic alliance" between GM and Daewoo Motor was the economic crisis. Daewoo Motor, not listed on the Korea Stock Exchange, had invested far beyond its means in building a new $1 billion plant in Korea, buying and rebuilding motor vehicle plants in Czechoslovakia, Egypt, India, Poland, Rumania, Uzbekistan, and Ukraine, and taking over the bankrupt Ssangyong Motor. Daewoo's sales indicated how hard the economic crisis was hitting. The company sold only 16,169 vehicles in January 1998, including 4,756 for export, as opposed to 45,106 vehicles sold in January 1997, including 22,211 for export. "It would be logical to presume that Daewoo is having financial concerns," said Perriton, but he denied that Daewoo had "asked us to come to the rescue." Ironically, as part of the agreement, Daewoo might rescue GM Korea, which had suffered from the economic crisis in quite a different way. GM's Korean distributor, Inchcape, a British company that had sold about 1,000 GM cars in Korea in 1997, shut down its Korea operation, and GM decided not to set up its own network amid sharply slumping sales of all imported cars. The agreement called for Daewoo Motor's aftersales network to service GM customers in Korea through ten outlets.[10]

The announcements by Daewoo and GM suggested, however, that Daewoo expected more from the talks than did GM. Although they had agreed on their press releases in advance, Daewoo omitted one sentence that GM had been careful to include: "The discussions would be exploratory in nature and did not presume a given outcome."[11] Lou Hughes, president of international operations for GM, confirmed on April 24, 1998, that he was in intense negotiations with Daewoo but refused to comment on rumors that GM wanted a majority stake in hopes of turning Korea into a regional base.[12] Gossip about Daewoo's talks with GM faded from the newspapers.

Instead, Daewoo might be the next conglomerate to face dissolution and possible bankruptcy of some of its companies. Stung by highly critical evaluations from foreign securities firms, Chang Byung Ju, president of the trading division of the Daewoo Corporation, said on November 20, 1998, that the group would wind up with a substantial profit for the year. Chang dismissed reports that Daewoo might have to sell subsidiaries to keep major affiliates from going bankrupt as "totally groundless." He blamed the group's troubles in the first half of 1998 on "additional loans" that the group had to obtain for Ssangyong Motor, which he said had lost 370 billion won, more than $300 million. As a result, the group showed losses of 90 billion won, $60 million, for the first half of 1998. Chang predicted, however, that the group's net profit for the year would come to 670 billion won, $520 million. Daewoo rested its hopes on profits from electronics, shipbuilding, and trading—enough to cover debts of two trillion won, $1.55 billion.[13]

Chang spoke out while the group was struggling to combat the impact first of an adverse report by Nomura Securities and second of brain surgery performed the same month on Chairman Kim Woo Choong. Kim's surgery appeared to be the lesser of the problems. Still only 62, he was soon on the way to recovery, as evidenced by his visibility as chairman of both the group and the FKI. While Kim appeared to have many years ahead, his illness raised the question of who might succeed him at Daewoo, which he had founded as a textile company in the 1960s. His eldest son had died years before in a motor vehicle accident, and he exercised tight one-man control.

The Nomura report was more difficult to gloss over. Headlined "Alarm bells ringing for the Daewoo group," the report said Daewoo had "survived solely on liquidity procured through the issuance of bonds" and was "likely to face huge difficulties in arranging equity financing." Daewoo's "only survival option," it said, was "through the sale of assets." The report, however, questioned whether Daewoo had "companies or assets attractive enough to entice buyers." Nomura no longer saw the need to provide information for clients on Daewoo Electronics or Daewoo Heavy Industries, two of the group's core companies. A sign of the group's difficulties was its decision to float $380 million in bonds in an effort at raising cash in place of credit that it normally received from banks. While rolling over Daewoo's loans, Daewoo's banks were reluctant to extend fresh credit. The group in the third quarter of 1998 had issued 9.2 trillion won worth of bonds, 27 percent of the country's bond issues, said the report.[14]

Samsung's lust for motor vehicles was, if anything, more perilous than that of Daewoo. On the ground floor of "Samsung Main Building," the Samsung headquarters near Seoul's Namdaemun or South Gate, was a car covered entirely with white cloth. Beneath the cloth was a prototype of the car that Samsung planned to introduce in March 1998. Lights flashed the number of days, hours, and seconds until the cover would come off, revealing this creation to the world. The car, however, was no mystery. It was a knockoff of the Maxima, a midsize sedan made with technology from Nissan. The mystery was how Samsung believed the car could survive. "We plan to export the car," said Cho Jang Won, a Samsung manager. "We have expected this financial crisis a long time. We have a crisis plan."[15]

Like Daewoo, Samsung yearned for a bailout. The chief executive officer of Samsung Motor, Lee Dae Won, flew to Detroit for "serious talks" with Ford Vice Chairman Wayne Booker on March 11. He hoped "to work out details of a strategic partnership" that might include production of "Ford's World Car through Samsung," "joint production and distribution of automobile parts," and research and development. "We shifted our strategy from taking over a foreign automaker to joining hands with a foreign company," said Samsung Motor Chairman Lim Kyung Choon, showing off Samsung's new car before its launch on March 28. Lim said Samsung would welcome a "strategic alliance" with Ford, Chrysler, or just about any major company except GM, since it was negotiating with Daewoo.[16]

The circumstances were dire. Korea's domestic market had fallen so sharply that Hyundai and Daewoo were both in trouble, and Kia, the number three car manufacturer, was in receivership. Samsung still hoped to produce 80,000 2-liter and 2.5-liter cars in 1998 but had to postpone plans to build a second plant needed to raise production to the point at which it could begin to make a profit in a fast-diminishing local market. By May, Samsung had sold about 20,000 cars. In a restructuring plan presented on May 6, Samsung Motor said it was negotiating to attract "foreign funds in strategic cooperation with foreign motor companies" and promised Samsung Motor would "actively cooperate with the government."[17] Nobody, however, wanted to pay a price approaching the $5 billion that Samsung had invested to produce its first car. While Samsung said it was negotiating with Ford, Ford remained aloof.

Ford was interested, however, in Kia, now in receivership, while Kia hoped Ford would increase its investment to as much as 51 percent. The dilemma at Kia mirrored that of a range of industries when I visited its plant on Asan Bay, 90 miles southwest of Seoul, two weeks after the signing of the IMF agreement in December 1997. The long line bearing the frames of Kia's latest model quietly stopped shortly before noon, and assembly line workers began playing table tennis, opening up newspapers, and chatting with each other. "The line is going half speed," explained Dan Dong Ho, a general manager at a modern complex capable of producing 650,000 cars a year. "The Korean situation is very bad. If the Korean market is better, the line speed is better."[18]

Often dismissed as barely alive amid mounting debts and pessimistic forecasts for the industry, Kia was still humming—albeit at a speed that raised daily alarms about its viability. Intermittent stoppage of the assembly line, carrying the Shuma, a sedan with 1.5- and 1.8-cc engines that Kia had unveiled a few days earlier, was commonplace in a period of tight money and a sluggish market. "We stop the line four times a day," said Dan. "Sometimes for 15 minutes, sometimes for 10 minutes, perhaps for lunchtime." When the line was not stopped altogether, it was possible to slow it to a snail's pace. A worker could relax for as long as 30 seconds before the next car moved up the belt.

In the complex on Asan Bay, built on reclaimed land by a natural port, the evidence of Kia's profligate spending was clear. Robots did more than 90 percent of the work once expected of humans in some of the individual plants, including one producing engines for all Kia vehicles. As production for 1997 declined to about 700,000, including vehicles from an older plant closer to Seoul, the dream lingered on. "By 2002 we will build a depot so we can ship our cars overseas from here instead of from Inchon," the port west of Seoul, said a blue-uniformed tour guide, briefing in an observation hall above the blue roofs and gray walls of the complex. "Kia's blueprint is to continue to explore overseas markets. By 2001 we plan to produce 1.5 million units a year"—a dream that no one believed impossible if only they got enough parts for today.[19]

For Kia, however, the fact that the assembly lines were moving at all was a miracle—so much so that Kia executives and managers talked of expanding production even though the group was in receivership. "Our major investment in infrastructure was completed two years ago," said Um Sung Yong, a director at Kia Motors' headquarters in Seoul. "Now is the time to reap all the fruits. All our lineups of new cars guarantee money coming in." The optimism reflected in part the survival of a group that had gone under court receivership five months earlier and now was 45 percent owned by the government's Korea Development Bank. Kia executives made no secret of plans for "restructuring," reducing the group's 18 companies to about five, including Kia Motors. They seemed far more interested, however, in getting new models into production and onto world markets than in downsizing.[20]

Not even bankruptcy dampened the vision of selling abroad. Founded in 1944 as a bicycle manufacturer, Kia showed no signs of shutting down even though the group was losing money as a result of poor investments in other industries, notably steel. "Last year the United States market bought 40-50,000 of our cars," said Kia spokesman Jeun Sang Jin. "We project 70,000 for this year. Not only Kia Motors but every manufacturer is putting more and more emphasis on the overseas market. It's export or die. That's what we've done ever since the 1960s." Jeun advanced the same arguments as those of Korean leaders and cabinet ministers and chaebol chairmen. "Look at our situation. Korea is a small market. We don't have natural resources."[21]

Much hinged, Kia managers acknowledged, on the outcome of the presidential election on December 18, 1997. Kia people believed that DJ might be more sympathetic to Kia's long-term survival if only because two of the group's major plants were in the Cholla region. Rather than get rid of the insolvent Asia Motors, in Kwangju, Kia Motors hoped to absorb the company and its 6,000 employees as a separate division. The other, Kia Special Steel, whose problems were blamed in large part for Kia's near-bankruptcy, employed several thousand people in Kunsan, South Cholla Province. It was now jointly administered by an unlikely alliance among Kia and rivals Hyundai and Daewoo.

Burdened with a debt that had ballooned to more than $10 billion as it built up its complex on Asan Bay and expanded into other industries, Kia got by day to day in a struggle for customers, parts, and ready cash. "For the past 50 years we have been fully committed to the development of technology in Korea," said Um Sung Yong. "People view Kia as a people's company. When we decided three years ago to introduce this new vehicle, the Shuma, we hoped for a best-seller." To the amazement of skeptics who believed that Kia was slowly being strangled to death by Hyundai and Daewoo, the Shuma was just one of nine new lines in varying stages of planning and production.[22]

The start-and-stop style of assembly line production, however, threatened all the models, new and old. The slowdown revealed not only Kia's troubles but also a tangled web of disintegrating relationships spreading through the parts industry on which all motor vehicle manufacturers depended. "The parts suppliers have money problems too," said Dan Dong Ho as the line ground to a halt with its row of glistening Shumas as well as Sportages and Sephias, all exported to the United States and other western countries. "They don't supply the parts on time. Today we are missing frames for the Sportage." Since the Sportage and the Shuma were rolling on the same line, both were delayed.[23]

An assistant general manager, Cho Jang Rae, said a gear was falling in short supply as well. Korea's two largest tire manufacturers, Kumho and

Hankook, were not delivering enough tires while raising prices and demanding payment in cash. Periodic bankruptcies of key suppliers heightened the uncertainty. Mando Machinery still supplied air conditioners and heaters—but for how long was anyone's guess. Cho was confident, though, that Kia would make it through the day—this day if not the next. "After lunchtime, more parts are coming," he said.[24] The plant still produced enough cars to justify double shifts, but the company could not pay overtime much longer and would no longer pay the customary year-end bonus.

Kia executives saw Ford in their future. Ford had dropped the Kia Pride, once sold as the Festiva in the United States, but Kia still produced the 1.5-liter Aspire under the Ford name for export to Australia, New Zealand, and Latin America. There was also talk about the impact of Samsung. "Samsung would like to buy Kia if it's got the money," said Karl Moskowitz, a business consultant, "but it's not certain it can afford to."[25] Kia hoped to raise one trillion won, approximately $570 million at the time, to keep the motor company alive while getting rid of most of its money-losing subsidiaries. Kia Chairman Jin Nyum, a former labor minister who took over as chairman when Kia's creditor banks filed for court receivership for the group in October 1997, appealed to Ford three months later to raise its stake. "It would be very nice for Kia Motors if Ford decides to expand," said Jin on January 22, 1998.[26]

Jin was eager to compromise with competitors. Rather than confront one another head-on, he suggested that Kia market its cars on the domestic market with those produced by Samsung. Alternatively, he said, Kia could market cars with Hyundai or Daewoo. Either way, he hoped that Kia would get a "one-trillion-won injection, half from domestic companies and half from foreign companies." His hopes rested in large part on the government's Korea Development Bank. A Korean court was expected to approve the application for receivership, which included a proposal for the bank to accept payment of loans in the form of shares in Kia. The plan would turn Kia into a company largely owned by the government, but Jin believed the company would soon resume making profits and with new investment would be free from government control.

"We hope other institutions will follow Korea Development Bank in converting loans into equity," said Jin. "By having this one-trillion-won injection, the profit margin will significantly improve."[27] Spokesman Jeun Sang Jin said the company's debt-equity ratio was 4.27 to 1 in 1996 but predicted it would sink to 2.88 to 1 by 2000 and would be about 2 to 1 by 2001. He offered no estimate, however, of the current ratio, believed to be at least 8 to 1. Total 1997 sales, worldwide, were 697,384 units, but sales slipped toward the end of 1997 from 62,913 in August to 46,731 in December. Kia blamed the slippage on the poor domestic market, a problem that had also hurt Kia's competitors. "I can guarantee you," said Jeun, "that Kia Motors will never go out of business."[28]

One reason for such confidence was Kia's determination to unload nine of its money-losing firms in addition to a dozen firms already shut down or sold off. The group would attempt to find buyers for Asia Motors, Kia Steel, and Kisan Construction. Both Daewoo and Samsung were possible buyers for Asia Motors, but neither seemed interested in Asia Motors without Kia Motors. In the end, Kia officials said Kia would retain only seven companies, including Kia Motors and its sales arm, Kia Motor Sales. The five others manufactured components used in Kia vehicles. Kia counted on a combination of seven lean companies focused on motor vehicles to deter Samsung.

All the motor vehicle companies at the time were in severe difficulty. Hyundai Motor suspended production on February 21, 1998, of all but a newly introduced minicar while awaiting tires from Korea's three tire manufacturers, demanding 15-80 percent price increases. Both Kia and Daewoo said they too would suspend production if price talks failed. Hyundai, Daewoo, and Kia employed 100,000 workers, and several times as many were on the payrolls of several thousand other companies, big and small, that supplied materials and parts. The problems of the tire manufacturers as well as Mando showed how the economic crisis might bring an industry to its knees. The tire makers, Kumho, Hankook, and Woosung, along with Mando, were all subsidiaries of chaebol suffering from fast-mounting debts. They all cited increases in the price of oil as the reason they had to charge more just to stay in business. Hyundai, accounting for nearly half the sales of Korean-made cars, trucks, and buses, sold 1,247,415 vehicles in 1997, including 645,573 in Korea and 601,851 overseas. As the crisis began to bite, Hyundai's domestic sales fell from 38,186 units in December 1997 to 17,677 in January 1998.[29]

The Kia group appeared likely to go into receivership when a court ruled in March 1998 on its application. The alternative would be court-ordered liquidation. "It is pretty obvious to us that the court will allow a reorganization program for Kia," said Yoo Seong Min at the Korea Development Institute. "Once the court approves the application, then Kia has to talk to the banks, and then the court has to approve their debt rescheduling plan. It is similar to the Japanese bankruptcy procedure."[30] Kia had initially battled applying for receivership, the process under which the court had named Jin Nyum to replace the previous chairman, Kim Sun Hong, imprisoned with three other Kia executives for falsifying the group's financial condition. Jin's optimism was forgotten, however, when he resigned in March to join DJ's cabinet as chairman of the Planning and Budget Commission. Now the group saw receivership as a reprieve. "Receivership can provide a lifeline," said spokesman Jeun Sang Jin, forecasting sales of 600,000 vehicles in 1998 despite a 50 percent drop domestically. "Ask anybody whether the court will grant receivership, and 99 out of 100 will say it will."[31]

While Kia looked for a lifeline, Hyundai and Daewoo were bolstered by Prince Al Walid bin Talala of Saudi Arabia, who flew in on March 15 as the biggest individual speculator in the crisis. The prince, meeting Daewoo Chairman Kim Woo Choong, invested $100 million in convertible bonds that could ultimately make him the Daewoo Corporation's largest individual investor with 10 percent of the holdings. (He already held Daewoo Corporation bonds valued at more than $50 million.) That done, he invested another $50 million in convertible bonds from Hyundai Motor after seeing Chairman Chung Mong Gyu. The prince, whose personal wealth of $12 billion made him the eleventh-richest man in the world, paid the ritual call on Kim Dae Jung before leaving on March 17.

As far as Hyundai Motor was concerned, Prince Al Walid could not have selected a more propitious time to call. On March 18, the company released statistics showing that Hyundai Motor's profits for 1997 had fallen to 46.5 million won ($31.6 million), down from 86.8 billion won the year before. Hyundai Engineering and Construction, the country's biggest construction firm, reported a profit of 14.9 billion won ($10.2 million) in 1997 after earning a net profit of 21.1 billion won in 1996. "We are facing a financial crisis," said Hyundai Motor spokesman Shin Hyun Kyu, noting that Hyundai Motor had suffered a precipitous drop in earnings even though total sales revenue had increased somewhat, rising from 11.5 trillion won in 1996 to 11.7 trillion won in 1997.[32]

As a sign of Hyundai's determination to bull its way through turmoil, the company on March 17 launched an entirely new series of Sonatas, its midsize car. Chairman Chung hosted a lavish reception introducing the 1.8-liter, 2-liter and 2.5 liter Sonata EF—for "Elegant Feeling." Hyundai Motor was sure to lose initially on the latest Sonata, to be marketed domestically for the first few months before being exported. The company also said it was shipping 50,000 of its Accent compacts to Rumania and would assemble 100,000 Accents in Rumania in collaboration with Automobile Dacia, a Rumanian manufacturer.[33]

While Daewoo negotiated with GM and Samsung talked to Ford, Hyundai saw conquest of Kia as the quickest way both to stave off the Daewoo threat to its number one position and to kill off Samsung Motor. On March 21, Hyundai Motor revealed tentative plans for the deal that would make Hyundai Motor one of the world's top ten motor vehicle makers. "We are seriously considering taking over Kia," said Hyundai's public affairs office. "But whether we will take over the entire company or only part of it is not certain yet." Hyundai Motor and the Hyundai Research Institute had a battle plan to acquire Kia Motors in an effort at forming a "globally competitive" company.[34] Hyundai companies, a spokesman revealed, had quietly purchased 6 percent of Kia Motors three years earlier.

A Hyundai takeover of Kia would be the type of "Big Deal" that Korean officials had said was needed to unload failing companies. The pressure had intensified several days earlier when Kim Dae Jung specifically cited Kia among bankrupt groups that he said were "smothering" the economy. Hyundai appealed to Korean patriotism as a reason for the takeover, declaring the Korean motor vehicle industry "could be destroyed in terms of technology, capital, and competitiveness if foreign firms take over management of local firms." Chairman Chung Mong Gyu told Korean reporters that his company, unlike its rivals, had "no intention" of negotiating with foreign companies in an effort at resolving financial difficulties.[35]

The Hyundai campaign to take over Kia turned into a bitter battle that illustrated the difficulties of mergers among major manufacturers. Kia attacked Hyundai's plan on March 23 with the fury of an aggrieved underdog. Portraying Kia as "the victim of this distorted economic structure," Park Je Hyuk, president of Kia Motors, blamed the chaebol in general and Hyundai in particular for attacking Kia "like hyenas." Kia, in an official statement, charged that the Hyundai plan "contradicts the efforts of the new government to restructure the economy" and also countered the promises of the largest chaebol to streamline operations.[36]

Kia managers promoted the image of the group as distinct from other chaebol in that its founding family had long since left the group, selling most of their shares and leaving control in the hands of professional managers. They argued that Kia was a relatively small organization of the sort that DJ had promised to encourage during his presidential campaign. "The government says it wants to prevent the chaebol from growing bigger," said spokesman Jeun Sang Jin. "It makes no sense for Hyundai Motor to swallow up Kia and grow still bigger."[37] Beneath the bravado, however, ran an undercurrent of pessimism as Kia awaited the court decision on the application for receivership. If the court granted the receivership as expected, Kia then would renegotiate its debts. The government's Korea Development Bank was expected to take over 35 percent of the shares in Kia Motors as payment of debt, and the Cho Hung Bank, Kia Motors' second-largest creditor, was likely to agree to a similar arrangement.

Hyundai officials said that Kia Motors, under receivership, would still have to agree to purchase of a majority of its shares by Hyundai. A portent of the takeover was that Kia Motors stock closed on March 23, 1998, at 7,080 won, up 750 points. Hyundai Motor stock was up 100 won, closing at 23,800 won. Kia executives saw the Hyundai takeover plan as an effort to frustrate Samsung. Kia's precarious struggle contrasted with the success of the Halla group, Hyundai's blood relative, in luring a "bridge loan," announced the same day, of $1 billion from Rothschild of New York at 12 percent interest. Halla needed the loan to meet debt payment deadlines and persuade foreign interests to invest in key companies, notably Mando Machinery.

A looming battle between Hyundai and Samsung over Kia might set the stage for a series of struggles over the remains of bankrupt enterprises. For Samsung, the contest might decide whether Samsung Motor could survive. For Hyundai, victory would mean that Hyundai Motor could not only recover the losses incurred in the crisis but climb at last into the "global top ten." Negotiations between Ford and Samsung or Kia, however, might end in a Ford-invested merger that would pose a serious threat to Hyundai as well as Daewoo.

On a larger scale, moreover, such talks portended the entry of scores of other foreign investors with roles to play in slimming down bloated industrial fields, bringing in expertise and capital, and reviving and restructuring the economy. That process was likely to go on for four or five years, said Robert Felton, director of the Seoul office of McKinsey and Co., a consulting firm that advised many of the chaebol. Felton predicted "basic restructuring, numerous mergers and acquisitions, foreign investment, and a lot of layoffs" but saw Korea "restabilizing its economy and returning to more attractive growth." The temptation existed, however, for Koreans "to wink at the IMF." With "business as usual," he warned, the gross domestic product would increase by no more than 3 percent a year while unemployment rose to 8 percent.[38]

McKinsey Global Institute on March 27 released a report that attacked the Korean performance in a wide range of industrial sectors, including motor vehicles. Although Korea ranked as the world's fifth-largest motor vehicle manufacturer, it said, "Korea lags significantly behind industry benchmarks in labor, capital and total factory productivity." The report lambasted the Korean motor vehicle industry for producing too many different types of cars, ignoring principles of manufacturing and assembly and failing to bring about labor efficiency. "These issues are a natural response to a captive domestic market," it said. Lack of competition from foreign manufacturers, which accounted for only 0.8 percent of the total motor vehicle market, "limits exposure to best practice and reduces incentive for productivity improvements."[39]

The question, however, was whether the scrap over Kia, whose executives were determined to stave off all attempts at a takeover, might indicate quite a different scenario. The chaebol might spend so much time, energy, and badly needed resources in fighting each other as to lose sight of the reforms needed for recovery. Clearly, they were not about to heed McKinsey's imprecations against unremitting emphasis on volume. For Hyundai, the ultimate dream was to manufacture 2.5 million vehicles, the total capacity of both the Hyundai and Kia motor vehicle plants. Responding to Kia's resistance to takeover and Samsung's drive to defeat Hyundai in the bidding, Hyundai set up a top-level team dedicated to winning a controlling stake in Kia. Issues of quality and productivity were secondary to the need to expand by eliminating competition.

In fact, the crisis was working against the very reforms that foreign experts, beginning with those who worked for the IMF, had recommended for laying the groundwork for long-term restructuring. Pleas for austerity meant that Koreans were more hostile than ever to foreign competition. "The government in its zeal to get the train back on track has pushed some of the wrong buttons," said James Tessada, president of Ford Korea, pitching Fords in an economy where motor vehicle imports had dwindled to almost none. "The market's been closed for so long, it's very difficult to change attitudes."[40]

The attitudes of workers, accustomed to labor practices that made layoffs difficult and also barred them from performing more than one task, were just as hard to change. Kia spokesman Jeun Sang Jin cited demands of the Kia union, representing 50,000 workers, 20,000 on the Kia payroll and another 30,000 working for parts manufacturers, as ammunition for Kia's campaign to remain on its own. "They are very much against Hyundai and Samsung," said Jeun. "They asked for guarantees for employment of all workers."[41] Workers, like executives, feared the imminent demise of their company after Kia Motors and its subsidiary, Asia Motors, were placed under court receivership on April 15, 1998, and the court appointed a new manager of Kia to look for buyers for both of them.

The news triggered a demonstration on April 16 through downtown Seoul by several thousand striking workers. "No takeover, no takeover" was the slogan, in Korean, shouted by the workers as they rallied in defense of Kia—and their jobs. The irony of the Kia strike, which shut down Kia's assembly lines, was that the workers said that they were fighting for their company and the men who now ran it. Several hours before parading through the heart of the city, about 400 workers surrounded the Kia headquarters, vowing to defend it with iron pipes and sticks. They made good on their threat shortly before noon when they prevented Yoo Chong Yul, appointed by the court to lead both Kia and Asia Motors through reorganization and sale, from walking into the building. Surrounded by menacing workers, Yoo said, "I am not an enemy of Kia Motors," before retreating to his waiting limousine.[42]

"He is an outsider," one worker shouted. Workers insisted that the Kia president, Park Je Hyuk, should have equal status with Yoo, vice chairman of a subsidiary of the Hyosung group, a secondary chaebol, in leading Kia through crisis. Among their worst fears was that Yoo would engineer a merger between Kia and Samsung Motor. Was Ford negotiating with both Kia and Samsung to form a single entity with Ford the major investor? "Please do not abandon us," said a leaflet distributed by workers, reminding readers that Kim Dae Jung, during his campaign, had promised to work for "the stabilization of Kia."[43]

Rows of policemen, wearing helmets and gas masks, lined the major arteries through the heart of the city during the afternoon of April 16 as the workers marched along the sidewalks, shouting slogans, beating drums and cymbals, and displaying leaflets. Fighting broke out intermittently, but for the most part the march was peaceful. The question was when the assembly lines would resume rolling. "Nobody is at the factories now," said a Kia spokesman, Kwon Yong Jun. "Nobody knows for how long. They are all on strike." Several hundred Kia managers resigned in protest against a takeover but then said they would return by April 20. "It was just a symbol," said Kwon.[44]

Yoo Chong Yul saw increased investment by Ford as Kia's salvation. He met Ford Vice Chairman Wayne Booker in Dearborn in May, returning with the impression that Ford hoped to raise its stake. As hundreds of policemen again surrounded the Kia headquarters on June 2, Yoo warned of the danger of labor action torpedoing the talks. "If the labor relationship is not resolved," said Yoo, "Ford will consider differently with regard to cooperation with Kia." Yoo's comment was an admission that Kia had nowhere else to turn. Acknowledging Kia's debt-equity ratio of eight to one, Yoo said Kia was ready to give a consortium led by Ford between 49 percent and 51 percent of the company. As Yoo talked, policemen in riot gear marched into position on nearby streets, standing five deep in front of the main entrance of Kia headquarters and filling the lobby of Asia Motors across the street.[45]

Two hours later, several thousand striking workers, bused in from the Kia plants outside Seoul, gathered outside, waving fists and chanting slogans, demanding bonuses and overtime that the company claimed the workers had agreed to forgo. The confrontation illustrated the dilemma of Korean companies and workers as both sides fought for survival. Foreigners, whom companies were beseeching for investment, were shying away while debts mounted steadily, one company after another pleading for a lifeline or sinking into oblivion as stock prices settled at their lowest levels in a decade. Ford "is now our hope," said Yoo, planning to go to Tokyo for more talks with Booker and executives of Mazda, Ford's investment partner.[46]

"Should Ford take over Kia, Ford will use Kia for producing Kia's cars under the Ford name," said Lee Chong Dae, president of the Kia Economic Research Institute and chief of Kia Motors' administrative office. "If Kia exports with the help of Ford, then Kia's export volume will increase drastically. We are discussing these issues with Ford now." Whatever happened, Kia had made clear its strong opposition to takeover by Hyundai, and Ford had broken off talks with Samsung. First, however, Yoo said the company had to "fix this abnormal relation where management has no right over personnel issues of production-line workers or the right to penalize them."[47]

The battle for Kia was about to begin in earnest. Korean and foreign car manufacturers squared off on July 6 for a bitter bidding war after Kia's main creditor, Korea Development Bank, set July 15, 1998, as the date to announce details for auctioning off the company by the end of August. Hyundai and Daewoo, as well as Kia's major minority shareholder, Ford, all were expected to bid in a test of Korea's view toward foreign investment in major enterprises. "All we can hope for is we have transparency and know what's going on," said Kenneth Brown, Ford's director of Asian markets.[48] Lee Keun Young, KDB president, said any motor vehicle company with the capacity "to normalize Kia swiftly" would be eligible. The bidder would also acquire Asia Motors. The bidding was to begin on July 27 with a briefing on the company's finances and problems. A critical factor could be the willingness of bidders to pick up some of Kia's debts.

Kia, in receivership, owed about $12 billion, but the debts of Asia Motors and other subsidiaries could drive the figure higher.

Ford wanted to turn Kia into a base for manufacturing and selling both Ford and Mazda cars around the world. "We've always been using Kia as a source for markets," said Brown, citing deals under which Kia produced the subcompact Ford Festiva and the Ford Aspire. "What we are trying to do is to protect the value of our equity and to protect our product. Anything that was good for Kia was good for us."

The prospect of Ford emerging as a major Korean manufacturer set off alarms at both Hyundai and Daewoo. Hyundai executives unabashedly cited the need to keep foreigners out of the Korean motor vehicle business. Kim Woo Choong at Daewoo even proposed that Hyundai and Daewoo together buy Kia. "We would like to establish a consortium with Hyundai Motor," said a Daewoo executive.[49] The eagerness of both Hyundai and Daewoo reflected the trauma of the industry. The Korea Automobile Manufacturers' Association forecast that domestic motor vehicle sales would decline by 45.8 percent in 1998 while exports rose only 8.6 percent. The industry seemed sure to fall below its ranking as the world's fifth-largest in 1997 with worldwide sales of nearly three million. Song Sang Hun, a research analyst at KAMA, said 120,000 workers might lose their jobs by the end of the year. The industry employed 281,000 people, including 102,000 at motor vehicle companies and 178,000 working for manufacturers of components.[50]

Kia opened its final struggle for survival with a campaign demanding that the winner of an auction be able to demonstrate financial and professional competence to run it. The campaign targeted both Hyundai and Daewoo as ineligible to take over. Kia executives expressed the fear that another Korean company, if it acquired Kia, would dismiss most of its workers and absorb its lines. Daewoo was interested in Kia's vans and sport-utility vehicles, which Daewoo did not produce, while Hyundai wanted to integrate technology and assembly lines. Kia confronted the government's ministry of commerce, industry, and energy after ministry officials said the company that submitted the highest bid would win. "The government wants to decide just on the basis of money," said Chang Sung Hyun, at the Kia Economic Research Institute. "Kia wants the bidding on other conditions, such as the financial status of the bidding company, the ability to operate Kia."[51] Kia's creditor banks insisted, however, that they would turn over the bids to an "independent" judge.

Kia's demands made clear it would greatly appreciate a push by Ford, whose executives indicated the company would enter a bid. Kia appeared to have won key support with a report that Kim Dae Jung had said that job security for the company's 18,000 workers should be a consideration. DJ's advisers said that the government should also consider which bidder would make Kia "internationally competitive" and should also be concerned about Kia's parts suppliers.[52] Kia mounted its campaign as Ford and Mazda considered bidding jointly. *Nihon Keizai Shimbun,* the Japanese financial newspaper, reported that Ford and Mazda might work with Itochu Corporation, the Japanese trading company, in formulating the bid. In addition to the Ford and Mazda stakes in Kia, Itochu owned another 2 percent, bringing combined equity to about 19 percent.[53]

Kia's argument against takeover by a Korean company rested in part on the dismal showing of both Hyundai and Daewoo in 1998. Research analysts projected Hyundai Motor's losses for the year would reach $100 million. Daewoo, still absorbing losses from its purchase of Ssangyong Motor, was expected to lose as much, and Samsung Motor would lose $170 million. Curiously, Kia might not be losing money at all under terms of the court-ordered receivership, which exempted it from having to pay interest on its debts.

All the companies in the bidding for Kia planned to ask the creditors to write off a large portion of the company's debts. Hyundai Motor's ratio of debt to equity had climbed above six to one while the Hyundai group now had a debt-equity ratio of five to one. Daewoo Motor, whose debt-equity ratio was about seven to one, still hoped to bid in a "consortium" with Hyundai Motor. Asked whether Daewoo would bid for Kia without Hyundai, a spokesman, Lee Chang Won, responded, "There is no 'if.' " Hyundai spokesman Shin Hyun Kyu said there was "a possibility of a consortium with Daewoo" but his company needed to know "the terms and conditions" of the bidding. Kia argued that a takeover by a Korean company would violate the terms set down by the IMF agreement. "If one of these Korean companies acquires Kia Motors, their debt-equity ratio will be much higher," said a Kia manager. "It will be very much against the government's policy of restructuring the chaebol."[54]

On July 15, 1998, one year after Kia had declared it could not pay its debts, Kia's creditors placed the company on the auction block. On the same day, a top-level Ford team defended Ford's qualifications. "Our vision for Kia is that it will be an independent company and will keep its own brand," said Ford Vice Chairman Wayne Booker before meeting with officials from Kia, the creditor banks, and the Korean government. "We will be bringing new technology to Korea, we will be bringing new cash. We will be bringing marketing know-how." The auction was emerging as the most indicative test so far of Korea's openness to foreign takeover of major companies. "One of the purposes of my trip is to determine how transparent this process will be," said Booker.[55]

Daewoo's idea of a "consortium" with Hyundai added to questions about the process, since the aim could only be to divide the remains, not to keep the company alive. Booker believed the Daewoo proposal was "to ensure they either eliminate a competitor," Kia, or to make certain that "Samsung can't take over." Booker, who arrived during a strike by more than 50,000 workers at motor vehicle plants and related industries, did not seem disturbed by labor problems. "We've been able to handle the labor situation in every country where we do business," he said. "I know it's volatile, but that's not too hard to understand. The country's going through upheaval, and workers must be wondering about their futures." Ford "would not have a plan to come in and lay off people."[56]

Lee Keun Young at the Korea Development Bank said that banks would have to write off much of the debt load of Kia Motors, 8.75 trillion won, and Asia Motors, 3.07 trillion won. He indicated that a crucial factor would be how much of the debt a prospective owner was willing to assume. Lee set ostensibly tough terms, saying bidders had to offer more than the current share price and be willing to acquire at least 50 percent of the stock. "I don't think there's any question that some write-off is essential," said Booker. At the same time, both Kia and Asia Motors would write off 90 percent of their capital and issue new shares. Under this plan, Kia would have 1.5 trillion won, $1.2 billion, as capital while Asia Motors would be left with 600 billion won in capital. Hyundai, Daewoo, and Samsung were all invited to bid along with any foreign motor vehicle manufacturer among the world's top 20. Bidding price and cash flow would each count for 30 of 100 points in deciding the winner, ability to guarantee jobs and market vehicles overseas for 25 points, and prospects for long-term development for the final 15 points.[57]

The auction process, however, was doomed from the start. The first attempt collapsed on August 30 amid reports the Korea Development Bank planned to cancel the sale the next day. Members of a committee that included Andersen Consulting and Banque Nationale de Paris halted the auction after Samsung appeared to have won and Daewoo had finished second. Kia Motors' court-appointed president, Lee Jong Dae, announced "the bidding was aborted" since the two leading bidders, Samsung Motor and Daewoo Motor, wanted write-offs of part of Kia's $12 billion debt. Bidders should have known, he said, they could not demand write-offs. Samsung, which had sold only 27,664 cars to date, claimed it had only suggested a write-off in an "amendment" to its bid and briefly threatened legal action. Hyundai and Ford were also sure to ask for write-offs but were disqualified for submitting bids below the minimum share price. Hyundai's bid was the lowest. Ford protested the entire process, claiming lack of transparency as Booker had feared, *Chosun Ilbo* reported on August 31, 1998.[58]

Cancellation of the auction, "by returning the whole issue surrounding Kia back to the starting point, puts the bankrupt carmaker's fate under uncertainty again," said the English-language *Korea Herald*.[59] Refusal of banks to write off loans was responsible for scuttling the auction, said Park Song Bae, in charge of the KDB's nonperforming loans. "We tried hard to explain to them, 'If you refuse to yield, you may lose everything.' They say, 'No write-offs.' Until now Korean companies are accustomed only to rescheduling debts but no write-offs." Park said he would try to persuade Kia's 33 other creditor banks to give up trying to collect the entire debt. Many banks, however, saw the write-off of billions of dollars as a prescription for failure. "It means they are forced to merge," said Park. "Every bank is having quite a hard time."[60]

KDB, which held 20 percent of Kia's debts, stood to incur the largest loss. If other creditors refused to budge on write-offs, all might wind up collecting nothing while Kia ran up more losses. Kia's court-appointed chief, Yoo Chong Yul, asking creditor banks to decide how much of the debt to write off, announced a new auction on September 11. Bidding would be open until September 21 and the winner announced on September 26. Write-off of more than 50 percent of Kia's debts would be essential to persuading any of the bidders to take over. At stake was not only the Kia debt but more billions of dollars owed by other bankrupt companies whose potential buyers also wanted write-offs.

The failure of the second Kia auction, on September 23, was more of a jolt to investor confidence than the first. Yoo Chong Yul announced the cancellation of the auction even though Samsung, as before, had submitted the highest bid and Hyundai and Daewoo had both submitted bids above the minimum per share. As for Ford, its chairman, William Clay Ford, Jr., personally pulled Ford from the bidding beforehand after examining Kia's debt burden. The failure "makes a mockery of the bidding process," said Peter Underwood, who specialized in motor vehicles as a business consultant. "It undermines investor confidence in Korea."[61] All three bidders were disqualified for attaching requests for write-offs beyond the $2.1 billion that Kia's creditor banks had reluctantly agreed on after cancellation on September 1 of the first attempt at an auction.

Cancellation of the second auction opened the possibility that Kia would have to sell off its facilities piece by piece, leaving most of its debts unpaid. Kia's debts were increasing by about $6 million a day, more than half for servicing previous debts. Kia had sold 325,084 cars through September 1998, a 36 percent drop from the first nine months of 1997. Kia's domestic sales, like those of its rivals, were less than half what they had been the year before. "The longer they wait, the less the value of the company," said Underwood, "but the worst thing is the message it sends to foreign investors that Korea isn't serious about writing off debts."[62]

Under strong government pressure, creditors said a third attempt at auctioning off the companies would set no minimum for write-offs. To sweeten the pot, the Korea Development Bank said on October 1 that bidders could state how much of Kia's $12 billion debt had to be written off. This time, said the bank, creditors would select as the winning bid whichever company offered the lowest write-off. Government officials viewed the shift in rules as a major concession to bidders, including Ford; Kim Dae Jung said the third auction would be the last. The government, exercising power over the banks through disbursement of reserve funds from the central bank, left no doubt that the result would be final.

To the shock of Kia as well as government officials who saw a Ford takeover as an omen for other foreign investment, Hyundai on October 19, 1998, was proclaimed the winner. Kia's president, Lee Jong Dae, said that Hyundai had "received the highest scores in the screening process" even though all bidders had asked for write-offs far larger than Kia's creditor banks had wanted to give. Daewoo finished second, and Samsung this time was third. Ford spokesman John Spelich said that Ford was "clearly disappointed" by the outcome, which gave Hyundai 51 percent of the shares and managerial control. Ford remained "the best option for returning Kia to global competitiveness," said Spelich, but Ford was disqualified for offering less than the minimum 5,000 won or $3.70 per share in Asia Motors.[63]

"Ford had the most to offer in terms of solving Korea's fundamental problems," said Underwood, but he was not surprised by its elimination. "I don't think Ford could play the Korean game as well as anyone else. I never thought a foreign competitor could win."[64]

Kia's creditor banks still had to approve the deal under which Hyundai had demanded a write-off of $5.6 billion in debts. "There is no other way," said Park Chun Sop, a deputy director at KDB.[65] Park Tae Young, commerce, industry, and energy minister, maintained the auction "was completed fairly and transparently" and "creditors will have to agree because they will end up with bigger losses if they don't." Hyundai Motor Chairman Chung Mong Gyu said, however, that Hyundai still hoped "to draw investment from major U.S. and European automakers."[66] Hyundai's need for foreign funding exposed the company's huge debt-equity ratio and the difficulty of shouldering Kia's debts even after the write-off. Hyundai had the right to conduct an audit to see if Kia's real debts totaled more than the $12 billion acknowledged by the company and the banks, but nothing would stop Hyundai from taking over Kia by March 1999.

Kia's absorption by Hyundai closed a chapter in Korea's rise as an industrial nation. Thrilled Hyundai Motor workers cheered as the result of the auction was announced on television before noon, and Chairman Chung Mong Gyu declared, "The most urgent thing to do is to put Kia and Asia Motors back on the track."[67] Acquisition of Kia, said Hyundai, would enable the company "to obtain 2.5 million units of production capacity, bringing its world ranking within top ten makers of automobiles."[68]

The numbers showed the possibilities for Hyundai Motor's recovery from the doldrums of 1998. Hyundai produced 416,575 vehicles from January through August 1998 while Daewoo produced 254,383 vehicles and Kia 238,891 in the same period. Asia Motors produced 45,952 vehicles, mostly trucks, buses, and vans, in 1997 and 16,384 in the first eight months of 1998.[69] Kia facilities by 1998 were outdated after two years of cutbacks on R&D and technology, but Hyundai had no doubts. Hyundai would retool or integrate Kia's facilities with its own, achieving "economies of scale" by using the same platforms and components for models marketed under the Kia name.

But what about Ford and Mazda, the major minority shareholders? Ford was not only bitter about the conduct of the bidding but mindful of the difficulties encountered in supplying knockdown sets for Hyundai to assemble into cars from 1968 to 1986. Ford in December 1998 cut its losses, selling its stake in Kia for $11.3 million to J. P. Morgan in New York. Nine months later, in September 1999, Mazda sold all its shares in Kia. Together, Ford and Mazda lost enormously on an investment that had cost a combined total of $150 million.[70]

Next out of the race had to be Samsung Motor, which had sold just 41,277 cars from a facility capable of producing 80,000 cars a year. On December 2, 1998, word leaked from the Blue House that Samsung was negotiating with Daewoo to exchange Samsung Motor for Daewoo Electronics, a producer of consumer electronic items. If the deal were to happen, though, it would be against strong resistance from both Samsung and Daewoo. Workers at Samsung Motor staged a sitdown strike when they got the news. At the Daewoo headquarters, a spokesman said that Daewoo Motor, after acquiring Ssangyong Motor a year earlier, already had "a full line of automotive products." As for letting go Daewoo Electronics, he said, "We do not have any trouble in that company, and we got a record profit overseas."[71]

An overall agreement on chaebol restructuring, reached at a meeting with Kim Dae Jung on December 7, 1998, confirmed the rumors. The agreement said that Samsung would yield Samsung Motor to the Daewoo group while Samsung Electronics got Daewoo Electronics, producing such consumer items as TV sets and refrigerators at a loss in 1998 amid declining domestic demand. The "Big Deal swap" of Samsung Motor for Daewoo Electronics would reflect the emphasis on "core areas" assigned to each of the top chaebol.[72]

Daewoo Motor President Kim Tae Gou signaled an agonizing negotiating process. There were "a lot of factors to consider," he said, when asked why Samsung Motor was ignored in an outline of Daewoo Motor's restructuring plan under which the company vowed to join the world's top ten motor vehicle manufacturers. To make the deal worthwhile, "we have to add some products." For the Daewoo-Samsung swap to work, a Daewoo spokesman said that Daewoo should acquire entities of Samsung Heavy Industries that produced commercial vehicles and components. Daewoo had not previously revealed its desire for Samsung's commercial vehicle division.[73] Financial Supervisory Commission Chairman Lee Hun Jai said, however, that Daewoo Chairman Kim Woo Choong and Samsung Chairman Lee Kun Hee had personally agreed on the exchange. Their groups were to set up a committee in which every creditor bank would be represented. The committee would make recommendations while an outside accountant evaluated assets and debts.[74] The hard part would come in haggling over the numbers.

Labor problems also clouded the deal. Samsung Motor's 6,000 workers and about 10,000 workers at Daewoo Electronics' two main plants walked off their jobs when the agreement was announced. Another 3,000 Daewoo employees demonstrated in front of Seoul Station, across a vast square from Daewoo group headquarters. Daewoo did not seem disturbed, citing the company's profits in recent years and predicting a similar profit for 1998 despite sagging demand. "The workers don't want to merge," said Lee Jeong Seung, a Daewoo spokesman. Still, he said, "the principle is to merge, and we will have to abide by the agreement."[75]

The merger, however, would never happen. On June 29, 1999, Samsung spurned the long-heralded "Big Deal" sale of its failed motor vehicle entity to Daewoo, choosing instead to ask a court to decide what to do with what was left of the company. The decision by Samsung Chairman Lee Kun Hee for court protection during liquidation marked a sharp rebuff to DJ's efforts to force heavily leveraged chaebol to "swap" some of their companies in order to get rid of the losers. While placing the fate of Samsung Motor in the hands of a court, Lee pledged to invest 2.8 trillion won, $2.4 billion, from his personal fortune to help pay off Samsung Motors' debt of 4.3 trillion won, $3.7 billion. He said he would raise the money by selling 4 million shares that he held in Samsung Life Insurance, one of the group's most profitable entities. After claiming repeatedly that Samsung and Daewoo were about to come to final agreement on terms for the takeover, officials were resigned to the fact that the groups remained far apart on terms. The FSC's Lee Hun Jai said the decision to apply for court receivership would "initiate a liquidation process for Samsung Motors," which had suspended production on December 7, the day the "swap" was announced, only to resume again four months later. By the time Samsung again suspended production in May 1999, it had sold just 45,091 cars.[76]

Choi Seung Jin, a Samsung spokesman, said the agreement to sell Samsung Motor to Daewoo Motor collapsed when the groups disagreed on how much the company was worth. An independent accounting firm "announced our asset value will be one trillion won," said Choi, "but Daewoo thinks our asset value is zero."[77] A crucial factor was that Daewoo ranked as the most heavily leveraged of Korea's top-five chaebol. Daewoo's debts at the end of 1998 were 59.5 trillion won, $52 billion, and they had increased by at least $5 billion since then. The two companies did not come close to agreeing on how much Daewoo could afford to pay—or how much of Samsung Motor's debts should be added to Daewoo's debt load. Nor could they agree on how much Samsung should pay for Daewoo Electronics, which Daewoo insisted was a money-maker.

The seeming demise of Samsung Motor heralded a period of maneuvering by Samsung to minimize its losses. Samsung Motor Vice Chairman Lee Dae Won, said his company had applied for court receivership "in the best interest of the national economy" in order to "settle this matter with transparency and honor." Daewoo Motor might still buy Samsung's manufacturing plant near Pusan—a political favor for DJ, worried about the impact of a shutdown of a major factory in an area where he had little solid political support.

Whatever Samsung got for its facilities, however, would hardly make up for its losses. Samsung's creditors were not happy about Lee Kun Hee's offer to "donate" his own shares in Samsung Life, which was not listed on the Korea Stock Exchange. They claimed the shares were over-valued and wanted repayment in cash. The ultimate threat was to try to call in the group's total debt, calculated at 42.3 trillion won, approximately $36 billion, at the end of June, but such drastic action was unlikely. To placate the banks, Lee Kun Hee in September decided to try to sell 400 million shares in Samsung Life to foreign investors—a risky scheme that would delay listing Samsung Life on the stock exchange.[78]

The banks had difficulty going aggressively after Samsung when Daewoo's troubles were far worse—not just for one or two companies but for the entire group, the chaebol system, and the Korean economy. On June 29, when the talks on the Big Deal for Samsung Motor and Daewoo Electronics ended, the presidents of 50 entities of the Daewoo group resigned en masse in order, said a Daewoo spokesman, to make it easier to slim down the group. That gesture fit in with Daewoo Chairman Kim Woo Choong's announced plan to reduce the group from 23 companies to 9, including Daewoo Motor and Daewoo Motor Sales, by the end of 1999. Kim Woo Choong by now was buying time, building up defenses against the wolves howling at his door. The game was up by July 19 when the FSC rammed through its stopgap rescue operation under which Daewoo's domestic creditors had to postpone demands for repayment of 11 trillion won.[79]

Driven into a corner, Kim Woo Choong was not about to surrender. Rather, he fought for a deal under which he might retain what he cared about most, Daewoo Motor. As part of a broad program for appeasing the bankers, he agreed to sell Daewoo Securities, one of the group's few profitable entities, as well as other companies, in whole or in part, but clung tenaciously to what he saw as his greatest legacy. Just to show the bankers he was serious, on August 6 he authorized yet another memorandum of understanding with General Motors on negotiations for reentering the partnership that was broken off in 1992.

Daewoo now was asking GM for $3.5 billion, 70 percent of the cost of its enormous buildup in the previous three or four years and more than 20 times what it had paid GM for its 50 percent share in 1992. Daewoo said GM would be welcome to run the company, even take a majority stake, an outcome that would be humiliating for Kim but preferable to losing out on motor vehicles entirely. Daewoo, however, was still asking too much. "They'd be lucky to get a billion," said consultant Peter Underwood.[80] There was speculation that GM wanted Daewoo's Polish plant, formerly Poland's state-run FSO manufacturer, for which GM had bid before Daewoo acquired it five years earlier. GM also was interested in Daewoo's plant at Kunsan in Korea. The Polish and Kunsan plants, each capable of producing 300,000 cars a year, might give GM valuable entrée into difficult markets in both eastern Europe and Asia.

Kim demanded the chance to succeed in automobiles as a condition for any agreement with creditors. Before the economic downturn in 1997, he had poured $5 billion in expanding and modernizing facilities in Korea and buying the plants in eastern Europe before taking over Ssangyong Motor. He could not give it all up. On August 16, he agreed to a sweeping deal with the government under which he would retain Daewoo Motor, Daewoo Motor Sales, Daewoo Capital and divisions of Daewoo Telecom, Daewoo Heavy Industries, and the Daewoo Corporation as a secondary chaebol. As for Ssangyong Motor, which had remained a separately listed unit on the Korea Stock Exchange even after Daewoo had taken it over, he would have to merge it with Daewoo Motor, never listed, while seeking foreign investment for the combined unit. All the rest of the group's 25 companies, including Daewoo Securities, had to go.[81]

In a book that was turned into readable English by Louis Kraar, author of flattering articles on Kim Woo Choong for *Fortune,* Kim advised, "Always think of the partner in a deal getting as much out of it as you do." Indeed, he went on, "If you impress upon people that they will not take a loss by dealing with you, your business will thrive."[82] At heart, however, Kim hoped to avoid having to bring in Generals Motors as an equal, much less a superior partner. He might lose most of the rest of the group, but he wanted motor vehicles to be his great contribution to Korean society. It would not be easy for GM and Daewoo, having broken up over Kim's wild expansion plans, to come together again. "There's a lot of work to do there, so it won't be anytime soon," said GM Chairman Jack Smith. While cutting production amid a shortage of parts from suppliers demanding payment, however, Daewoo did not appear to have a strong hand. For ten days, Daewoo Motor had to stop its assembly lines, resulting in a reduction of 14,000 units, 70 percent of them for export, in September.[83]

The rumors got better. One of the best was that GM would take over not only Daewoo Motor and its offshoot, Ssangyong Motor, but also Samsung Motor—a deal that would divide the Korean industry between GM and Hyundai. Ssangyong and Samsung, however, might be liabilities rather than assets. The future of both companies seemed to lie largely in the hands of bankers wrangling over the advantages of swapping debt for equity, liquidating assets or seeking foreign buyers. Samsung promised to cover some of the debt of its motor vehicle company while applying for court receivership—and protection. Political considerations were always paramount. Samsung Motor on November 1, 1999, resumed production, turning out about 2,000 cars a month at a snail's pace for sale around Pusan, a bastion of support for Kim Young Sam and the opposition Grand National Party. The company, no longer part of the Samsung group, entirely in the hands of creditor banks, had to prove to potential foreign buyers that it was "in good operation," said Seong Jun Ke, one of the few managers still at work in the gleaming facilities.[84]

The takeover of Kia, the failure of Samsung Motor and the troubles afflicting the Daewoo group, including Daewoo Motor, could only strengthen the position of Hyundai Motor—but not Chung Mong Gyu. In a sad subplot of the drama of the Korean motor vehicle industry, Mong Gyu was to lose the company that his father, Chung Se Yung, had nurtured for him from infancy, his and the company's. The group founder, Chung Se Yung's eldest brother, Chung Ju Yung, remained the real power. Ju Yung's oldest surviving son, Chung Mong Ku, was not only group cochairman but also chairman of Hyundai Precision, which had produced utility vehicles in Ulsan since 1991, and Hyundai Motor Service, which marketed Hyundai cars in Korea. Having been the major influence behind the manipulations that had led to the Kia takeover, Mong Ku wanted it all. What better way to appear to be "restructuring" and "downsizing" than to put everything in a single company? Soon after the Kia deal was announced, he made himself acting chairman of Hyundai Motor. Mong Gyu was demoted to vice chairman.

The humiliation of Chung Se Yung and Chung Mong Gyu was not over. In the final act of the family feud, Chung Ju Yung, 83, barely able to walk but still the Confucian boss, in early 1999 ordered the resignation of Chung Se Yung as honorary chairman of Hyundai Motor. Hyundai officials confirmed the move on March 2 after Seoul newspapers made it front-page news. The family feud erupted at a stockholders' meeting at the end of February in which Chung Se Yung, called "the godfather" of the industry by the media, challenged big brother's authority by attempting to name his own protégés to top posts and advance the power of his son. The public display of internecine strife in a chaebol family shocked the business community. A cartoon in *Chosun Ilbo* depicted "King" Chung Ju Yung, his arm around Mong Ku, asking, "Who the hell is against my son?" as a doctor, muttering, "Impossible to cure," carried Chung Se Yung away on a stretcher.[85]

The shift came as the company prepared to expand on its dominant position over the Korea motor vehicle industry after losing money in 1998 for the first year since producing its first car in 1966. Hyundai Motor was about to take over bankrupt Kia at the end of March by paying 1.7 trillion won, $1.4 billion, while assuming debts from Kia of 6.3 trillion won, nearly $6 billion. Hyundai in all of 1998 had sold 868,904 cars, down from 1,247,424 in 1997, and suffered a loss of 33 billion won, nearly $30 million. With the Korean economy showing signs of recovery from economic turmoil, however, sales increased 52 percent in January over January 1998 and 9.5 percent in February.[86]

For three more days after their fall from grace, Chung Se Yung and son Chung Mong Gyu remained Hyundai Motor's largest stockholders, with 8.2 percent of the stock. Chung Ju Yung and Chung Mong Ku, however, brandished the ultimate weapon, their control of other Hyundai companies, which owned a majority of the shares. On March 5, Chung Ju Yung forced Se Yung and Mong Gyu to trade their shares for Mong Ku's large stake in Hyundai Industrial Development and Construction. Thus SY and son bowed out of the motor vehicle business, leaving Mong Ku to run an expanded HMC as he saw fit. Chung Ju Yung had betrayed his brother and nephew, then saved a little face and a great deal of money for them by giving them a unit they did not want. SY, swallowing the bitter pill, said at a press conference that it was all "in the best interests" of the company over which he had held sway for more than 30 years.[87]

For reasons that had nothing to do with the shift at the top of Hyundai Motor, Korea's slimmed-down motor vehicle industry was on a roll after the shakeout of companies in 1999. Many Koreans, having forsworn new cars for more than a year, were emerging from their cocoons, taking stock of recovery and deciding to buy. The figures for Hyundai Motor were all the more impressive since the company had not had to cope with a debilitating strike, as in the summer of 1998. Hyundai sales zoomed to 107,166 in August 1999, an increase of 750.9 percent over August 1998. Domestic sales increased from 8,694 in August 1998 to 48,651 vehicles in August 1999 while exports shot up from 3,900 to 58,515. In all 1999, Hyundai sold 1,271,471 vehicles, up 40 percent from the 898,208 vehicles it sold in 1998. The 1999 figure exceeded the 1,247,424 vehicles Hyundai sold in 1997 but was still below the HMC record of 1,366,508 vehicles sold in 1996, the last full pre-crisis year. Kia, under Hyundai ownership, improved still more, selling 834,084 vehicles in 1999, an increase of 76 percent from the year before. Daewoo Motors, despite slowdowns and stoppages caused by financial troubles, finished 1999 with sales of 945,572, up five percent.[88]

Paradoxically, however, Hyundai Motor as a result of the Kia acquisition was in trouble. On September 16, Hyundai Motor revealed that it had sold $500 million worth of stock to foreign investors for 15.85 percent less than its actual share price after failing to find enough buyers—"the strongest evidence yet," Bloomberg reported, "that investors are concerned about Korea's debt-ridden industrial groups." The group by the end of June 1999 had a debt-equity ratio of 4.5 to 1 with total debts of 64.9 trillion won, $54.4 billion, up from 61 trillion won at the end of 1998. It was a measure of Hyundai's desperation for cash that Hyundai Motor was willing to sell shares overseas at discount prices. Overall, Hyundai said that it had raised $1.84 billion abroad in 1999, including the half billion dollars in Hyundai Motor sales.[89]

Nearing the end of the race, the survivors had to wonder if the struggle had been worth the costs. Hyundai, with Kia, faced an uncertain future, as did Daewoo, with or without GM. Motor vehicles might be glamorous symbols of national success but were uncertain money-makers.

Both the Samsung and Ssangyong groups were better off then they had ever been as motor vehicle manufacturers. By the time the finalists reached the finish line, there might be no place for also-rans.

GM, on December 14, 1999, presented a formal offer, reportedly for $5-6 billion, to the governor of Korea Development Bank, Daewoo's lead creditor, for all Daewoo's passenger car facilities in Korea, including Ssangyong, and most of its foreign operations, notably the Polish plants. GM wanted "exclusive" bidding rights, said GM President Alan Perriton, warning that a protracted auction process "would be long-term destructive for the industry in Korea."[90] Hyundai Motor President Lee Kye Ahn countered that Hyundai was "especially interested" in having the plant in Poland for itself—and warned that GM would undermine the entire Korean motor vehicle components industry while using a Korean operation as a takeoff point to get into the China market. Lee had the full support of the FKI, which claimed that GM's price strategy would "drive Hyundai Motor out of business."[91]

Hyundai, with 70 percent of the Korean motor vehicle industry, lusted for more—along with membership in the "global top ten." Ultimately, in an elimination contest against a dreaded foreign competitor, Korea, Inc., perceived a threat not just to Hyundai but to basic national interests and sovereignty.

# Chapter 9

# Selling Off/Selling Out

T he failure of Ford to take over Kia proved a point: logical though
foreigners might have seemed as Big Deal partners, they were
often excluded from buyouts. Window-shopping before everything
from motor vehicle companies to chemical giants to high-tech spin-offs,
foreigners were reluctant to enter fields where the sellers set the terms.
"There's been an unprecedented surge of interest," said Tom Pinansky, an
attorney who had spent ten years advising foreign companies in Korea.
"I've never seen anything like it. A whole lot of companies have a whole
lot of questions. People are getting all excited by M & As. It's the flavor
of the month." Pinansky, however, saw everything as "very complicated
and difficult because of all the regulations."[1]

An example was real estate, a market that the government was liberalizing. Companies now could only own the land for their plants, but what
if they wanted office buildings or tracts for future expansion? "Most of the
real estate revisions they talked about haven't occurred," said Pinansky.
"It's like peeling the onion. They're down one or two layers, but that's all."
As the real estate market opened to foreigners in mid-1998, prices
remained sky-high. Owners hated to think the land on which they had
placed a premium, often as collateral for more loans, was not worth nearly
as much as they had paid for it.

Foreigners, their lawyers, and a host of consultants cited a litany of
obstacles, beginning with ingrown bureaucratic prejudices and ending
with rules and regulations that were changing only slowly under the
weight of Korea's need for hard foreign currency. For all the difficulties,
however, investors and consultants alike had the sense that the business

landscape of Korea had changed since December 1997. Investors should exploit the discounts now on the market, in the view of Robert Broadfoot, managing director of Political and Economic Risk Consultancy of Hong Kong. "The door is going to be closed to many industries that have been closed to outside investment. Korea has to be aware it's not in isolation. This is a terrific opportunity."[2]

Broadfoot had in mind mostly small and medium-size enterprises, but great opportunities should have been available among some of the nearly bankrupt giants as well. The prospects for the Big Deal were just as hot in oil as in motor vehicles. Previously limited to 50 percent ownership of energy companies, foreigners as of April 1998 could own up to 55 percent and might eventually bid for 100 percent if the new government saw no other way to rescue them.

Deals that appeared to make sense, however, had a way of souring. Hanwha Energy, Korea's fourth and smallest oil company, a member of the 20-company Hanwha group, the ninth-largest chaebol, had lost at least $30 million in 1997. Rumored to be the next to fall into bankruptcy, the Hanwha group had to borrow $300 million to keep Hanwha Energy afloat. The Hanwha group had a debt-equity ratio of about nine to one, far above the national average of four or five to one. Financial pressures dictated negotiations between Hanwha Energy and foreign companies anxious to get in on a potentially huge market by selling oil at cut-rate prices. Hanwha pressed for a deal with Texaco, but Texaco showed little interest. "Foreign companies want to buy our company at too low a price," said Ahn Hun Mo, a manager in sales at Hanwha Energy Plaza, responsible for marketing Hanwha Energy products.[3] By 1999, Hanwha was absorbed by Hyundai Oil.

Ssangyong Oil Refinery was similarly disappointed. A core company of the Ssangyong group, ranked sixth until it sold Ssangyong Motor to Daewoo Motor, Ssangyong Oil was anxious for Aramco to increase its 35 percent share to 55 percent. Aramco saw little to gain from the investment. Finally, in March 1999, Ssangyong Cement, the mainstay of what was left of the Ssangyong group, struck up a deal to sell its 28.41 percent stake in the oil company to the SK group, which owned Korea's largest refinery.[4]

The government in 1998 faced a tough battle in overcoming skepticism among foreigners despite what appeared to be bargain-basement prices. Foreigners get "scared off when they learn of the debt ratio of local companies hiding behind complex financial sheets," said Kim Dong Woo of Shinhan M and A Technology. They are "stunned by their actual mountain of debts."[5]

Preconceptions and prejudices against foreign penetration were intimidating. What happened to Hangul and Computer after it tried to make a deal with Microsoft revealed a mind-set that foreign investors most feared. For Hangul founder Lee Chan Jin, a 32-year-old computer whiz described in his official biography as "leader of the software industry in Korea," the choice initially was simple. Either he yielded to the protests and let his company sink into oblivion or he ignored aggrieved citizens' groups and sold out to the American giant. "If not, our company will go bankrupt," said Park Soon Baek, vice president of Hangul.[6]

The dilemma confronting Hangul in mid-1998 epitomized that of scores of companies that appeared to welcome foreign investment but faced deep-seated nationalist sentiment against it. In the 18 years since Lee had founded Hangul, the company had come to represent the ultimate in successful, entrepreneurial venture capitalism. Until the onset of the economic crisis, it seemed to be thriving on one product, the software for Hangul, the Korean writing system that embodied the deepest strains of Korean patriotism. Now, however, with a Korean market share of more than 60 percent, Lee wanted to withdraw from that side of the business and give it all to Microsoft Korea, which had about 30 percent of the local market. In return, Hangul would accept $20 million from Microsoft, which it could then invest in other products. The deal, agreed on before Microsoft chief Bill Gates visited Korea for three days in June 1998, generated a storm that no one had anticipated.

The implications were disturbing to a government committed to persuading foreign investors to bail out a wide range of companies. "I was with Bill Gates when our president asked him if he was aware there was a movement regarding Hangul," said Park Jie Won, the presidential press secretary. "Mr. Gates said he was aware of the sentiment. He said he might find some other way to invest if that became an issue." Park, making clear Kim Dae Jung's distaste for the protest, said flatly, "We know that foreign investment is the only way to revitalize our ailing economy."[7] Such arguments, however, were lost upon the critics. "The Committee to Save Hangul Software will fight to save Hangul and Computer," said a radio talk show commentary. "If Hangul gives up Hangul software, it will be a tremendous loss for the country. The entire business will be taken over by Microsoft, and people will have to learn MS Word."[8]

The message was that Hangul was guilty of lack of patriotism for selling a Korean birthright, its unique lettering system, to foreign interests. The fact that Microsoft did not demand operating control and would limit its investment to 19 percent of the company's value did not impress the critics. A "Hangul Venture Company Committee" was set up to raise the funds needed to keep the company alive without Microsoft. "Lee Chan Jin and his staff must fight until the end," said Lee Min Hwa, committee chairman. He claimed the cost of retraining people on Microsoft software would exceed Hangul and Computer's debt of about $14 million, five times the available capital in a company valued at about $50 million.[9]

For Lee Chan Jin, however, the campaign to save his company was unrealistic. The reality was that 80 percent of the software in Korea was pirated, said a Hangul aide, Kim Jung Soo. "We can't make any profit by selling software anymore. Without any kind of investment from an outer source, it will be really difficult to do anything."[10] To the critics, any foreign rescue plan marked an ignominious defeat for Lee, whose biography confidently described his company as "the largest software developer group throughout the entire nation." Abandoning Hangul Software would be as wrenching for company workers as for the citizens who wanted Hangul and Computer to stay afloat on its own. "We love our product," said Kim. "The public is angry about the deal with Microsoft, but the public doesn't buy the software. If they had bought the software, we wouldn't have this problem." She denied, however, that Microsoft managers would meddle in Hangul's affairs. "Microsoft will never touch Hangul and Computer's management. It will be a separate entity, but they will send people to work with us."[11]

Three weeks later, the deal was off. Hangul and Computer reversed its position and spurned a bid by Microsoft on July 20, 1998, to take over its business and merge it into Microsoft's own Korean software. Instead, the company accepted an offer of $10 million from a group of Korean investors and donors to enable it to pay off some of its debts and invest in research and development needed to revive its fortunes. Lee Chan Jin and Lee Min Hwa announced their startling rebuff to Bill Gates at a crowded press conference that caught Microsoft unawares. Lee said he was "very happy" that he had decided to save the software program. "The money is not enough," said Lee Chan Jin as Lee Min Hwa beamed approvingly, "but the Hangul program will survive."[12]

The Hangul case was a caricature of the problems that foreign investors perceived, in many different forms, in approaching Korean companies. In the face of an alarming downturn in direct investment in the first few months of 1998, the government on September 25 announced an international "road show" to persuade foreigners to place their money on Korea. The show, with stops from Tokyo and Hong Kong to New York and London, reflected the worries in the aftermath of other much greater debacles.

These included not only the failure of Ford in bidding for Kia Motors but also cancellation by AES Corporation of Arlington, Virginia, on September 23, 1998, of an agreement with Hanwha Energy to purchase a 1,500-megawatt thermal power plant for $874 million. The collapse of both deals deepened the doubts over whether the chaebol could carry out the tortuous process of restructuring while burdened with at least $150 billion in debts to foreign creditors and domestic debts several times that amount.

The bottom line was that foreign direct investment by the end of October 1998 was $5.53 billion, down from $5.85 billion for the first ten months of 1997 but up in September and October. The biggest investor in 1998 was BASF of Germany, which pumped more than $1 billion in petrochemical facilities, including about $600 million for the purchase from Miwon, a chaebol on the brink of bankruptcy, of a plant producing the food additive lysine. The largest single deal with a foreign firm in 1998 was the sale of the heavy equipment division of Samsung Heavy Industries to Volvo Construction Equipment for $720 million. Other major deals with foreigners included British Telecom's investment of $400 million in LG Telecom, a $270 million investment by Commerzbank of Germany in the Korea Exchange Bank, and a $250 million investment by Interbrew of the Netherlands in the Doosan Brewery, part of the Doosan group.

"In past years before the Korean crisis, Korean companies found money by borrowing from outside," said Moon Jae Woo, director of foreign direct investment at the ministry of finance. "They were afraid of losing control if foreigners invested." Attitudes were different now, he said, citing a poll that showed 87 percent of the people viewed foreign direct investment as "very positive" to the economy. "Our economic outlook is better, and we have stability in securities and foreign exchange. Now many companies think this is the time to make investment. Investors see a strong will to keep up the reform process." Moon predicted that foreign direct investment in business and industry in 1998 would exceed the 1997 total of $6.9 billion.[13]

To convince the doubters, the government on November 17, 1998, introduced incentives under the newly passed Foreign Investment Promotion Act. Besides simplifying procedures for obtaining government licenses and opening offices, the act provided a ten-year tax holiday for investors in plant facilities along with discounts on rental fees for real estate, one of the biggest barriers to foreign investment. Except for telecommunications, publishing, and broadcasting, certain defense industries, and agriculture, there were, on paper, no restrictions on 100 percent foreign investment. By 1999, foreigners theoretically could even buy Korean newspapers, once seen as off limits to foreign influence, much less control. Foreigners still complained, however, of protracted efforts at penetrating veils of secrecy. Financial consultant Hank Morris cited a company with 40 employees that claimed its payroll was below the equivalent in Korean currency of $500,000 a year. "For 40 employees, you need double that amount. When we noticed that, we knew it didn't make sense."[14] The worst problems occurred in trying to assess a company's debts.

Another major area of disagreement was the value of the land on which a company was built. Koreans still viewed real estate as a treasured commodity, often using it in the years before the crisis as overly valued collateral for wildly accumulating loans. Although the government in 1998 opened the real estate market to foreign investment, potential investors lost interest after hearing the prices. "You have a gap in how things are priced," said Sam Hageman, a director at the Seoul office of John Buck, a Chicago-based real estate firm. "Legally the market is completely open, but it's unproven." One question: "Once you buy a property, how do you get out if you want to?" He knew of no cases in which a major property had been resold. Still, he surmised, "Eventually something is going to happen—the question is when."[15]

The government also wanted to lure foreign tourists. An advertising campaign in 1998 showed Kim Dae Jung in a TV commercial popping up in an airline cockpit offering a guarantee that he hoped would prove irresistible. "I'll be sure you have a safe and pleasant journey," DJ promised in Korean, translated into three different languages in versions for Japan, China, and English-speaking countries. The image of the 73-year-old president was just one of the lures for an industry that had become a major source of income. Hong Too Pyo, president of the Korea National Tourism Organization, estimated the country would earn $3 billion in 1998 from foreign visitors. Footage of champion golfer Pak Se Ri, Los Angeles Dodgers pitcher Park Chan Ho, a couple of Koreans on Japanese baseball teams, and conductor Chung Myung Hoon were interspersed along with shots of dancing, singing, and scenic sights.[16]

There was one kind of foreign penetration, however, that Koreans, from factory workers to chaebol chairmen, could hardly stomach—foreign competition on the Korean market. While government officials said repeatedly they did not support an anti-import drive, foreigners saw everything from bureaucratic influence to corporate payoffs to cultural patterns behind the efforts of consumer groups to discourage the influx of foreign goods. The specter of an invisible hand quietly seeking to staunch imports had long haunted foreigners in the new era of "globalization" and "market-opening," two terms from the Kim Young Sam presidency. Did consumer groups reflect just a desire for self-discipline, or was the campaign anti-foreign? Businessmen were uncertain whether the barriers were officially sanctioned or the unofficial legacy of anti-foreign habits buried deep in history. The main concern, U.S. diplomat Richard Christenson told Korean business and political figures in June 1997, was "not the austerity drive but the erection of new barriers against the common principles" of the OECD and the World Trade Organization.[17] Michael Brown, president of the American Chamber of Commerce in Korea, observing that "the frugality campaign is problematic" for importers, asked authorities to guarantee that officials "do not misinterpret austerity policies and impede in the distribution or sale of imported products."[18]

Raucous demands of consumer groups for spurning import items ranging from tobacco to liquor to cars struck foreign businessmen as secretly inspired by high-level mandarins. John Alsbury, president of British-American Tobacco in Korea, saw a high-level connection in the efforts of the government-owned Korea Tobacco and Ginseng Corporation to hold down the sale of foreign tobacco—not legal at all in Korea until 1988. "They don't leave many fingerprints," said Alsbury. "They're hiding behind civic groups. You can never prove anything."[19] Consumer advocates, however, revealed the complexes. "This is just a developing country," said Park Chan Sung, chairman of the Citizens' Movement Center for Anti-Overconsumption, financed by donations from religious organizations, Protestant, Catholic, and Buddhist. "It's too early for us. Average spending is almost as high as advanced countries."[20]

One organization that ran afoul of the rules was Amway Korea. The Michigan-based company had built up a force of 144,000 part-time door-to-door sales "representatives" in four years since an Amway executive, David Ussery, was jailed for nine days in 1993 for violating a law against "direct" or "network" sales. The law was lifted, but the National Council of Consumer Protection Organizations and the Korea Soap and Detergent Association claimed Amway reps in 1997 had violated another law barring companies from comparing their products with those of rivals. "The campaign is not against foreign products," said Kim Kyu Hong, the Korea Soap and Detergent Association's director. "It is only against the Amway method." Amway, whose sales in Korea leaped to $350 million for 1996, filed a complaint with Korea's Fair Trade Commission. "Our business is suffering," said Brian Chalmers, president of Amway Korea. "The impact of demoralization of our distributors is difficult."[21]

Motor vehicles made the easiest target. Even though foreign motor vehicle sales in 1996 came to only 25,000 among total domestic sales of more than 1.5 million, foreign cars were still too visible for the critics. "We're hearing reports of the police stopping imported vehicles just because they're imported," said Wayne Chumley, president of Chrysler Korea, which sold 2,200 units in Korea in 1996. He also cited threats of tax audits of owners of foreign cars, once not seen at all in Korea. "Customers have a perception, if they buy our import, they'll get audited," he said.[22] Officially, however, the tax office told bureaucrats to call off the audits and claimed importers could sell all they wanted as long as they paid the 8 percent tariff and got each vehicle inspected at a government facility.

The frustration of foreign car dealers rose in proportion to the decline of foreign car sales in the IMF era. Foreign sales had been falling since mid-1977, when the economy was showing the first signs of serious problems. "Our customers are afraid of driving their new cars because people persecute and harass them," Choi Byung Kwon, chairman of the Korea Automobile Importers and Distributors' Association, said in March 1998. "They accuse them of selling out the nation."[23] Ford Korea sold 260 vehicles in January and February 1998, making Ford the leading foreign dealer with 23 percent of sales.[24]

The anti-import campaign meant that owners of foreign cars were not safe from policemen flagging them down for tickets as well as other drivers hurling insults, and sometimes eggs, at them and their vehicles. Restaurants and nightclubs were known to deny parking space for foreign cars, gasoline stations occasionally refused to serve them, and owners found nails in their tires and antiforeign warnings smeared on windows and doors. One woman, said Yoon Dae Sung, executive director of the foreign car dealers' group, was "grabbed by the neck and abused" when she got out of her imported car. Yoon said the woman decided to sell it rather than risk injury and damage to the car.[25]

Such protestations had minimal effect. "There are basically no imports here," said Chumley in September 1998. Chrysler in the first eight months of 1998 led foreign car companies, selling 467 vehicles in Korea as opposed to 1,650 vehicles in the same period in 1997.[26] Ford sold 401 cars from January through August 1998, and GM sold none.[27] Foreign manufacturers still hoped for drastic cuts in the tariffs and taxes that added about 85 percent to the cost of their products, but U.S. negotiators encountered a bleak response. After three days of talks in September 1998, a negotiating team threatened trade sanctions if the Koreans failed to come to terms by October 19. On October 20, the United States and Korea signed an agreement that appeared to pry open the market, but there was no chance of equal access. The agreement eased the testing of motor vehicles entering the country but did not lower taxes and tariffs of 8 percent.

South Korean officials were hardly sympathetic with U.S. complaints in a period in which sales of South Korean cars on the domestic market were plummeting and exports were down because of lack of funds for parts and materials. "They ask us to lower tariffs on vehicles from 8 to 4 percent," said Kim Ho Shik, deputy finance minister. "We don't have any intention to accept that." He maintained that 8 percent was "already low" compared with tariffs of 10 to 12 percent levied by European countries. As for U.S. complaints that local taxes posed a still greater burden, Kim said, "We admit it, but it is equal for local products as well."[28]

Korean exports to the United States, however, had risen out of all proportion to imports since Korea had gone to the IMF. Overall, Korean exports to the United States earned $17.4 billion in the first nine months of 1998, up 2.4 percent from 1997, against U.S. exports to Korea of $10.7 billion, down 43.4 percent. Korean industry earned $1.32 billion from steel exports to the United States in the same time frame, up 68.4 percent. American and European steel manufacturers complained that banks, propped up by government funds, were subsidizing weak Korean specialty steel companies through low-interest loans while Pohang Iron and Steel was exporting iron and steel at below-market prices.[29]

The Koreans, though, were driven by a certain desperation. The Bank of Korea on November 25, 1998, revealed that Korea's gross national product had plunged 8 percent in the third quarter of 1998 from the same period of 1997 while heavily leveraged companies struggled to pay interest on loans. The report showed Korea in the midst of its worst recession since the final period of the Korean War in 1953, the last period until 1998 in which the economy had shown negative growth for three quarters in a row. The Bank of Korea blamed "increased interest payments to the rest of the world" in part for the dip. Korea's foreign debts now totaled $154 billion. The gross domestic product dropped 6.8 percent for the third quarter from the third quarter of 1997, equaling the year-on-year decline reported for the second quarter of 1998.[30]

The Bank of Korea gave rising reluctance of Korean consumers to spend money along with a regimen of stern cost-cutting by manufacturers as reasons why the economy remained in recession. "Dragged down by a persistent reduction in household spending," said the report, "private consumption showed a 12 percent decrease." Major factors were 7.3 percent unemployment and 21,000 bankruptcies in 1998. "Fixed capital formation registered a 28.3 percent decrease," reflecting a 46.3 percent decrease in investment in machinery and equipment and a 15.8 percent drop in construction.[31] How could Korea completely recover while battling to pay off debts and keep pace with competitors in Japan and the west? Korean manufacturers relied on import of machinery from abroad to turn out products for export at prices below those of more advanced rivals.

Korea's construction industry, mainstay of the economy through two generations since the Korean War, was hit hardest, falling 12.9 percent in the first nine months of 1998 from the year before because of "the steep decline in residential and nonresidential construction." Manufacturing was also hard hit, falling 7.9 percent, but the Bank of Korea noted the decline was "less severe than in the previous quarter," when manufacturing dropped 10 percent year-on-year. The report cited "increased production of the electric and electronics sector" as the reason for the smaller decline in manufacturing but said that "manufacture of wearing apparel, industrial machinery, transportation equipment, and nonmetallic mineral products continue to shrink sharply." Services fell by 6 percent largely because of contraction of wholesale and retail trade, transport and storage and finance and insurance. Agriculture, forestry, and industry fell 5.6 percent "on account of the shrinkage of almost all subsectors except animal husbandry."[32]

The numbers added urgency to the demands for the chaebol to downsize and for companies to accept foreign investment, even foreign control and ownership. Like other countries in the IMF era, however, Korea was weighted down by companies whose raison d'être was the pride of owners who did not want to admit failure by selling out to domestic rivals, much less to foreigners. On December 4, 1998, a year and a day after the signing of the IMF agreement, Sohn Byung Doo, executive deputy chairman of the Federation of Korean Industries, made an extraordinary appeal to foreign businessmen. "The government listens to what the IMF and what the foreign people say," said Sohn. "You can tell them that restructuring takes time." From there, Sohn made another plea that seemed even more amazing, given the dominant position of the chaebol. "If you tell a child to walk quickly, if you scold a child, it will never be able to take a first step," he said. "So we are asking you to help us to walk properly."[33]

While the top five chaebol had replaced the all-powerful office of the chairman with "restructuring offices" and pledged their resolve to streamline, they made no secret of their disdain for the pressure exerted by policymakers viewed as amateurs. "Now I am afraid all these companies will be programmed," said Kim Duc Choong, an economist and older brother of Kim Woo Choong, chairman of both Daewoo and the Federation of Korean Industries. "It is time to re-review, to see what these companies did" that had made them world-class industrial giants. As for the "swap" of money-losing companies demanded by the government, he said, "These are government-designed programs in the name of private enterprise."[34]

Such rage showed the risks of twisting the arms of chaebol leaders. In a hierarchical society in which the president was the ultimate authority figure, Kim Dae Jung had to bring unwilling conglomerates into line the way his dictatorial predecessors had done it: by telling them what to do. His task was to convince the chairmen of the top five that they had to cut down to three or four core industries apiece or face penalties. The government had moved gingerly to use the kind of dictatorial power that Park Chung Hee as president had taken for granted, namely authority to order state-owned banks to extend or withdraw credit. The government no longer owned most of the banks, but Kim Tae Dong, senior secretary for economic policy planning, said there were "many ways the government sector can take for correctional purposes" if the chaebol misbehaved. "First, the creditor banks may refuse to roll over their existing debt. Then, the creditor banks can ask the chaebol to get rid of their shares of commercial banks and other financial institutions."[35]

Kim Dae Jung demanded "voluntary" action, voluntarily or not. On December 7, 1998, he summoned the chairmen of the top five chaebol, including Kim Woo Choong, still recovering from brain surgery, for "a brainstorming session." After weeks of acrimonious negotiations, the five, under his gaze, made a deal that DJ believed was final. The government "finally came to a complete agreement with the tycoons," he said, surviving "sabotage and resistance" that had made restructuring impossible. "The family-run chaebol will revamp their organizations to make them fit in better with a market-oriented management system," said Kang Bong Kyun, chief economic secretary. The chaebol had "agreed to share the pain" endured for more than a year by smaller companies and workers laid off "under the IMF era."[36]

The language of the agreement matched the mood. "Concrete results that fall out of restructuring of the top five chaebol is the key to bringing Korea out of the current economic crisis," said the agreement, ending "growth-driven strategies based on an excessive number of affiliates and intra-chaebol transactions." The agreement also called for the chaebol to cut the number of subsidiaries from 264 to 130 and work out restructuring with their banks to raise about 20 trillion won, nearly $18 billion, enough to cover 15 percent of their debts. The chaebol promised, as before, to cut debt-equity ratios to two to one by 2000 and to give up cross-guarantees of credit and subsidies of companies within the same chaebol. The owners of the top five, assumed to have secretly banked billions overseas and in Korea, had until December 15 to say how much they would contribute from personal accounts.[37]

Just as Park Chung Hee had seen no point in big companies' killing each other off by selling the same products, DJ insisted on mergers for the same reason. Hyundai would focus on motor vehicles, construction, ship-building, finance, and electronics. Samsung would keep electronics, finance, service, and logistics. Daewoo would have motor vehicles, ship-building, and trade, LG would be in chemicals, electronics, and communications, and SK in information, telecommunications, and energy. The agreement also listed the number of companies that each of them had to unload. Hyundai would go from 63 to 30 companies, Samsung from 65 to 40, Daewoo from 41 to 10, LG from 53 to 30, and SK from 42 to 20. There would still be competition. Hyundai and Daewoo would compete in motor vehicles, shipbuilding, and construction, for instance, while Hyundai and Samsung competed in DRAM chips after Hyundai Elec-tronics merged its semiconductor entity with LG Semicon.[38]

The Korea Stock Exchange responded with a flash of confidence to the realization that the chaebol might cooperate. The stock market on Decem-ber 7 shot up 23.81 points, 4.6 percent, closing at 514.52 with trade vol-ume peaking at a record 328.2 million shares. "Most securities issues surged to the new permissible high of 15 percent," said Yonhap. "Con-struction, wholesale/retail, and other manufacturing sectors also gained more than 10 percent."[39] The won held steady against the dollar, which closed at 1,208 won, down from 1,214 on December 4. There were doubts, however, as to whether all the reforms would come to pass. "There's going to be a strong temptation on the part of the conglomerates to appear to be downsizing," said Richard Samuelson at Warburg, Dillon Read. "Until there's real evidence of families removing themselves from power, you have to be suspicious."[40] The chaebol said they would cut cross-guarantees of credit among companies in the same group, but the Fair Trade Commission said cross-ownership, a more reliable indication of tight control, had increased over the past year.

FSC Chairman Lee Hun Jai liked to compare cowing the chaebol to the last roundup. "In order to tame the wild horses, you have to set up fences strong enough and high enough so the wild horses could not escape" was one of his lines. When Daewoo appeared uncooperative in talks over the "swap" of Daewoo Electronics for Samsung Motor, Lee suggested, "Maybe a critical bank will take action" in retaliation for "breaking the promise." Daewoo, as always, had an answer, claiming its restructuring, "in line with the economic reform programs and policies" of the government, was designed "to maximize Daewoo's business and management strengths in four core operations," motor vehicles, heavy industries, trade and construction, and finance and services. While reducing its subsidiaries from 41 to 10, Daewoo expected to attract $2.7 billion in foreign capital and slash its debt-equity ratio to just under two to one by the end of 1999.[41]

The "swap" between Hyundai Electronics and LG Semicon would be difficult, maybe impossible. On December 11, 1998, Lee said he "would not expect the merger of these two chipmakers will be successful." They "refuse to cooperate," he complained. "Both sides could not agree how to evaluate their companies."[42] Hyundai Electronics was battling LG Semicon over which would be the majority shareholder and who would manage the new semiconductor company they had agreed to form. The clash between Hyundai and LG was a test of the willingness of the chaebol to bury their differences in mergers and of the ability of the government to force them. Hyundai Electronics, with a debt-equity ratio of eight to one, insisted on 70 percent of the assets of the new company, rejecting a proposal for 50 percent ownership. LG Semicon, which had a smaller debt-equity ratio, wanted to manage the merged company.

John Dodsworth, senior IMF representative in Seoul, warned against excessive optimism as the major chaebol entered a new period of tough negotiations on Big Deals. He lamented that there had been "only limited gains" in coping with two of the chaebol's biggest problems—downsizing industrial capacity and reducing debts. "The Big Deals can certainly be part of the solution," he said, but "the underlying problems can only be solved over an extended period of the next several years."[43]

While the chaebol balked, the government did all it could to bring in foreign direct investment. Kim Dae Jung, hoping for $15 billion from abroad in 1999, promised on January 28 to cut all regulations "hindering foreign investment" by the end of the year. In his first 11 months as president, he claimed to have cut by half the "11,000 regulations" that had long made life difficult for foreign companies in Korea. "This year we will make sure the regulations are all abolished," he said.[44]

Kim underlined the shift in outlook toward foreign interests that he had sought to bring about. "Koreans in the past have had a very negative attitude toward foreign investment," said DJ. "That is because we are a homogeneous nation." He maintained, however, "We were able to convince

Korean people" of the need to view foreign business positively while foreign direct investment rose in 1998 to $8.85 billion, 27 percent above the 1997 level of $6.97 billion. Foreign business people warned, however, of overconfidence while the country coped with unemployment and the reluctance of the top chaebol to restructure or downsize. "We have the risk of declaring success prematurely," said James Rooney at Templeton Investment Trust. "Korea's credibility rests on recognizing the huge restructuring task that lies ahead. There's no clear focus on delivering growth or shareholder profit." The greatest concern was that "we will lose competitiveness by premature strengthening" of the won, now pegged at 1,173 to the dollar at the end of January 1999.[45]

On February 25, 1999, visiting Seoul on the first anniversary of DJ's inauguration, Lawrence Summers, U.S. deputy treasury secretary, warned against "complacency" that could "put a brake on necessary restructuring." Summers mingled criticism with praise for Korea's "effective crisis response" in a year in which the currency had stabilized, foreign exchange reserves had risen from almost none to more than $55 billion, and the gross national product was forecast to be going up by 2 percent. He was anxious, however, to puncture the balloon of inflated self-praise floated by Korean leaders. "There is an important difference between recovering from a heart attack," he said, "and changing your lifestyle to be sure you never have another one."[46]

Summers's remarks struck a responsive chord among analysts worried that big business would resume its old habit of covering up the figures, relying on easy credit to prop up unprofitable entities and repress competition. "Corporate restructuring stands in the way of a resurgence," said a report by Stephen Marvin at Jardine Fleming, "and the government stands in the way of corporate restructuring." Marvin charged that "policy-makers have consistently shielded the incompetent management and unviable operations of big companies from market forces, inflicting serious damage on healthier competitors and the financial sector." The chaebol, he wrote, still suffer from the familiar problems of "too much capacity, too much debt and too many people."[47]

Summers coupled his warning with a reminder of a matter of immediate American concern—that the government was supporting companies suspected of dumping their products on foreign markets with easy credit and other breaks. "In the steel and semiconductor sector," he said, "it will be necessary to put to rest any doubts that subsidies are continuing." Officials acknowledged that steel exports had risen by nearly 60 percent in 1998 even though output had dropped by more than 12 percent as a result of sharply declining sales to local companies. In a year in which semi-conductor sales worldwide had fallen dramatically, semiconductor exports from Korea rose slightly.

The latest statistics underlined the mood of optimism. Mindful of the anniversary of DJ's inauguration, the National Statistics Office announced that industrial production had risen 14.7 percent in January 1999 from January 1998. True, January 1998 had been one of the worst months of the downturn, but the fact was the increase was the largest year-on-year monthly jump in three and a half years. At the same time, retail sales increased 2.8 percent year-on-year—the first such jump since the country had had to go to the IMF. A range of other statistics indicated the economy had hit bottom. One hopeful sign was that investment in new machinery, needed to build exports, was up 8.8 percent year-on-year in January 1999 after dropping 34 percent in all 1998.[48]

Michel Camdessus, IMF managing director, captured the mood in a talk on February 24, 1999, in New York. Less than three months before, the fund had predicted a 1 percent drop in economic growth, he said, "but the people of Korea were so courageous, and the program worked so well, that we have already had to correct our forecast" to a 2 percent increase for the year.[49]

Analysts, however, feared a relapse if the chaebol did not move faster. "Korea will not be secure against future economic crisis without structural reform of finance, enterprise and labor markets," warned economist Hilton Root. "Reform of the Korean finance system will stop at recapitalizing banks and solving bankruptcies," he pessimistically predicted, while "enterprise reform will not be sufficient to restrain family control" over the chaebol. As a result, "prospects for future sustained growth are fragile."[50]

Indicative of the problem were the doubts raised about pledges by the top five chaebol to reduce debt-equity ratios to less than 200 percent by the end of 1999 from levels in February of more than 400 percent. The reductions largely reflected readjustments in the valuation of assets, said Lee Soo Hee at the FKI's Korea Economic Research Institute. Other factors included an influx of foreign investment and sale of large numbers of shares on the local stock market. "In a fundamental sense, the chaebol are at the starting point," he said. "They'll have to do more to downsize."[51]

The chaebol, however, were resisting. Hyundai Electronics and LG Semicon remained $3.5 billion apart on the terms under which Hyundai Electronics was finally to take over LG Semicon. Samsung and Daewoo had yet to agree on the swap of Daewoo Electronics for Samsung Motor. "Is the political will strong enough to get these guys to come to terms?"

asked David Young, in charge of the local office of Boston Consulting. "My big fear is things get better, and the pressure reduces." Or, as Summers put it, "If what Korea had to fear in 1998 was fear itself, what it has to fear in 1999 will be the lack of fear itself."[52]

The fears of ordinary workers were real. The government on February 26, 1999, the day after the first anniversary of Kim Dae Jung's rule, announced the highest unemployment figures in 33 years. As the National Statistics Office revealed that a record 1.76 million people, 8.5 percent of the labor force, were out of work as of the end of January, Kim warned that "attack from the developed countries," notably the United States, "will close markets" and destroy attempts at opening economies worldwide. "Without globalization," he said, such a campaign "will threaten us."[53] DJ himself suggested the western countries, notably the United States, might be working against Korea's interests by threatening strong measures against Korean exports. He introduced this element of criticism into a glittering World Bank conference that appeared largely to celebrate the struggle of South Korea and other Asian countries against the economic crisis that had swept the region.

Kim's remarks countered the warning from Summers against South Korean subsidies of exports of such highly competitive products as semiconductors and steel. "In the United States, super 301 is being revisited," said DJ, referring to legislation invoking stern measures against countries engaged in unfair trading practices. One reason, he said, was the decline of the U.S. steel industry—blamed by Americans on unfair competition from a number of countries, including Korea. Kim defended Korea's efforts at increasing exports in keeping with the longstanding view of both government and business that Korea had to export its way out of duress.

Concern about growing criticism in the United States of mounting Korean exports was driven in part by the steady rise in the number of people out of work. Problems would mount as 300,000 new graduates of universities and secondary schools poured into the labor pool in the spring. Some officials predicted that official unemployment would cross the two-million threshold. The Korean Confederation of Trade Unions stated that real unemployment, including workers who moved from job to job and were not eligible for unemployment benefits, was already more than four million.

Foreign companies, meanwhile, reported increased difficulty in exporting to Korea. The decline was attributed both to the inability of industries to purchase new equipment from abroad and the reluctance of consumers to spend. Exports in 1998 totaled $132.31 billion against imports of

$93.28 billion, a decline of 35.5 percent from 1997, for a surplus of $39 billion, down only 2.8 percent. U.S. exports to Korea in 1998 fell to $16.5 billion, down 34 percent, while Korean exports to the United States in 1998 climbed 3 percent, to $23.9 billion.[54]

The American Chamber of Commerce in Korea became the butt of criticism for exerting pressure for market opening, including removal of restrictions on financing the purchase of foreign vehicles that still had a market share of 0.2 percent.[55] The chamber's recommendations amounted to "interfering with our domestic affairs," observed *Maeil Kyungje,* Seoul's leading financial paper.[56] "You can say we are exerting pressure," Jeffrey Jones, the chamber president, responded, "but we believe that liberalization and opening of Korean markets will be the only way for Korea to gain international competitiveness."[57]

U.S. Commerce Secretary William Daley took up the cudgels for American business. The U.S.-Korean "trade relationship is not what it was" before the onset of economic turmoil, he told local businessmen.[58] "A surge of imports" of steel had resulted in antidumping cases against three countries, he reminded Korean leaders. Although Korea was not among them, he warned, "We will enforce our trade laws if there is illegal dumping going on in our country."[59] Koreans were in no mood to compromise. "We feel the United States is being too harsh on Korea," Lee Soo Young, chairman of Oriental Chemical, told Daley, accusing Washington of "undue pressure."[60] Korea's rising trade surplus, besides upsetting American and European trade partners, was symptomatic of the crisis. Unless major industries increased exports, experts said, industry would go into long-range decline and be unable to compete effectively worldwide.

In fact, the country suffered the worst drop in annual growth in 1998 since 1953, the last year of the Korean War, falling by 5.8 percent in the first full year of the IMF era. In real terms, the gross domestic product in 1998 was $321.3 billion. The economy dropped in all industrial and agricultural sectors, with the most precipitous drop in construction, which fell by 9 percent because of cancellation of thousands of building projects. Overall, manufacturing fell by 7.2 percent and income from farms, forests, and fishing fell 6.3 percent.[61]

The sharp decline was the first annual drop since 1980, when the economy slipped by 2.7 percent amid a violent change in governments, massive demonstrations, and, in May, the Kwangju revolt. One sign of the extent of the current crisis was that investment in facilities had plummeted by 38.5 percent while thousands of companies retrenched or went bankrupt. Average individual income fell to $6,823 in 1998, about two thirds the $10,397 reported for 1997. The Bank of Korea said the Korean gross domestic product in 1998 had fallen from eleventh worldwide in 1997 to seventeenth in 1998. Per capita, income dropped from thirty-third to forty-second. The economy had reversed course after having grown 5 percent in 1997, when signs of weakness were clearly visible, 6.8 percent in 1996, and 8.9 percent in 1995.[62]

Other statistics verified the official view that the crisis had "bottomed out." The current accounts surplus now stood at a record $40 billion, a reversal from the end of 1997 and 1996, when current accounts were measured in deficits—minus $8.17 billion in 1997 and minus $23 billion in 1996. Another good omen was that the gross domestic product in the last quarter of 1998 had dropped only 5.3 percent from the year before after having fallen 7.1 percent in the third quarter and 7.2 percent in the second quarter.[63]

Stockholders' meetings showed the rules of the game, however, had hardly changed. Minority shareholders shouted, heckled, and asked annoying questions but faced frustration in their battles on March 20, 1999, against five of the country's most prestigious companies. After holding forth at a nine-hour annual meeting of shareholders at Samsung Electronics, Jang Ha Sung, the Korea University professor who had spearheaded the drive, admitted having lost a battle but not the war. "We were very close to a substantially good agreement" that would have given minority shareholders the power to elect a member of the board of each of them, he said, but "our plan was aborted."[64]

Banded together in a group called the People's Solidarity for Participatory Democracy, Jang's followers were defeated in shareholders' meetings at all the companies they targeted—Hyundai Heavy Industries, Daewoo Corporation, LG Semicon, and SK Telecom as well as Samsung Electronics. The People's Solidarity said the group had targeted those as the most important of more than 500 companies that staged annual shareholders' meetings in March. "There were also hopeful signs," Jang claimed. "The managements were responding sincerely. Overall both minority shareholders and the company side were trying to change their attitude. Both did equally well."[65]

While gaining little in practical terms, minority shareholders raised issues that the companies would have preferred to ignore. One of the most troubling was why Samsung Electronics had been the largest single investor in Samsung Motor, a project that had cost about $5 billion. Jang cross-questioned Samsung Electronics executives about Pan-Pacific Industrial Investments of Ireland, a minority shareholder permitted to sell back most its shares in the motor company. He claimed that Samsung Electronics and other Samsung companies stood to lose $270 million from just that aspect of the deal when Samsung Motor was sold to Daewoo Motor. "The shareholders should not stand for it," he said, rejecting the board's apology.

Minority shareholders also shouted out questions about overseas investments and the support given by strong companies to weak ones in the same chaebol. In vain, they protested that none of the tycoons who controlled the groups bothered to attend. A motion demanding no pay for Daewoo Chairman Kim Woo Choong as a penalty for nonattendance was quickly voted down at the Daewoo Corporation meeting. Minority shareholders did, however, win a few concessions. Samsung Electronics, for instance, promised to add more outside directors. As proof of progress, Jang noted that the Samsung meeting in March 1999 was nearly four hours shorter than the one in 1998, which had lasted thirteen and a half hours.[66]

Kim Dae Jung for his part was unhappy about the pace of chaebol reform. A week after he warned Daewoo and Hyundai of "stern measures" in the form of cutoff of credit, they again publicized elaborate plans. Daewoo on April 19, 1999, promised to "concentrate its strengths on the automobile, trade, and finance industries" and sell off the shipbuilding and commercial vehicle divisions of Daewoo Heavy Industries, the Seoul Hilton, "and other high-profiting companies." The objective was to raise $6 billion and "thereby achieve a financial structure as that of advanced corporations." The news included price tags: $4 billion for shipbuilding, $900 million for the bus/truck/engine division, $450 million for a cement plant and three overseas mobile phone subsidiaries, $240 million for the Seoul Hilton, $966 million for the Diners Club of Korea and five other companies, and $528 million for stock in local telecommunications companies. Chairman Kim Woo Choong would sell "personally owned stocks amounting to $240 million," all to go into Daewoo Motor. Daewoo would be left with eight companies, dominated by motor vehicles.[67] Kim put some of his most successful units on the block in order to reduce the group's debts of 59 trillion won, approximately $50 billion, by half. Daewoo, like all the other chaebol, vowed to cut its debt-equity ratio from its current level of 354 percent to less than 200 percent by the end of 1999.

What would happen, though, if Daewoo failed to find the buyers it hoped to get for the entities it said it would sell? Japan's Mitsui, rumored to be interested in the shipbuilding unit, professed no knowledge of the deal. "We're open to any opportunity," said Kim Tae Gou, the Daewoo

Motor president in charge of group restructuring. "That's our first priority—to sell."[68] Skeptics suggested, however, that Daewoo had deliberately set prices too high, hoping to be able to claim that it had made a good-faith effort but found no buyers. "Smoke and mirrors" was the wry assessment offered by consultant Hank Morris. "Clearly Daewoo is in a tight spot," said Edward Campbell-Harris at Jardine Fleming. "We have to see if their price tags are realistic. From an investor's point of view, we want to see if this isn't hot air to boost the stock prices. The problem with Daewoo is no one knows what's going on. It's all in Chairman Kim Woo Choong's head. No one can get to the root. You have to piece up a huge puzzle."[69]

The same applied to the Hyundai group, which might be the most difficult chaebol to curb. Hyundai executives had said on December 4, 1998, that they planned to break up the group into mini-chaebol, but they would remain under the six surviving sons and one nephew of founder Chung Ju Yung. Like Daewoo, Hyundai had to wrestle with the fact that its debt had ballooned over the past year and a half even though it had resolved to slim down. Hyundai on April 21, 1999, said it would sell subsidiaries and stocks to comply with demands from its main creditor, the Korean Exchange Bank, 30 percent owned by Commerzbank but a government-dominated entity in which the government's Bank of Korea and Export-Import Bank of Korea together held as many shares.

Officials, however, had also questioned two of the group's top executives, including Park Se Yong, chief of restructuring and chairman of the Hyundai Corporation, about share manipulation. Investigators believed they had elevated the value of Hyundai Electronics stock in 1998 by having Hyundai companies purchase them at inflated prices in order to lower the company's debt-equity ratio, more than eight to one. Hyundai Electronics stock more than doubled as a result.[70] The Fair Trade Commission promised to broaden the investigation to include other chaebol, but no one raised the topic when Park announced the Hyundai restructuring plan on April 23.

Hyundai would concentrate on motor vehicles, electronics, construction, heavy industries, and financial services. In each field, the goal would be a place in the "Global Top Ten," a kind of slogan. The group would "spin out all companies outside the five core fields," selling off assets in petrochemicals, power generation, aerospace, and the production of railroad cars, slashing the number of companies from 79 to 26 by December 31, 1999. The details were exact—13 to be "separated," 13 "sold," 15 "merged," 4 "liquidated"—and 8 Kia companies also to "merge" with

Hyundai Motor. The group's burgeoning debt load of 79.3 trillion won, about $80 billion, 449.3 percent of equity, would fall to 45 trillion won, $40 billion, less than 200 percent of equity. The company would raise 14.2 trillion won by selling or "separating" entities, but the biggest source would be real estate and new stock issues, which together would bring in 19.65 trillion won. Another $4.5 billion would come from foreign investment, $1.76 billion by year's end. "By the end of 2003," said the announcement, "the entire Hyundai group will be divided into five separate entities"; the announcement did not say that they and other Hyundai companies would remain in Chung Ju Yung's family.[71]

By no coincidence, Hyundai Electronics completed the deal for LG Semicon the same day, April 23, 1999, agreeing to pay 2.56 trillion won, about $2.2 billion. The announcement promised "no layoffs" with all employees to be treated "on merit."[72] Implicit in LG's cooperation was fulfillment of its dream of gaining control of Dacom, with Korea Telecom one of Korea's two fixed-line phone companies. Sure enough, on May 6, the ministry of information and communications lifted the 5 percent limit on LG's stake in Dacom. LG could now acquire the 28.6 percent it wanted for about 540 billion won, $456 million. The price included Hyundai's 5.25 percent stake in Dacom, which Hyundai was giving as partial payment for LG Semicon. The decision overrode a protest from Samsung, which now owned 24 percent of Dacom after hastily buying up shares.[73]

The incipient takeover of Dacom by LG raised another question, this one about the rights of foreigners. The government made much of the passage of legislation in April 1999 to allow foreign investors to buy up to 49 percent of telecommunications companies, including SK Telecom and Dacom, starting in June. Previously held to 33 percent, foreign interests still had to surmount backstage deal making by government and chaebol in their drive to open up Korea's $12 billion telecommunications market. The politicians who opposed the bill feared such competition would bankrupt local firms.

The leaders of the top five chaebol, however, claimed less was more when it came to the number of entities or the size of the assets in their empires. "The focus is on selective core business," said Lee Chong Suk, senior executive vice president of the LG group, on the day that Hyundai announced its restructuring plan and takeover of LG Semicon. "Our goal is to integrate our resources."[74] Those words could apply to any of the chaebol as they submitted, under prodding, to the reality that they no longer had the unlimited credit on which they once relied. "We have to change the system to maximize the efficiency of our group," said Choi Eui

Jong, vice president for restructuring of the SK group. "We will focus on business which has proved core-competitive."[75]

LG on May 18, 1999, won the largest single direct investment so far in Korea's economy—the sale for $1.6 billion of 50 percent of its liquid crystal display unit, LG LCD, to Royal Philips Electronics, Europe's biggest manufacturer of consumer electronics products. LG Electronics had formed LG LCD as an independent entity in early 1999 as a magnet for the type of deal offered by Philips. LG group Chairman Koo Bon Moo predicted the new joint company would hold a competitive edge against Samsung Electronics as well as its Japanese competitors. John Koo, chairman of LG Electronics and Chairman Koo's uncle, called the deal "an excellent example of a win-win strategy." Cor Boonstra, Royal Philips president, said the deal "exactly meets Philips' criteria" even though LG had had to sell more than $6 billion of assets to pay off crushing debts since 1997.[76]

Kang Yoo Shik, president of the LG group restructuring team, said the investment from Philips would reduce the group's debt-equity ratio, now about 350 percent, and also strengthen its core areas of electronics, information and communications services, and chemicals. He linked the sale to LG's efforts to buy controlling stakes in both Dacom and Korea Life Insurance. "The new funds provide LG the financial power to pursue these projects," he said.[77] The LG deal with Philips again showed it was possible for a chaebol to make good on its promises to bring in foreign money. LG in the past year had also signed deals for $320 million with Dow Chemical, $500 million with Caltex, and $400 million with British Telecom and several hundred million in a joint venture with Nippon Mining and Metals.

Foreign investors wondered, however, if Koreans had suddenly shown a reluctance to accept foreign investment now that the economy had improved from the depths of 1998. The chaebol, when in need, welcomed the foreign investment, but management control, even in the case of LG's venture with Philips, remained in the hands of mysterious family groupings whose inner workings and power relationships, rivalries, and trade-offs were largely hidden.

The issue of family ownership crystallized tragically around the nation's flagship carrier, Korean Air. More than 800 people had died in Korean Air crashes, including 228 in the crash of a Boeing 747 on a hill in Guam in August 1997 and 269 aboard a 747 shot down by a Soviet MiG fighter after straying into Soviet airspace in September 1983. Korean Air in 1998 contracted with Delta Air Lines for a team of ten pilots to conduct a safety study. Delta, upset by what the pilots found, canceled its flight partnership with Korean Air on April 16 after an MD11 cargo jet crashed in Shanghai on April 15, 1999, killing three crew members and six people on the ground.

Kim Dae Jung on April 20, 1999, blamed the airline's troubles on "mismanagement by the owners of the company and the owners' management style." His attack pinpointed the top-down approach of Korea's wealthiest families—and the family ownership of all the chaebol. The airline was the leading company of the Hanjin group, Korea's sixth-largest chaebol, an empire with interests ranging from shipbuilding to construction. The Korean Air president, Cho Yang Ho, eldest son of the Hanjin chairman and founder, Cho Choong Hoon, had succeeded his uncle, Cho Choong Keon, as president in 1992, while Yang Ho's three brothers ran other companies in the group. DJ cited Korean Air as "an example" of the problem of ownership by extended families. "Instead of making the best efforts to acquire skilled pilots," he said, the owners "concentrate too much on profits."[78]

Korean Air had just announced a net profit of $246 million for 1998 that it said was "a turnaround after two years of losses," including $397.5 million in 1997.[79] Until then, the Cho family had profited immensely from a venture that was hemorrhaging badly before Hanjin acquired it from the government in 1969 at the behest of Park Chung Hee. Now the world's thirteenth-largest carrier with 112 planes, Korean Air was the only airline with its own manufacturing division—an aspect that critics charged also distracted owners from safety. Like other chaebol, the Hanjin group had grown recklessly, piling up a ten-to-one debt-equity ratio before the economic crisis. The group planned to combine its lucrative shipbuilding unit, Hanjin Heavy Industries, with money-losing Hanjin Engineering and Construction while counting on Korean Air to stay profitable.

In response to DJ's criticism, Korean Air on April 22 announced an executive shakeup that indicated little had changed. Shim Yi Taek, chief executive officer since 1994, replaced Cho Yang Ho as president with a pledge "to ensure safety even if it means reducing flights and cutting frequencies." The shakeup, however, had the aura of a ritual. Cho, noted for an autocratic style that brooked no criticism, was named Korean Air chairman in place of his father, Cho Choong Hoon, 79, who grandly accepted "the entire responsibility for Korean Air's recent aircraft accidents" but remained Hanjin chairman. Shim, aged 60, insisted he would take "full responsibility" for Korean Air operations while Cho Yang Ho, ten years his junior, provided counsel "due to his vast aviation experience." In a simultaneous histrionic gesture, 29 Korean Air executives submitted resignations en masse "to take responsibility." Sitting at a table flanked by several of them, however, Shim said it was "up to the company" to decide whether or not to let any of them go.[80]

After the U.S. Department of Defense barred military personnel and Pentagon civilians from traveling on the airline on official business, Korean Air on May 4 announced a series of safety measures. Battling a seniority system that placed too much emphasis on the military backgrounds of former Korean air force pilots, the airline said they now had to log 4,000 rather than 3,000 hours to make captain. The head of flight operations had "stepped down," said the airline, accepting "responsibility"—a favorite word—for "recent accidents."[81] Lee Yo Yul, at the transport ministry's international air transport division, hoped the Pentagon ban would not hurt Korean pride. "We should focus on the company, not the country," he said.[82]

The stock market by now was doing so well, however, that the chaebol, those that survived, might lose whatever incentive they had to follow through on the reforms they had been promising. On May 6, 1999, the raging bull went on a rampage, soaring above the 800 level minutes before closing as programmed buying by foreign investors pumped in a final 20 billion won, nearly $20 million, in the last five minutes of trading. Korean stocks had become so hot that government officials and analysts were afraid that investors might forget the country had only recently begun to recover from the economic crisis that sent stocks below the 300 level for much of 1997.

"It's overshot," said Richard Samuelson. "Even bull markets don't go on forever."[83] The market was powered in part, said analysts, by an assurance from Chon Chol Hwan, governor of the Bank of Korea, that the government would not boost interest rates by tightening the monetary market. The central bank decided, however, to keep the local call rate at 4.75 percent rather than go on lowering it, as it had been doing for more than a year. Fearful of a disillusioning downturn, Korea's finance minister, Lee Kyu Sung, warned investors not to "jump on the bandwagon."

What accounted for the upturn of an economy still in recovery mode? For months both Korean and foreign investors had been flooding the market with cash, some of which might otherwise have gone into bank accounts when they were producing high interest rates. "We have an enormous buildup of liquidity and a heavy inflow from abroad," said Stephen Marvin at Jardine Fleming. "That's because domestic rates are low and foreigners are bullish on the economy." He warned, however, of the possibility of a "massive correction" if the chaebol failed to follow through with restructuring.[84]

As blue chips soared, few investors were worrying. The market closed on May 6, 1999, at 810.54, up 39.37 points or 5.11 percent. Companies owned by the top five chaebol led the surge. Goldman Sachs in Hong Kong raised its rating for Samsung Electronics from "market performer" to "market outperformer." With Korea's domestic economy and the worldwide semiconductor industry "on a recovery path," said the announcement, "Goldman Sachs is positive about Samsung Electronics' earning growth prospects for 1999 and 2000."[85] One factor: demand for dynamic random access memory chips, which accounted for half of Samsung's earnings, was coming off a cyclical low.

Korean financial institutions had as much to do with the surge as did the inflow of foreign investment. "They're putting their money in the market rather than the real economy," said James Rooney at Templeton. "There's a lot of money chasing around Korea looking for something to do." Failure of the top chaebol to restructure, however, could slow the market. "It will need to be driven by fundamental changes in the way Korean companies do business," he said. Otherwise, "We will have a stock market that fails to realize its full potential and a stock market that's underperforming."[86]

The failure of Korean companies to pay billions of dollars borrowed in the wild days of nonstop expansion before the economic crisis was luring major foreign investment firms in bidding for the right to collect the debts themselves. The vehicle for the bidding war was KAMCO, the government's Korea Assets Management Corporation, which in mid-1999 held loans with a face value of 43 trillion won, $35 billion, and was likely to nearly double that figure as it purchased bad loans from banks. "We are going to have to resolve 50 percent of these loans in three years," said Park Jae Ho, manager in KAMCO's capital market department. "In five years, we have to resolve 100 percent." The government set that time frame for KAMCO to sell the loans, 60 percent of them accrued by chaebol that had had to apply for court intervention after declaring their inability to pay their debts to banks. The rest were piled up by small and medium-size enterprise, which, unlike the chaebol, put up real estate as collateral.[87]

KAMCO officials planned to hold auctions every few months for the next few years for packages of loans with a face value of several hundred million dollars apiece. Their confidence reflected in part their greatest success—the sale at an auction in June 1999 of a bundle of 8,100 nonperforming loans to the Lone Star Fund of Dallas, Texas, for $455 million, 50.6 percent of the face value of the loans. Lone Star barely edged out such competitors as Goldman Sachs, Lehman Brothers, Bankers' Trust, J. P. Morgan, and Morgan Stanley in the bidding, the fourth and largest auction held by KAMCO since December 1998. "They're certainly off to a good start," said Bradley Geer, vice president of Samjong Houlihan Lokey. "They've shown the resolve to get this thing done." Geer viewed Lone Star's winning bid as "an extraordinary price" to pay for loans that Lone Star now had to try to collect. The bidding, organized by Samjong Houlihan Lokey, was "very, very competitive"—and a bonanza for KAMCO, which had previously sold loans backed by real estate for collateral at about 35 percent of face value and settled for 15 percent on chaebol loans, not backed by collateral.[88]

Charles Larson, regional representative for Houlihan, Lokey, Howard and Zukin, the American partner in Samjong Houlihan Lokey, rated KAMCO's record as far superior to efforts elsewhere in the region at selling off nonperforming loans. In Thailand, government entities had sold off small portions of nonperforming loans that were probably as high as those in Korea. Discipline, however, remained incredibly lax. "It would be unheard of to do in Thailand what they have done here," said Larson. "You start with a different political profile, different banking systems." As for Japan, banks there had sold off a small percentage of nonperforming loans estimated as high as $700 billion, but there was no equivalent of KAMCO.[89]

Adding to the interest in the auction was uncertainty about the response of debtors to efforts at persuading them to settle their debts at more than 50 percent of the face value of loans. For a winning bidder, the chance to earn money from the transaction derived from how well it could do in getting debtors to pay more than the price it had paid to take over the loan. "One of the unknowns about this whole thing is how discounted payoffs will work out in Korea," said Geer. "This is something new here. It will add to the process of resolution of loans." Although any winning bidder would like to resolve loans through negotiations with the debtors, bidders were counting on Korean authorities, through the courts, to back them up if they had to foreclose on properties. Collateral in the latest bidding included stores, office buildings, and a hotel as well as private residences.

Steven Lee, president of Lone Star Advisers' Korea, responsible for the winning bid, said he expected virtually all the loans would be settled through negotiations. "People have a tendency to try to work things out," he said. Underlying his optimism was that "the currency is stable and the government here is committed to reform." No one was certain, though, how much Korean companies borrowed that they would never be able to pay back. "KAMCO holds under half of what is there," said Harrison Jung, managing director of Samjong Houlihan Lokey. "The total is in the range of $125 and $150 billion."[90]

Those figures, however, included only loans made by banks—not loans extended directly from one company to another. Nor did they include another $10-15 billion loaned by insurance companies, many of which were on the brink of failure. Yet another major area of nonperforming loans included loans made by Korean banks to entities in Southeast Asian countries. Between five and ten percent of Korean bank loans fell into this category, including about $10 billion loaned to now bankrupt companies in Indonesia.

The chaebol with the worst loan problem was Daewoo, whose troubles now emerged in stark clarity. In the face of a government-engineered campaign to keep it alive until the end of the year, Daewoo in July 1999 appeared likely to collapse unless rescued very quickly. The group, unable to significantly lower its debt load of $57 billion, faced a seemingly impossible struggle to sell off some of its best and worst units in order to become financially viable. It needed to sell off the best units in order to pay off debts. It had to jettison the worst ones to keep them from dragging the group still further down.

The truth was, the group had been unable to find a buyer for one of its better units, the shipbuilding division of Daewoo Heavy Industries, and now faced the prospect of having to sell Daewoo Securities. Once Korea's largest securities firm, now rated number three behind Hyundai Securities and Samsung Securities, Daewoo Securities was the only company in the group that one could be certain was making a profit. The main obstacle to reform remained the resistance put up by founder Kim Woo Choong. Kim until recently had opposed government efforts to restructure the economy, repeatedly calling instead for postponing new programs. In the meantime, Daewoo, unlike other major chaebol, showed few signs of recovery even while reducing the number of companies from more than 40 to 22 in fields ranging from motor vehicles to electronics to trade and finance.

Kim liked to say that he would quit when the group recovered but was resisting pressure to leave its disposal to professional managers. "There seems to be a big struggle between Daewoo and the government," said Lee Keun Jang, chief operating officer at Jardine Fleming Securities in Seoul. The central question was whether the collapse of Daewoo, a victim of unbridled overinvestment at the behest of Kim, would be a blessing or a setback for Korea's efforts at recovering from crisis. Some analysts saw the prospect of the collapse of Daewoo as a major reversal for DJ's reform program—and one reason why the local stock index fell from 1024.58 on July 19 to 965.11 on July 21. "It's going to be a big blow to the financial sector here," said Lee. "Daewoo group is in technical default. The impact is serious for the Korean economy."[91] Others were less pessimistic. "As a group, Daewoo's finished," said Jeffrey Jones of the American Chamber, "but I don't think it will have that much impact." If anything, said Jones, a lawyer who had worked on complex financial transactions involving Korean and foreign companies, the demise of Daewoo "will bring in more foreign investors" and buttress the fortunes of some of its domestic rivals.[92]

The crisis of Daewoo was so fraught with dangers for the economy that government officials clearly placed it in the category of "too big to fail." While 11 other chaebol had slipped into bankruptcy in 1997 and 1998, the impression all along was that none of the "top five" was in danger of falling apart. Officials worried that failure of Daewoo would jeopardize the jobs of many of its 90,000 employees in Korea and also saddle the government with having to cover much of the unpaid debt, as had happened with other bankrupt chaebol. The government hoped that domestic as well as foreign companies would buy up chunks of the Daewoo empire, but Daewoo so far had sold only a few relatively small units—not nearly enough to cover debts.

The fear of the implications of the demise of Daewoo explained why the government pressured Daewoo's 69 domestic creditors on July 19, 1999, to roll over seven trillion won, $5.9 billion, in loans falling due almost immediately. They had the right to refuse, but the government's majority stake in the largest commercial banks insured they would do as told. Investment trust companies, which held nearly three-fourths of Daewoo's loans in corporate bonds and commercial paper, balked at extending credit but also had to comply. "It's up to them, they have a choice," said Cho Won Dong, director-general for economic policy at the finance ministry. "If they want repayment of their loans, they might be successful if they wait." But if the group's liability "becomes an insolvency problem, if it goes bankrupt," said Cho, "there is little chance to get the loans back."[93]

Rollovers alone, however, would not give Daewoo the breathing space it needed to recover from excessive investing. "The Daewoo group is requesting another four trillion won in loans," said Lee Keun Jang at Jardine Fleming. "There's no bottom." Lee likened Daewoo's appeal for life support to the series of efforts to save Kia, which had gone into bankruptcy exactly two years earlier.[94]

Investors showed their distrust in Daewoo's future on July 21 as the stock of 9 of the 12 listed Daewoo companies declined markedly. The Daewoo Corporation and Daewoo Heavy Industries, mainstays of the group, both lost ground along with Ssangyong Motor and Daewoo Motor Sales. Government officials, however, were hopeful that Daewoo could overcome its problems, partly by selling off entities. "This is a matter of survival for Daewoo," said Chung Duck Koo, commerce minister. "The creditor banks will take action." By focusing on motor vehicles and trade, he said, "We believe Daewoo will contribute to Korea's economic growth." Meanwhile, "Many of Daewoo's affiliates that do not have competitiveness will be sold to Korean and foreign investors."[95]

Investors, however, did not share his confidence. The Korean stock market on July 23, 1999, skidded 71.7 points, the biggest single-day drop in the history of the exchange, a record 7.3 percent, as investors panicked over Daewoo's future. Acrimony among domestic creditors fueled the drop amid reports that they still were unable to agree on extending the additional four trillion won, $3.4 billion, in credit needed to stave off bankruptcy. Investors questioned whether Daewoo's 69 creditors had come to terms as reported on fresh credit for Daewoo after agreeing to roll over 7 trillion won. "The market seems to be in a panic because of Daewoo," said Lee Keun Jang. "Many investors have asked investment trusts to return their money. That money belongs to the investors, not the investment trusts. Investment companies cannot dispose of their portfolios." The FSC proposed that creditors convert Daewoo's credit into equity, but "it may not happen," Lee said. "Everybody's concerned whether it's workable."[96]

The dropoff on July 23 was the most precipitous of a regional decline in which stock prices fell on every exchange in Asia except Taiwan. An external driving force was the fear that U.S. interest rates might rise after U.S. Federal Reserve Chairman Alan Greenspan vowed to act "forcefully" to fight inflation. Rising interest rates in Korea also played into the volatility of the local market. "It's Daewoo and it's interest rates," summarized Adrian Cowell at Kleinwort Benson. "Daewoo, of course, is the big problem."[97] The problems of Daewoo—and the Korean stock market—jolted the economy just as Korean leaders were saying the country had virtually recovered from crisis. The stock market began sliding downhill after hitting 1052, its highest level in three years, on July 12, more than triple the lows to which the market had plunged toward the end of 1997 and early 1998.

Down, Kim Woo Choong was not yet out. Stung by mounting pressure to force him to jettison most of the group's remaining companies, Kim spoke publicly on July 25 for the first time since the group's creditors had agreed six days earlier to roll over more than $8 billion in loans. He would "spare no effort to normalize the status of Daewoo," after which he would "resign honorably," he pledged at a press conference staged before an emergency meeting at which he feared government finance officials would tear his group out from under him. As ever, he sought to give an impression of cooperation, promising that Daewoo would "stabilize liquidity problems" by "cooperation with creditors" and "complete second-half restructuring plans through transparent and objective procedures." Kim calculated domestic debt at 49 trillion won, $40 billion, while overseas debts were $6.84 billion, including $2.71 billion due within six months. Daewoo, he said, was "not burdened by excessive foreign debts" and would never declare a debt moratorium.[98] Sales peaked at $71.5 billion in 1997 when assets were $44 billion, said a spokesman, but the group did not reveal the latest statistics. "We are under restructuring," he explained. "Every day the figures are changing."[99]

Kim Woo Choong's accounting conflicted with that of the government, which said Daewoo's foreign debts came to $9.94 billion. Besides the $6.84 billion borrowed by Daewoo companies incorporated abroad, the group's companies in Korea had borrowed another $3.1 billion from foreign creditors, which Kim included under domestic debt. Kim made no apologies for the debts, domestic or foreign, much of which had financed expansion at home and investment in motor vehicle plants in eastern Europe. Rather, he said, "I am going to build confidence in the market and will provide transparency in management." Seeking to silence the critics who accused him of stonewalling efforts to dismantle the group, he said, "I will be free from avarice."[100]

Immediately after the press conference, the government's top financial guns, FSC Chairman Lee Hun Jai and Finance Minister Kang Bong Kyun, intensified pressure on Daewoo in hopes of restoring market confidence. Intent on Daewoo's survival, Lee said the government would provide emergency loans totaling as much as three trillion won for investment trust companies, which had threatened to blow the plan apart by saying they could not extend the due dates for Daewoo's commercial paper and bonds. Thus Lee said the government hoped to eradicate "instability in financial markets" caused by Daewoo's troubles but warned that Daewoo shareholders could "inevitably" find the value of their shares going down as creditors sold off assets. At the same time, Lee and Kang said the government would pressure Daewoo's creditors, as they exchanged Daewoo's debts for equity, to begin selling off the assets.[101]

The IMF, once the advocate of high interest rates in order to reduce over-borrowing, now had to deny pressing for higher rates as a device to keep inflation in check. With the Korean market rebounding on July 26 and 27 after a week of turmoil, John Dodsworth, the IMF representative in Seoul, said "fears of inflation and rising interest rates are not well founded." Financial analysts had "misread" a warning in the IMF's latest review on Korea "to be careful about inflation" as a suggestion that the fund believed interest rates should go up. Dodsworth sought to soothe concerns over inflation partly to calm the jittery stock market, which jumped 6.2 percent from July 26 to close on July 27 at 928.85. "It's right they don't raise interest rates until they see some inflation," said Dodsworth. "There are no signs of inflation." Tight fiscal policies rather than higher interest rates, he believed, were the antidote to inflation that he expected to increase by just 1 or 2 percent over the rest of the year. Meanwhile, he saw the Daewoo problem as the proverbial blessing in disguise. "This has given some new impetus to corporate restructuring," he said. "Everybody's going to look very closely at Daewoo for the next six months. People are calling it a wake-up call."[102]

Dodsworth talked about the future of Daewoo and the Korean economy as Daewoo's domestic creditors appeared finally to have united on a plan to keep the group alive. The FSC on July 27 said that domestic creditors rather than Daewoo would be responsible for initiating and carrying out the program for dumping assets. Thus the FSC overcame the reluctance of creditors to roll over debts and extend another $4 billion in credit by promising emergency funds of more than $2 billion. Kim Woo Choong in turn was forced to put up 10 trillion won, $9 billion, of assets as collateral, including 1.26 trillion won, more than $1 billion, of what were described as his "personal assets." The FSC said Daewoo's four major creditors, Korea First Bank, Hanvit Bank, Cho Hung Bank, and Korea Exchange Bank, all of them at least partly owned by the government, would lead the program for selling off the holdings.[103]

Faced with a deadline of August 11 for making a firm commitment to get rid of assets, Kim Woo Choong was rapidly being forced into submission. No longer appearing in public, he had his headquarters state on August 11 that it would sell Daewoo Securities but resisted creditors' demands for swift completion of a deal. Rather, a Daewoo spokesman said, "We are still in final discussions with our creditors." Daewoo also backed down on a commitment to sell its huge construction operation, part of the Daewoo Corporation, the group's central company. A man identified as a Daewoo official, talking in a disguised voice to the state-owned Korea Broadcasting System, said he doubted if anyone would want to buy the construction unit. He also complained about pressure from creditor banks. "If we sell construction, we want to sell it when we want to sell it," said the man, whose identity was not revealed either by KBS or by Daewoo.[104]

The resistance to sale of key units showed the type of fight that Daewoo was likely to wage against impossible odds. Kim had repeatedly said he would restructure the group, cutting it down to just nine companies, but he put off the inevitable. Kim doubtless entertained flickering hopes of clinging to Daewoo Securities as the conduit for bringing cash into the group, which he still wanted to hold together largely as a motor vehicle manufacturer. "They might have to think this thing through a little more," said Edward Bang, branch manager of Kleinwort Benson in Seoul. "I don't think anybody will want to get rid of a cash cow"—a reference to Daewoo Securities. Still, he said, "They will have to come to a hard decision."[105]

The news of no news sent the stock market plunging 24.6 points to 944.88 on the same day, with Daewoo entities taking some of the biggest losses. Shares in the Daewoo Corporation, refusing to sell its huge construction unit, fell 9.3 percent while shares in Daewoo Heavy Industries, under pressure to sell its shipyard, dropped 8.2 percent. "Daewoo Securities is the only thing they've got that's making money," said financial consultant Peter Underwood. "Their dilemma is, how else can they make money, and what else are they going to sell?" With most companies in the group hemorrhaging badly, Daewoo no longer had room for maneuver.[106]

Kim Woo Choong's stubbornness was an affront and a challenge to Kim Dae Jung, miffed by reports that economic reform was faltering and fearful that he might appear weak. On August 15, fifty-fourth anniversary of the Japanese surrender and the demise of Japanese rule, DJ vowed to be "the first president" in the history of his country "to reform chaebol and straighten our economy." His mission, he said, was to enable the economy to grow "on the basis of the middle class."[107] DJ "is highly sensitive to the fact that three months ago we were all saying the reforms are slowing down," said Graham Courtney, executive director in charge of research for Warburg, Dillon Read in Asia. "The government is delighted to tear Daewoo apart."[108]

The next day, August 16, Daewoo agreed to the deal under which Daewoo was to retain just 6 of its 25 companies. Kim Woo Choong, still at the helm, made plain his unhappiness. He did not comment, and a Daewoo spokesman said he did not know where he was. His fadeaway contrasted with the sycophantic publicity he had long sought in foreign and local media as he built the empire that he had founded as a textile trader with the blessing of President Park Chung Hee, who had studied under his father in Taegu. One thing was clear: under the agreement, the group's major domestic creditors, mostly the government-controlled banks, were to acquire Daewoo Securities. The FSC said Daewoo's creditors would finish taking over Daewoo Securities by the end of the third quarter and then seek a buyer. The agreement also called for division of two major Daewoo companies, the Daewoo Corporation and Daewoo Heavy Industries. The group agreed to sell the construction unit of the Daewoo Corporation, whose trading and administrative arm would remain the headquarters for the group, as well as the shipbuilding unit of Daewoo Heavy Industries while retaining the latter's machinery unit. The group also would retain most of Daewoo Telecom and a small financial entity, Daewoo Capital.[109]

While giving the group until the end of the year to get rid of three quarters of its assets, the agreement was vague about what would happen if Daewoo failed to find buyers. Also unclear was how much the selloff would earn—and how much debt would remain. Officials said they still expected the group's debt-equity ratio to shrink from its present level of 5.8 to 1 to 2 to 1, but that goal was not realistic. The stock market responded uncertainly, going down by 2.3 percent on the morning of August 16 before recovering somewhat on the news of the agreement and closing at 907.28, off 10.19 points for the day. Daewoo's troubles were far from over. Several companies that relied on Daewoo for business had to ask for court receivership, and Daewoo Motor and Daewoo Electronics temporarily shortened hours after suppliers slowed down shipments of parts.

Daewoo was in no hurry to follow through on its commitment to sell beleaguered units. Increasingly impatient, the FSC on August 26 placed the Daewoo Corporation, Daewoo Electronics, Daewoo Motor, Daewoo Heavy Industries, and eight other Daewoo companies on a "workout program" under which creditor banks, not Daewoo executives, had final say on restructuring and paying off debts. While the group had not been "nationalized," as Reuters reported, the companies under the program no longer had the power to act on their own. Since Daewoo's "short-term liquidity situation has not been completely resolved" and "can further spread to Daewoo's subcontractors," said the FSC, creditor banks "will be able to play a more proactive role in Daewoo's restructuring in a more predictable and stable environment."[110]

"Workout," however, had a downside that might plunge the economy deeper into trouble. The companies under the program did not have to pay interest or principle on their debts for three months. Investment trust companies, reluctant partners in the initial agreement for rolling over Daewoo's debts, now faced still more problems. They needed more than twice the 10 trillion won the government had promised investors wanting to redeem their bonds. What would happen if investors panicked and demanded all their money back? The bond market could collapse while many more companies went bankrupt. Creditors had to move quickly to avert disaster.

Selloff was difficult—and not always that remunerative. In order to take over Daewoo Securities, creditor banks paid approximately $300 million for Daewoo's 14.85 percent stake. They then had to devise a plan for paying another $333 million to increase their ownership to 33 percent, enough to give them the power to tear it away from the group and keep it from financing Daewoo enterprises.

There was talk of an agreement with Walid Alomar & Associates, a California firm, to pay $3.2 billion for 80 percent of the assets of Daewoo Electronics, whose debts exceeded $4 billion, but that also fell through. Good deals were never certain. After announcing plans in June to sell the telephone-switching division of Daewoo Telecom to an American firm, Laves Investment, for 400 million won, $340 million, Daewoo in September said talks on the deal had ended. The Seoul Hilton, meanwhile, was auctioned to a Hong Kong group for $228.5 million after the collapse of an agreement to sell it to a Luxembourg firm, but only a made-in-heaven sale of Daewoo Motor would provide a real reprieve.

For Kim Woo Choong, once so critical of his foreign partners while charming correspondents with "exclusive" interviews when times were good, there would be no quick fix. As the stock market dipped precipitously amid unease about the repercussions of Daewoo, the founder was left with little choice. For him the swiftest rhetorical solution was to "resign," in the Korean tradition, accepting "responsibility" for the troubles that Daewoo had inflicted on the country and the economy. First, on October 9, he quit as chairman of the Federation of Korean Industries even though he insisted other chaebol chieftains had begged him to stay. Then, on November 1, he and a dozen other executives of Daewoo companies handed in their resignations to the creditors who now had control over the group's future. Again, the operative word was "responsibility," which a group spokesman said they had "fully accepted" in a joint statement issued on behalf of all of them. Naturally, none was available to take on the added responsibility of talking individually, much less answering questions.[111]

Forced to surrender his shares to Daewoo's creditors, Kim was loathe to bow out entirely. At the time he resigned, he was visiting Frankfurt, looking after his European investments—a convenient ploy for delaying acceptance of his resignation by the corporate restructuring coordination committee charged with overseeing restructuring of the group. As the only one privy to many of the group's top secrets, Kim may have dreamed of exercising leverage over its future. Besides, at age 62, he saw himself as too young to abandon the struggle in a society conditioned to equating age with wisdom. There was always the hope that major entities, having restructured, would miraculously regain value, surviving alone, merging with other companies, or at least polishing themselves up as trophies with enough sparkle to sell at pre-crisis prices.

Kim's career seemed to have ended, however, when, on November 25, 1999, his name was formally removed from the rolls of the companies that he had nurtured to greatness. In a sad letter to Daewoo employees from his European hideaway, he apologized for the "errors of judgement, mistakes and negligence" that had brought his empire to its knees. Daewoo's Korean creditors, meanwhile, battled to a new agreement, deferring payment on more than $16 billion owed by the Daewoo Corporation and another $3.5 billion owed by Daewoo Motor as foreign creditors howled in protest against Korean demands that they accept 18 cents to the dollar for their loans to the former and 33 cents for loans to the latter.

Through it all was the discomfiting sense that foreigners were different. The feeling was that Korea Inc. should not sell out to them even while rushing to protect its own institutions. The battle of Coors Brewing for a Korean company that bore its name tested Korea's openness to foreign investors. Coors hoped to recover Jinro Coors, which it had formed in 1991 as a joint venture with the Jinro group, investing $22 million for one-third of the shares and supplying much of the technology for a new brewery. Coors in December 1997, at the height of the economic crisis, exercised a "put option," remitting its shares with little hope of recovery after Jinro went bankrupt, but then offered to buy Jinro Coors outright. Jinro's creditors, rejecting Coors's offer, chose instead to have an auction.

In the midst of the bidding, in July 1999, Coors made an emotional demand that a rival Korean-backed bidder be disqualified. George Mansfield, regional managing director of Coors Brewing International, said Oriental Brewery, the brewer of OB beer, the leading Korean brand, should be "red-carded and disqualified" for having compromised the bidding by submitting a second bid after discovering the contents of the Coors bid. How did OB discover the contents of the supposedly confidential Coors bid? Was OB in collusion with the Jinro creditors responsible for the auction? Doosan group, which had controlled OB until selling 50 percent of it to Interbrew of Belgium in 1998, responded that OB "never broke the international rules of the bidding."[112]

While foreigners often complained privately of difficulty penetrating the bureaucracy or understanding rules and regulations, they rarely laid their cases before the public. In vain, Mansfield said that Coors would not bid again unless OB was ruled out of a contest whose rules not only called for confidentiality in bidding but also barred any company from resubmitting a bid. The battle of the breweries provided an unusual glimpse into the kind of disagreements that often arose between Koreans and potential foreign investors. Coors never had a chance. There was no way South Korean officials would side with foreigners against their own system. In response to Mansfield's protest, the auction was postponed and opened for a new round—with OB participating and Coors refusing to participate. Coors's withdrawal was just what Doosan wanted. On July 31, after a court ruled against Coors's protest, the same court proclaimed OB the sole bidder in a second auction. OB's "winning" bid was 454 billion won, $380 million, about 7 percent more than Coors had bid for what would have been its largest overseas investment.[113]

For all Kim Dae Jung's efforts to encourage international investment, resistance to foreign control of Korean companies remained strong. In some cases it was the foreign investors' methods that rankled Korean authorities. On July 6, 1999, the FSC warned Tiger Management, the U.S. investment fund, against "excessive interference" in South Korean companies if it wished to invest in the country. The order was prompted by a battle between Tiger and SK Telecom over a planned rights offering that would dilute Tiger's holding in the company. Tiger was ordered to pledge not to repeat such practices if it wanted approval of a new $150 million fund in Korea.[114] Tiger had been contesting a decision by SK Telecom, flagship of the SK Group, to raise 1.5 trillion won, $1.28 billion, through the largest special rights issue ever held in Korea. Tiger, contending that the rights issue would reduce the value of SK Telecom shares, sold 140,000 shares of SK stock after SK Telecom's board ignored protests by minority directors and decided to go ahead with the rights issue in the last week of July. SK Telecom also ignored a plea by Robert Dole, the former U.S. senator, who called on SK Telecom's president, Cho Jung Nam, in June.[115]

"They see the potential for dilution of the value of their shares," said Richard Samuelson at Warburg, Dillon Read. "They're afraid the rights issue will bring the price down." Samuelson observed that boards of American companies would have been more receptive to Tiger Management's view. "The decision process is different," he said. "Here it's very difficult for foreign management to make the decision. It's pretty obvious what is the reasoning here. It's to secure control."[116] Overseas investors were limited to one-third of the shares in telecommunications companies until July, when the limit was raised to 49 percent. SK Telecom's board had a mission to deprive the foreigners of a serious policy role.

Tiger's campaign assumed the aura of a crusade. In a filing with the U.S. Securities and Exchange Commission on August 4, 1999, Tiger said that it would call for the ouster of SK Telecom Chairman Son Kil Seung at the next shareholders' meeting. While his removal "would be largely symbolic," said Tiger's letter, "it sends an important message that management must observe the protocols of internationally recognized standards of corporate governance."[117]

Tiger's main complaint was that SK planned to divert funds from SK Telecom to buttress other companies in the group—a familiar practice in Korea. In hopes of increasing the numbers of shareholders who might oppose top management, Tiger also wanted a 50-to-1 stock split along with the appointment of one more independent director. Tiger had 4 of SK Telecom's 11 directors on its side. The company's three independent directors, along with a fourth director representing Korea Telecom, the state-run telephone company, SK Telecom's second largest shareholder after the SK group, had all opposed the rights issue, which SK discounted in its rush to find buyers and reduce Tiger's stake.

For all the ferocity with which Tiger carried out its campaign, Tiger played into the hands of SK's top management by continuing to sell shares. In papers filed with the U.S. Securities and Exchange Commission on August 23, Tiger said that it had sold 363,914 shares to the SK group on August 23. The price was 1.25 million won, $1,031 per share.[118] Two days later, on August 25, Tiger sold its remaining 259,574 shares, 3.1 percent of the company, in the United States. Tiger did not say how much it got in the second sale, but total earnings for both sales of what had been a 7.5 percent stake exceeded $600 million. SK group companies now owned 36.5 percent of SK Telecom—a guarantee against takeover by foreigners, who held 29.4 percent of the shares without Tiger. SK Telecom's top executives might have wanted Tiger's money but not if Tiger hoped to use it to undermine their power. The fewer shares this trouble-maker held, as far as they were concerned, the easier it would be for them to run the company any way they pleased.

SK executives now had no trouble at a shareholders' meeting on August 27 at which the SK Telecom board made certain that a majority defeated motions both to get rid of the company chairman and to split the stock 50 to 1 so that more people could buy it. Jang Ha Sung, leader of the shareholder rights' movement, asked pointed questions and introduced motions, but the outcome was preordained. The only positive result was that the company did agree to make room on the board for one more "independent" director.[119] Koreans might sell off as a last resort, but they would resist any challenge to their authority. There was not much difference, in the Korean view, between selling off and selling out, especially when foreigners were the buyers or investors.

Tiger won a moral victory by publicizing SK's traditional style, but the SK group had won a far greater victory by turning back a difficult foe, proving that Koreans did not have to sell out after all. Tiger pulled out only after concluding there was no point in playing the role of Don Quixote, tilting at the windmills of the Korean system. There was no reason, however, to shed tears for Tiger. By unloading before the shareholders' meeting, the fund maximized profits. As Tiger managers knew better than the Koreans, selling out in the end could be advantageous. That was a lesson the chaebol found extremely troubling.

# Chapter 10

# Framing a Peace*

T wice a day, a decrepit shuttle bus rattled over the old Japanese-built bridge across the Yalu River between Sinuiju, the second-largest city in North Korea, and Dandong, China. On board were privileged North Koreans, ready for another round of deal making, shopping, and spying. "They eat, they clothe themselves here," a Korean-Chinese businessman who ran a textile factory in Sinuiju told me. "When they come here, they can buy anything they want. Some of them come here often. They have connections. Their sons and daughters work for state trading companies. They are a few chosen people."[1] Dandong, a hustling enclave of glass-and-steel towers, concrete office blocks, and crowded apartment buildings, provided North Korea's best access to the outside world beyond the tight-knit circles in which they moved in their own starving society.

Several hundred North Korean officials had set up permanent residence in Dandong, said Korean-Chinese businessmen who commuted to small factories they operated across the bridge in Sinuiju. North Koreans also crossed legally into China at several other points on a frontier that wound more than 300 miles along the Yalu and then down the Tumen River on the east. None, however, matched Dandong, a city of more than a million people, as a window on the world. In Dandong, the North Koreans, operating from companies and shops fronting as stores for Korean-Chinese merchants, coordinated their trade along the border with China, by far their largest trading partner and source of support. From here they also shipped products through a fast-modernizing Chinese port several miles downstream at the river's mouth on the Yellow Sea. In Dandong, said the Korean-Chinese who did business with them, the North Koreans found an opening without which their impoverished economy could not function at all.

On the esplanade along the banks of the Yalu, along the narrow streets leading to the central business area, the North Koreans wore one badge that set them off in the crowd. "You can tell them by their Kim Il Sung pins," said a South Korean businessman, up from Seoul for a few days of trading with the Chinese. All North Koreans had to wear the lapel pins, emblazoned with the head of Kim Il Sung, the "great leader" who had died in 1994 but was still North Korea's "eternal president." So privileged were some North Koreans that the Chinese government provided them housing and other perquisites—rewards not for politics and diplomacy but for their value as business contacts. "A lot of Chinese have gotten rich here off the North Koreans," said the businessman. "They come in here with western currency, American dollars. They also change their own currency at the legal rate into Chinese currency and buy products with Chinese money."[2]

As North Korea sank deeper into poverty, however, business had slumped. Increasingly, the deal making was done by barter. Fish and vegetables were among the most common items imported from the other side of the river, but the quality and quantity were declining. "For every ten tons of vegetables, three tons are rotten," said a South Korean woman whose husband often dealt with North Koreans. "You never trust what you are getting. It's hard to get money this way. You have to check everything." Rows of fishing boats lined the banks on the North Korean side, rusting and falling into disrepair. A wisp of smoke curled from one of several smokestacks rising from old factories beyond the bank. In a children's park, a Ferris wheel rested motionless. "The North Koreans don't have enough oil for their boats," said Kim Byung Gon, up from Seoul to expedite cargo shipments from South Korea to China. "They don't catch a lot of fish, so the price is high."[3]

The brimming markets of Dandong contrasted with the poverty of Sinuiju, said those who had seen both sides. Sinuiju, population 750,000, now was a ghost town where the lights were never on and people, most of them jobless, with barely enough food to stay alive, conserved their energy by staying at home. For North Koreans lucky enough to get to Dandong, there was no better place for filling routine needs. "They have nothing over there," said one of the Korean-Chinese businessmen. "One hundred percent of their necessities, like soap and toothpaste, are from China." North Koreans cut or scraped off any labels indicating South Korean origin, though Chinese and even Japanese labels were acceptable at home. In the Hong Kong Coffee Shop, run by a South Korean woman who had spent many years in Hong Kong, local businessmen sometimes met North Korean contacts. South Korea's Korea Broadcasting System, received by cable, was constantly on. "The North Koreans don't pay attention to the television," said a waitress, from North Korea. "Or they say it's terrible, but there is nothing they can do."[4]

Those who came illegally, however, were in for a different reception. Ensconced in a comfortable chair in the coffee shop, the Chinese police commander got a call on his cellular phone, then rushed out to investigate a report of six North Koreans just picked up on the Chinese side. A few minutes later he was back, assured the North Koreans were in custody and awaiting deportation to Sinuiju. "Many people are coming over, I see them often," he said.[5] The fate of the North Korean refugees, however, might not be as bleak as a few years ago. In those days, North Koreans were herded back publicly, an object lesson for others who might try to escape. Quite often, they were led by wires stuck though their ears and noses, according to those who had witnessed such macabre scenes.

The Chinese commander, who often visited the North Korean side, said escapees no longer faced execution on return. "They used to execute many people, but not anymore," he said. "They will be put into prison for three years. They escape for hunger so North Korea gets a little softer on them when they go back. If they have a connection, they will only stay in prison for 30 to 40 days." The North Koreans stopped piercing ears and noses, he added, after Chinese complained that the practice created a bad impression. "You still see them sometimes on trains or trucks," said a South Korean businessman. "They tie them with rope. If they are not politically related, they are treated harshly."

The relative leniency reflected the reality of famine so widespread in North Korea that everyone, including officials of the Workers' Party and the government, understood the need to cross the frontier in search of food. The refugees slipped across the Yalu by boat in the lower reaches, sometimes by foot as it wound through the mountains upstream. To the northeast, they forded, swam, or, in winter, walked over the ice of the narrow Tumen River, crossing into China or, near the mouth, into Russia south of Vladivostok. Most disappeared into a Korean-Chinese community of more than two million people. Chinese officials said between 150,000 and 200,000 people were crossing the border each year, including those who came legally on business. As many as 50,000 North Koreans crossed illegally into China in 1998.[6]

No one had definitive statistics, but the number was increasing as hunger worsened. Chinese police in 1998 estimated 100,000 North Koreans lived illegally in China. Other estimates ranged from 200,000 to 500,000. Hundreds of North Korean girls served as bar hostesses, many of them led into prostitution by Korean-Chinese who they believed would help them find jobs as waitresses. "They are hired, and then they sell them into the countryside," said a Korean-Chinese. "They are here to survive," said a Chinese official who commuted regularly between Dandong and Sinuiju. "We don't want to arrest them. We know they will have a hard time if they go back." Many more Korean girls were sold into marriage by Korean-Chinese go-betweens. The trade in wives reflected a shortage of young women in a society where abortion of female fetuses was widespread. Girls, less likely to become wage earners, were viewed as burdens.

Northeast, along the border formed by the Tumen River, refugees from the starving North Korean countryside described a wave of summary public trials and executions, usually by shooting but occasionally by hanging. In cities and towns across the border in North Korea, the pattern of executions as recounted by North Koreans was the same. The prisoner, half dead from beatings, bound and sometimes gagged, was dragged in front of a security official. While several thousand people watched silently, the official read the charges over a loudspeaker and asked if the culprit admitted guilt. After he nodded or mumbled, "I admit," he was shot.[7]

Sometimes the executions were for crimes that would warrant the death penalty in many countries—murder and cannibalism. Stories of the latter, said refugees, were heard everywhere as members of a disintegrating society battled day to day against famine and starvation. More often, however, refugees said the executions were for crimes that would merit no more than prison sentences elsewhere. Typical capital offenses were attempting to smuggle bronzeware across the border for sale in China, breaking down TV sets and selling the parts, and stealing equipment from trains and vehicles.

Whatever the crime, North Korean authorities were conducting public executions in towns and districts throughout the country, according to refugees interviewed near the Tumen River border. Refugees often were reluctant to speak openly amid a stepped-up Chinese campaign to arrest them and send them back to North Korea. Those who talked, however, described the same routine. "They announce them on loudspeakers before the trial," said Park Chul, a 30-year-old farmer who had come to the village of Dongsung, about ten miles from the Tumen River, in search of food and shelter among Korean-Chinese living there. "Everybody knows the time and place. Many people gather," including children marched to see a spectacle designed to terrorize a populace that might otherwise rebel. "For one prisoner, there are three riflemen," said Park. "They line them up and shoot at the same time. One shoots three times in the head, the other three times in the chest, the third three times in the legs."[8]

Interviewed in a farmhouse near where he was working, Park Chul said that he had seen an entire family of five shot to death in Heidong, a district not far from the Tumen River. The father and mother had been luring small children into their house, drugging them with wine, then strangling them, chopping up their bodies, and mingling the flesh with pork, which was then sold on local markets. Executed along with the parents were their three sons, the youngest age 12—the youngest person Park ever saw executed. Park said seven other people were executed at the same time, all for economic crimes.[9] A 19-year-old girl from the same district said she had heard of the execution of the family but had not seen it. "They shoot smugglers and cannibals," she said. "I heard other times babies were killed too."[10]

Representatives of aid organizations reported similar stories from starving refugees. "The executions are widespread," said Park Ji Hun, coordinator for the Korea Buddhist Sharing Movement, dispensing food and clothing in the area from its headquarters in Seoul. "We know of a person who cut electrical wire to exchange for food and was executed. We heard many such stories."[11] South Korea's National Intelligence Service said the government was using executions "to try to stop robbery and bribery."[12]

Without exception, North Korean refugees reported having witnessed several, in some cases a dozen, executions. The number had risen steadily since the one-year period of mourning that followed the death in July 1994 of Kim Il Sung, who had ruled ever since the founding of the Democratic People's Republic of Korea in 1948. "For one month after Kim Il Sung's death there were no executions," said a woman who fled from the east coast port city of Hamhung in 1997. "For a year there were very few." Then word spread that Kim Il Sung's son and heir, Kim Jong Il, wanted to "hear the sound of gunshots again" and stepped up the killing.[13]

Those were just the executions held in public, to establish the criminals as examples in order to discourage rampant crime. "Then there are closed executions," said Kim Myung Goo, 42, who served ten years in the North Korean army, then became secretary of a youth group but was denied membership in the Workers' Party when it was discovered his father had owned land during the Japanese colonial era. "Political prisoners are accused of slandering people and the party, so it's done secretly. With those political criminals, it's not only he or she who goes to prison or is executed but all the family."[14]

The increase in death sentences revealed Kim Jong Il's mounting inability to control his starving people by other means. The fact that the regime kept political executions a secret also indicated its fear of open defiance—a fear borne out by reports of an abortive coup staged by the army's VI Corps in 1995.[15] "You still cannot say bad things," said Kim Myung Goo, who first left the North in 1998, then crossed the Tumen to see relatives and returned to China early in 1999. "In general the whole public has a bad feeling toward the government, especially educated or retired people or those who are dismissed from the army. People think something needs to be done."[16]

Kim Myung Goo talked about the executions as if they were such ordinary events as hardly to merit discussion. Asked to describe one, he said that "hundreds of people watched on the execution grounds"—in his case in Chongjin, an industrial port on the northeastern coast. "They wanted as many people as possible to see it. Policemen signaled passersby. Children were marched from school. There were old people as well as young. There was a special section for prisoners from the local jail, but they were separated from the crowd." Finally, eight condemned men were hauled in aboard a truck.[17]

"They were tied around the neck, waist, and leg," said Kim. "After the charges were read and the sentence announced, they were asked if there was anything you want to say before you die, but they were gagged so they couldn't say anything." The condemned men were tied to eight stakes in a row. "This time there was only one rifleman for each man, but each rifleman fired a number of shots." Only once did Kim hear a condemned man try to speak. "I saw his mouth moving, but it was just a moaning," he said. "He was already mostly dead."[18]

Most people had stories to tell about cannibalism—some just rumors, others based on certain knowledge. "I heard many times that people kill people and eat the bodies," said a 16-year-old boy named Lee Han Seung, living in the home of a Christian religious worker in Yanji, a Korean-Chinese center about 30 miles west of the Tumen River border. "I know of a murder case in which a man was shot for killing his nephew." Lee said he had seen his first execution about two years earlier when his teacher in his village school near the northeastern free trade zone of Rajin and Sonbong led his class to the execution ground. "The teacher said, 'Do not commit any crime—come to school and study hard.'" He had last seen an execution on February 1, 1998, the day before he fled across the Tumen on the last of three attempts to escape to China. "There was a man and a woman," he said. "They were executed for escaping to China three times and being returned to North Korea."[19]

Nowadays, authorities had many more serious crimes to worry about than fleeing to China. "When Kim Il Sung was alive, when people died, the government provided a coffin," said a man named Kim Chul Soo, who worked in a factory near Hamhung until all the factories were shut down. "Now people are just put in the ground. At night people go to the burial of a recent corpse, dig it up, cut it apart, cook it, and eat it." When word spread that a dumpling with fresh meat was on the market, "it sells out in ten minutes," he said. "I saw dumpling with human flesh inside. North Koreans don't care. They just eat it."[20]

Runaway North Korean children, their growth severely stunted by lack of food, recounted their experiences while begging. "I left my home on September 28, 1998," said a girl named Kim Yun, who looked about 8 years old but was actually all of 12. "That day, my sister and two other boys crossed with me. My sister drowned. I was nearly drowned, but a Chinese rescued me. The two boys ran away." Kim, about four feet, two inches tall, slightly pudgy off the handouts she received, survived by asking for money on a walkway beside a bridge across the Tumen linking the Chinese town of Tumen with the North. She said her father had died several years ago, probably from starvation, and she tried not to think about her mother, slowly starving to death. "What's the use," she countered, when asked if she missed her mother. She recalled the day when a child wrote "Death to Kim Jong Il" on a wall by a toilet at her school. "They had a long investigation," she said. "The child and his whole family were sent to prison."[21]

A bored Chinese policeman stared from the bridge while a few tourists, most of them from South Korea, photographed the bridge and small signs with Chinese and North Korean flags proclaiming "Eternal Chinese-Korean Friendship" in Korean and Chinese. There was no traffic, no sign of life among the closed factories on the North Korean side. Down the river, big characters in Korean lettering, painted in white on a hillside, read, "Let's Brighten the Youth by Lighting Up Kim Jong Il's Era."

Suddenly, half a dozen more children emerged from nowhere, grinning, hands out, waiting for coins or snacks from nearby food stands. Unlike North Korean adults, who preferred to hide for fear of arrest, deportation, and harsh punishment at home, the children counted on their eager smiles and heart-rending stories to bring sympathy from local Korean-Chinese, who sometimes provided clothing and shelter. They all told of their struggle to live without parents, teachers, or guardians. "My father went away,

I don't know where he is," said Choi Han, who at about four and a half feet tall looked 10 or 12 but was actually 16. "I've been arrested three times and sent back across the river. The last time I came back here was three days ago. I walked on the ice. I had 15 yuan"—less than two dollars—"and hid it on the soles of my feet. I used it to buy some bread."[22]

Beside him, Yoo Sun Eung, also 16 but an inch or so shorter, smiled, revealing that he had lost two or three teeth beneath a scarred upper lip. "I was struck by a soldier because I ran to China," said Yoo. "It happened after they sent me back in August, but I came back in November. Since then I have been arrested three times. I came back yesterday." Yoo wore a red warm-up jacket, given him by a South Korean tourist, with the words "Chicago Bulls" inscribed above a Bulls symbol. He could not read western letters and had never heard of Chicago or the Bulls. His main concern was keeping North Korean guards from seizing his clothes when they captured him. "Once they capture us, the army tries to tear the clothes apart," he said. "They say they want to keep me from selling the clothes for money. Sometimes they take all our clothes. They always take our money if they find it."[23]

All the children said they had seen people executed but were too hungry to worry about what they had witnessed. Yoo grinned as he imitated the response of a condemned man before the firing squad. "First the man stares at the riflemen in front of him," said Yoo, whose teacher had led his class to watch a number of executions. "Then when the guard yells 'Fire,' the man closes his eyes. Then his head snaps back when the bullet hits him." Choi Han, 15, said he had seen one hanging as well as a number of shootings. "The hanging was for a bank robbery," said Choi, talking quickly. "The reason for hanging was to give the criminal time to reflect on his sin before he died. About three years ago, execution was not so common. Now it happens almost every day. More people are starving, and more are executed."[24]

Refugees from the North, children and adults, counted on church members to shelter them and to offer occasional meals and advice. At a church in Yanji, members of the congregation urged a couple of children to return home and go back to school. Han Sung Chul, 16, looked down and mumbled. "Some teachers don't even come to school because they are hungry too," said Han, 16, about four feet four and a half inches tall. "It is useless. Please don't make me go back."[25] One of the church leaders, Nam Bok Ja, said she had gone to North Korea four times in the past year to spread the word of God as the answer to suffering. She said she spoke to North Koreans in basements, 10 or 12 of them at a time, praying for God's help to free them from the daily struggle. "I tell them, despite all their difficulties, God leads the way," said Nam. "I get a connection with them through God."[26]

Nam conducted missionary work in the North despite the certainty of severe punishment, possibly death, if caught. She counted herself as one of a rising number of Christian missionaries, all Chinese citizens of Korean descent, bringing the word of God to a people crying for spiritual support. Almost all of them were Protestants, members of congregations that the Chinese government had permitted since easing its own rules against religious worship in the 1980s. Nam and other missionaries had a simple excuse to offer North Korean authorities when they entered the country, open to Chinese. They carried rice and clothes, which, they said, they were bringing to friends and relatives.[27]

Bearing such gifts as these to a starving people, missionaries formed the nuclei of small congregations. Many North Koreans, they discovered, had never heard the word "God," much less seen a prayer book or Bible. Those old enough to remember the pre-communist era of Japanese rule, when missionaries still had congregations in what is now North Korea and Pyongyang was known as "a city of churches," had long since been silenced. The two Christian churches now in Pyongyang were closed to all but a few members of the elite.

As long as no one betrayed her to authorities, however, Nam found a ready audience for a simple message. "I tell them how rich are China and South Korea," said Nam, who had last visited the North in January 1999. "I tell them that only North Korea, in between South Korea and China, is getting worse. I say, it's because of God's will. Without God's will, nothing can be done. Then I tell them how to pray." She insisted her message was nonpolitical; she never suggested that anyone to whom she spoke should be disloyal to Kim Jong Il. Never, she said, did she raise ideological or political issues.[28]

"Even though they don't talk about Kim Jong Il, they all know the system and the leader are bad," said Nam. "The people don't believe in anything. So it's time for the church to spread the word." She had no illusions, however, about the punishment that awaited those who preached the gospel. "There was a case when a Korean-Chinese went to North Korea to teach about God," she said. "When the North Koreans found out, they cut out his tongue and hit him on the head. Then they let him go. Now he can't remember anything and can't talk." As for North Koreans discovered to be engaging in religious worship, "They are sent to prison where they cannot survive."[29]

For the few North Koreans who made their way to the South, however, life could be riskier than hiding in China—or even staying at home.

"Come see your brothers and sisters in the South," said a typical South Korean propaganda leaflet dropped in the North. "Live in peace and freedom. We promise you money and job opportunities." The reality, recounted by North Korean defectors to the South, was different. "We have no money, no jobs," said Han Chang Kwon, president of an association of North Koreans in Seoul. "We demand the South Korean government stop this propaganda. It is totally misleading."[30]

The disillusionment began within minutes after a typical defector got to Seoul by a circuitous route through China or Russia. "When we arrive at the airport, we are given flowers and have a photo session for television and the newspapers," said Han, in Seoul since 1995. "Then we are taken by van to an interrogation center. They strip off our clothes. They insult us with crude remarks. They pour questions so you are confused, dumbfounded. They beat you like a dog with sticks."[31]

Such charges infuriated the powerful National Intelligence Service, responsible for interrogating defectors to see if any were spies. The intelligence service, responding to a hyperbolic report by an Australian freelance in the *Sydney Morning Herald,* said the defectors made their claims "in a desperate attempt" to extract more "settlement money."[32] Interviews, however, indicated that many were mistreated—and were still unhappy even though they obviously were more comfortable than before.

Typically, defectors had to spend most of their settlement money to buy the small apartments in which they lived. Thrown into the hard-driving capitalist society of the South, most of them survived through menial, temporary jobs. "I would never have come if I had any idea what life is like here," said Hong Jin Hee, who had bribed North Korean border guards to cross the Yalu into China in 1993. He got a taste of what awaited him when he visited the South Korean embassy in Beijing and was told to leave. "They said they would call the police if I didn't get out," said Hong, who had to see South Korean officials in Hong Kong to defect. "The whole world knows how terrible is life in North Korea, but it's not that much different here. The police and National Intelligence Service agents still come to my apartment and beat me up occasionally." He slowly lifted his right hand to demonstrate permanent injury to his arm.

Nonetheless, Hong described life in the North as worse. "Every day you hear shooting," he said. "They execute people openly. They threaten to kill family members. They come in the night, catch people, and drag them off like dogs."[33] Refugees tended to forget such realities while coming to terms with new lives far from home. "They're limited in their ability to adapt," said Oh Hye Jung, a Roman Catholic nun who worked for a church committee on unification. "There is a huge cultural gap between the two Koreas. They're like oil floating on water. They find it extremely difficult to get along with South Koreans, so they have this emptiness. They feel isolated, alone. They feel like aliens here. They are not one of us. Some of them even want to commit suicide. Many of them want to move to other countries, but the government won't let them."[34]

While the beatings largely stopped under Kim Dae Jung's presidency, the issue pointed to a larger problem. How could South Korea absorb thousands of North Koreans who might want to come to the South when the government still did not know how to deal with the 751 North Koreans who had defected to the South from the end of the Korean War up to February 1999? "The problem lies in both lack of policy and lack of determination of defectors to adjust to our society," said Oh Hye Jung. "The government's support program is inadequate. This problem should be seen from a long-term view. The government just gives them a small amount of settlement money, and that's all. That way, the problem cannot be solved."[35]

Researchers at the Korea Institute for National Unification, a government think tank, said they were working on contingency plans for a flood of refugees if North Korea collapsed. "But we are not quite serious about such planning," said Choi Jin Wook, a research fellow. "We do not want immediate unification. We prefer coexistence." He justified tough interrogations as the only way to find out the real backgrounds of some of the defectors. "Many of them might have been involved in crimes in North Korea. They do not want to report exactly what was their position in the North. They inflate their rank in the workplace. They say they were managers when they were just workers."[36] Such arguments only added to the fury of Han Chang Kwon, the president of the North Korean refugees' association, as he contemplated suing the government for damages. "If the government has this mentality, even if the two Koreas are united, it is quite likely there will be another Korean War."[37]

Fear of a second Korean War was never far from the Korean consciousness—and was very much on the minds of forces along the demilitarized zone that had divided the two Koreas since the signing of the truce that ended the first one on July 27, 1953. "I always feel there could be a second Korean War," Lieutenant Colonel Chun In Bum, standing on a craggy outcropping in the northeastern corner of South Korea known as Hill 911, facing the mountains of the North, told me in May 1997. "If I didn't feel that way, the whole purpose of being in uniform would be lost." The possibility assumed urgency amid reports of North Korean military exercises along the DMZ, the four-kilometer-wide strip that wound 155 miles across the peninsula, below the 38th parallel on the west, above it on the east. South Korean defense officials warned that more than half of North Korea's 1.1 million troops—nearly twice as many as South Korean forces—were within 30 miles of the DMZ. The worst-case scenario was that the North Koreans might attack "like a rat in a corner," in a common Korean phrase, in a desperation engendered by a severely worsening food crisis.

Chun was also on guard for the opposite scenario—a rush of defectors. "Our troops are trained on what to say if somebody shows up in front of a guardpost," he said. "They know what to do if they're military or civilian." The view from the GOP (general observation post) atop Hill 911— 911 meters, nearly 3,000 feet, above sea level—added to the sense of impending drama. It was only a few meters from the barbed-wire fence that marked the southern edge of the DMZ. "The only foreigners who've been here are officers from the UN Armistice Commission," he told me. "Otherwise you're the only foreigner who's been here since the Korean War. The North Koreans would kill to know what's here. We don't know about them, and they don't know about us."[38]

South Korean, not American, troops pushed the North Koreans far beyond the 38th parallel in this sector in the first months of the Korean War—and fought back across the parallel again after the Chinese entered the conflict. "We lost a lot of men," said Chun, on a jeep ride along the barbed-wire barricade. "It's the roughest terrain in the country." As a result of the war, South Korean territory on the eastern side of the peninsula jutted up to about 40 miles south of the 39th parallel. The armistice, certifying a jagged line across the peninsula, was signed at Panmunjom, in the DMZ south of the 38th parallel on the west. Chun proudly observed that South Koreans were alone on Hill 911. Most of the 37,000 American troops—including the 2nd Infantry Division south of Panmunjom—were more than 100 miles away.[39]

Sunk into a shoulder of Hill 911, an M4 tank of World War II vintage, rusted and broken, its 85-millimeter cannon not fired since 1953, testified to the fighting that had once echoed through these mountains. Down the next ridgeline, inside the DMZ, the UN flag flew over a South Korean guardpost. Through the barbed-wire mesh of the demarcation line, the shallow waters of the Nam River dividing North from South were visible far below as it meandered to the "East Sea"—the "Sea of Japan"— glimmering several miles to the east.

A North Korean army sergeant had made it across and defected in October 1996. Chun talked to him before rotating up to Hill 911 with his battalion. "He was really sure of himself," said Chun. "He certainly wasn't starving, and he wasn't lacking in training. They conduct training for infiltration right in the DMZ. He gave a picture of how great is their military." The sergeant lent credence to the widespread South Korean view that the North fed its soldiers while civilians starved. Why, then, did the sergeant cross the line? "He was discontented with a late promotion," said Chun. "He was also in deep trouble. He had bought some ballpoint pens while on a mission in search of supplies in Pyongyang. It turned out the pens were made in South Korea." The defector had left in daylight, when he knew the North Koreans shut down the power on two electrified fences that shielded their side of the DMZ. South Korean soldiers heard shots being fired as North Korean guards tried to stop him. Hiding inside the DMZ, he held up his hands in surrender before a South Korean guard-post the next morning.[40]

Since then, two North Korean civilians had made it to the South along the ten-mile stretch covered by South Korea's Thunderbell Division, including Chun's battalion. A 26-year-old woman weighed 80 pounds when she showed up one day in front of a South Korean bunker. "Now they seem to have fortified their defenses," said Chun. "They're working all the time. We see trucks and people. They have artillery nearby in tunnels, right in the Zee." How tough would it be for a sizable force to make its way through the heavily forested valleys below Hill 911 before the South Koreans could stop them? The defenses looked thin, the mountains dauntingly steep. "My mission is to delay and warn," Chun responded. "We don't have electricity running through our fences like the North Koreans, but we have lights and everything. If they come, they'll pay a heavy price."[41]

Down in Seoul, the rumor factory carried suspicions to sensational extremes. Word was that Washington had known all along about forays of North Korean submarines into South Korean waters before a 325-ton North Korean Shark-class submarine foundered off South Korea's east coast on September 18, 1996. The Americans, according to the rumor making high-level rounds, had not told their South Korean "friends" about the incursions for fear of ratcheting up North-South tension.[42] President Clinton "has every reason to keep the status quo, just keep them at bay so as not to cause any problems," reasoned SaKong Il, a former finance minister who now ran his own Institute for Global Economics. "South Koreans are immediately threatened. We would like to see Americans come forward more strongly. There are different perspectives, so there seem to be uncertainties between allies. Seoul is within artillery range of the

North. The DMZ is only the distance from Dulles Airport to the White House."[43]

South Korean distrust of the United States was rooted deep in the alliance. South Koreans believed that the United States had sold them out initially by agreeing with Moscow to divide the Korean peninsula into Soviet and American occupation zones after the Japanese surrender in 1945. The Americans then compounded the offense, in the Korean view, by doing a deal with the North for dividing the peninsula roughly on the same 38th parallel at the truce talks in Panmunjom in July 1953. The United States was eager to get out of a war that had gone back and forth between American and Chinese forces since the first year of hard fighting up and down the peninsula. Dwight Eisenhower, the former World War II general, won the 1952 presidential election partly on his "I will go to Korea" campaign pledge—a promise to get out short of "total victory."

The image of that sellout at Panmunjom—on a piece of paper that South Korea's President Rhee Syngman refused to recognize—endured in the "agreed framework" signed at Geneva on October 21, 1994. The accord called for a consortium to provide money and expertise to build two nuclear reactors in the North while the North forswore efforts to produce a nuclear warhead. The United States pressured the South into agreeing to supply 70 percent of the $5 billion for the reactors, while Japan agreed to supply only about 20 percent, and the United States shipped 500,000 tons a year of heavy oil until the reactors were up and running. Two years later, still under Kim Young Sam, Korean officials were unhappy about having to send food aid to the North while the North sent propaganda and spies to the South.

The standoff over the submarine meant the whole peace process might fall apart. The Americans wanted to prevent such a disaster while trying to talk the North into agreeing on such contentious issues as nuclear nonproliferation and the search for several thousand Americans still missing from the Korean War. South Korean officials under the Kim Young Sam administration saw the standoff as inevitable, at least until the North collapsed. Some South Koreans believed the best they could do was speed up an inevitable crisis by adopting an implacably tough line. Americans were unwilling to risk a second Korean War when Washington had enough trouble in the Middle East and the Balkans.

At South Korea's ministry of unification, bureaucrats in the Office of South–North Korean Dialogue acted out scenarios for "reunification talks" as if they were characters in a stylized drama. One bureaucrat played the North Korean heavy, screaming out impossible demands, while another played his South Korean opposite number, parrying demands with demands of his own, stating and restating the government's position. "We are working on various scenarios," said Kim Young Min, a ministry staffer. "It's a matter of what kind of offer can we suggest, what might they give us. We do simulation of actual dialogue." Since July 1995, simulated dialogue had been all the negotiators had been able to do. It was then, for the last time, that ministers from both sides had met in Beijing— only to break off in a dispute over terms for holding talks on a regular basis as both sides had agreed to do in 1991.[44]

In the aftermath of the submarine incident 14 months later, U.S. officials were trying to get the process on track again. Somewhere through the fog of distrust, the Americans perceived a glimmer of hope that representatives of the two Koreas, the United States, and China might sit down for the "four-party talks" suggested by Clinton and Kim Young Sam on Cheju Island, off South Korea's southern coast, in April 1996. The sense was that negotiators would come up with a solution to the puzzle: how to get the North to say something that the South could accept as an apology for the submarine. Tensions had been rising ever since the incident. Only one of 26 crew members and saboteurs was captured alive; one escaped and the others were killed, a number of them executed by their own mates, while five South Korean soldiers and several civilians were killed in the manhunt. YS had made the issue of an apology and a promise of "never again" a key point of prestige and power.

After the submarine incident, optimism generated by talks in New York between the Americans and North Koreans contrasted with measures and countermeasures that appeared to be driving relations in the opposite direction. South Korea was holding back on contributions to KEDO, the Korean Peninsula Energy Development Organization, formed to oversee the reactor project. The South also banned all travel by South Korean businessmen to North Korea, delayed on a program for refugees, and then outlined vague plans for a camp capable of holding only 500 people.

Several elements were on the table in New York. Most important was the nuclear freeze agreed on in Geneva and progress on "canning" the nuclear fuel processed at the North's nuclear facility at Yongbyon, 90 miles north of Pyongyang—that is, storing it in canisters. The United States for its part had to deliver heavy fuel oil, a low-grade residue from refined oil used largely for heating and power production, despite opposition in Congress to appropriations; the South had to send a survey team to begin the site study for the first reactor at Kumho on North Korea's east coast about 100 miles north of the DMZ.

The cancellation of permission to go to North Korea threatened a small but promising flow of goods between North and South—not overland but by sea, mainly between Nampo on North Korea's west coast and Inchon. The North stood to suffer far more than the South from the reduction, largely because it had a favorable trade balance, most of it in the form of cheap clothing produced by small textile factories. South Korean interests, anxious to open up a potential market and a source of cheap labor, saw no chance of success in the near future. The worst sign: a Daewoo team, after investing $5 million to build several factories to produce toys, golf bags, and shirts and blouses in Nampo, had returned to Seoul. South Koreans were equally pessimistic about the chances of investment in a special economic zone in northeastern North Korea, an area first developed under Japanese rule before World War II.

What alarmed the South most of all in late 1996, however, was the feeling that whatever was said in New York, North Korea was building up militarily. The defense ministry accused the North of increasing artillery pieces near the DMZ from 300 to 500, of doubling multiple rocket launchers to 280, and of raising sea and air infiltration troops from 20,000 to 40,000 and special force troops from 100,000 to 110,000. At the same time, the North raised the call-up age of reservists from 40 to 45, adding another 500,000 reservists.[45]

Why, South Korean officials asked, should the South send food to the North while the North refused to apologize for the submarine? "The North can say, 'Even though the submarine thing was an accident, we feel sorry about it,'" said Ro Jae Bong, executive director of the Korea National Committee for Pacific Economic Cooperation. "If they can say some kind of words which can be interpreted as an apology, Kim Young Sam can take it. If North Korea has real difficulty and is at a level of a food security problem and they need quick action from Japan or the United States, they may say something."[46] On December 29, 1996, the North did just that, expressing a begrudging "deep regret" and promising that "such an incident will not recur."[47]

Still, South Korea's policy toward the North did not change until Kim Dae Jung introduced his "sunshine policy" of "engagement." Whatever offenses the North would commit, the South under DJ would perceive sunshine through the clouds. "Engagement" began right after his inauguration with word that the South was on the verge of lowering the barriers blocking South Korean companies from investing in more than minor projects in North Korea. The government "plans to radically simplify administrative procedures on investment in North Korea," Yonhap reported on

March 1, 1998.[48] The government hoped for increased commerce with the North, and DJ renewed his call for exchanging envoys between North and South and holding reunions between families separated by the Korean War. North-South trade for 1997 was $308,339,000.

A sharp rise in North-South commerce was seen as a crucial first step. "The government is talking about step-by-step deregulation of economic contacts with North Korea," said Selig Harrison, a Washington-based expert who had visited North Korea a number of times. "If they implement what they say they are going to do, that will make a big difference." No question, DJ was approaching the North in a style very different from that of YS, who had discouraged more than minimal contact. "The question is how fast Kim Dae Jung can move," said Harrison. "He wants change early on in the economic field. He believes in separation of economic concerns and politics."[49]

South Korea revealed its plans for sharply improving trade with North Korea on the eve of talks in Seoul on March 2, 1998, with a U.S. team led by Charles Kartman, in charge of U.S. diplomacy vis-à-vis the North. Kartman was in Seoul to discuss the next round of four-party talks in Geneva on March 16 among representatives from the two Koreas, the United States, and China. DJ's aides had indicated they would not object if the United States wished to improve its own ties with the North, possibly by recognizing the North Korean government in Pyongyang and easing trade sanctions. The government might more than double the $5 million limit on investment by South Korean companies in the North and also ease restrictions on business travel. The $5 million ceiling, set by Kim Young Sam in 1994, had discouraged all but a handful of investors. Another measure that might encourage investment in North Korea was a proposal for compensating business interests for losses there.

As delegates from the four principal protagonists of the Korean War gathered in Geneva on March 14, there were hopes, said a U.S. diplomat, of moving "beyond sterile debate and into matters of substance." Whether or not negotiating teams from North and South Korea, China, and the United States could turn rhetoric into action depended on how the North viewed the policies of Kim Dae Jung. "What we're waiting for is for the North to decide it's ready for genuine dialogue," said the diplomat. "It's a matter of waiting for them to make up their minds." As far as the Americans and South Koreans were concerned, the talks would be a success if they got the North to budge on a single issue that had frustrated attempts to negotiate a lasting peace ever since the death of Kim Il Sung—the North's refusal to negotiate directly with the South.[50]

South Korean and American negotiators delineated clear goals for the talks. Negotiators might discuss the "peace mechanism on the Korean peninsula," including steps for reducing forces, while North-South talks should cover economic cooperation, exchanging envoys, and reuniting families separated by the Korean War. U.S. and South Korean negotiators expected the usual rhetoric from Kim Gye Gwan, North Korea's vice foreign minister and leader of the North Korean delegation, in the opening minutes of the first session on March 16. No one doubted that Kim would reiterate, as he had in the first round of the four-party talks in December 1997, the two central points of North Korean policy. First, the United States had to withdraw its 37,000 troops from the South, and second, it had to negotiate a bilateral peace treaty concluding the Korean War, in a state of suspension since the 1953 ceasefire. The United States might offer the bait of easing sanctions, but skeptics were not hopeful. "The United States and China are the main players," said a Korean foreign ministry official. "South Korea is kind of playing along."[51]

The breakdown on April 18, 1998, of the first direct North-South Korean talks in nearly four years, however, meant that the North might adopt as tough a line toward Kim Dae Jung as it had toward Kim Young Sam. The South Korean team was still optimistic about keeping up the dialogue after getting home on April 19 from nearly a week of meetings in Beijing with North Korean delegates, but nobody had any idea when the talks would resume, much less what issues they would cover. "There is a constant sense of frustration in dealing with the North," said Gerald Segal, Asia specialist with the International Institute for Strategic Studies in London. "It wouldn't take much for them to take the basic step, but they refuse to do so."[52]

The leader of the North Korean team, Chon Gum Chol, based his refusal to keep talking on the South's insistence on linking a deal on the North's demand for fertilizer to a plan for reuniting millions of families divided by the Korean War. The South's chief delegate, Jeong Se Hyun, called for establishing a center for that purpose at Panmunjom. "Although we need fertilizer, we will not exchange it for our independence," Chon warned before leaving Beijing. He advised the South not to assume the North was so desperate for aid as to agree to another meeting before the South dropped "political conditions."[53]

The North's response raised doubts as to whether it was willing to pursue the dialogue on any terms but its own. The South suspected North Korea stalled on family reunions for fear family members would tell their relatives from the South how terrible were conditions in the North.

Richard Grant, head of the Asia program at London's Royal Institute of International Affairs, suggested that the North was "testing the waters" to see if DJ's position was really softer than that of YS. "The North is trying to sound out how much Kim Dae Jung is willing to move from the previous government," he said. "The room for maneuver is not all that great."[54]

Matching South Korean diplomacy, Kim Jong Il called on April 29, 1998, for "a wide-ranging, nationwide dialogue" in a drive for reunifying the Korean peninsula. All Koreans, "North, South, and abroad, must visit one another, hold contacts, promote dialogue, and strengthen solidarity," he was quoted as having said.[55] In Seoul, Kim Dae Jung told visiting Japanese editors that he believed that eventually, "in the course of South-North contacts," he and Kim Jong Il would hold summit talks.[56]

Significantly, Kim Jong Il's plea, in an "open letter" in the North Korean press, was dated April 18, the day that talks in Beijing had broken down between North and South Korean vice ministers. His remarks synthesized North Korean statements since February, when the North had called for "dialogue" between groups from North and South. In the letter, Kim Jong Il suggested participation by "representatives of all political parties and social organizations, including the authorities and the figures from various walks of life in the North and South and the overseas compatriots." The fact that Kim Jong Il, who rarely spoke in public and issued few statements, attached his name to the letter showed the desire of the North to come up with a response to DJ's proposals.

Basically, however, analysts perceived no change in North Korea's policy. "The important thing is they show us by actions, not words," said Lho Kyung Soo, professor of international politics at Seoul National University. "If we can't agree on cooperation, where are we?"[57] The statement, Kim Jong Il's first on North-South relations since he was named general secretary of North Korea's Workers' Party in October 1997, was issued around the fiftieth anniversary of a meeting of social and political groups from North and South in Pyongyang. There was an irony in the North's observance of the anniversary. North Korea in 1948 was promoting a popular revolt in the South as a prelude to the Korean War. Kim Il Sung had risen to power in the North that year.

There was no stopping, however, the drive from the Blue House to encourage rapprochement through dialogue, if not between governments, then between groups and individuals. Thus it was that a Unification Church song-and-dance troupe called the Little Angels sat through eight hours of briefings at the ministry of unification on the do's and don'ts of how to behave in the North as a prelude to a ten-day mission beginning May 2, 1998. "Nobody can give a small gift, especially a gift of food or money," warned Yun Sang Sop, international manager of the church-owned Korean Cultural Foundation, which operated the group. "Nobody can go out without a guide." The ensemble took off May 1 for Beijing en route to Pyongyang as the first private South Korean cultural group invited to the North. More significant, the mission served as a conduit for economic ties as the Unification Church explored investment possibilities.[58]

In an act of timing that did not appear coincidental, the South Korean government on April 30, the day before the Little Angels took off, lifted the $5 million ceiling on investment in the North. The government also did away with the ceiling of $1 million that South Korean companies previously could spend on machinery sent to the North to build factories. The new rule permitted South Korean companies to engage in any type of business in the North except strategic defense industries, including electronics, aeronautics, and computer science. The government also decided to grant multiple permissions to go to the North for people with business interests there. "Our business dialogue with the North will resume," said Pak Bo Hi, right-hand man of Unification Church founder Moon Sun Myung. The South sent messages to the North through Pak describing its new rules on investment there and conveying its blessing for private initiatives even if the North rejected diplomatic overtures.[59]

No one was more enthusiastic about the possibilities in the North than Pak, leading an entourage that included 38 performers, all girls aged 9 to 14, and 30 adults. Some of the grown-ups would join him in talks having far more to do with business than culture. Pak cited fields ranging from machinery to soft drinks to tourism for investment by the Tongil group, the business arm of the Unification Church in Korea. "On our behalf, I will talk about business possibilities," he said.[60] The mission provided special entree for the church, which owned factories and a chain of small stores in the South under Tongil, the Korean word for "unification." The wealth of the church was a secret, but the group ranked about thirty-fifth among the chaebol.

Pak hoped the trip would give Tongil a tactical advantage in the North over the Hyundai group, which had sent a team to Pyongyang in April to talk about opening up the region of Mount Kumkang, Diamond Mountain, to tourism. Hyundai founder Chung Ju Yung, whose native village of Asan was the near the east coast 20 miles north of the area, had first proposed exploiting the region for tourism in a pioneering mission to North Korea in 1989. Chung, now 83, hoped to revisit the North in a few weeks, but

Pak would be competing for much the same business. The Unification Church, through its companies in Japan, already owned the hotel in Pyongyang where the Little Angels would stay.

The fact that Pak, who had helped Moon found the ensemble in 1962 and the *Washington Times* as a conservative newspaper in the American capital in 1982, could go on such a mission symbolized the shift in South Korea's outlook under Kim Dae Jung. The shift was all the more significant considering that Pak had angered South Korea by attending the funeral of Kim Il Sung in 1994 and meeting with Kim Jong Il. Threatened with arrest under South Korea's national security law for unauthorized contact with the North, he did not dare return to Seoul from his residences in Tokyo and Washington until mid-1997 after receiving assurances that he would not face charges.

North Korea hosted the group despite the failure on April 18 of North-South talks in Beijing. As a further sign of the significance that South Korean officials attached to the mission, they permitted live broadcasts of the Little Angels' three Pyongyang concerts on at least one South Korean radio station and a televised version after the group returned to Seoul on May 12. "The doors of South Korea are open right now," said Pak. "It's a matter of North Korea accepting the degree of contact they want."[61]

More than 30 South Korean firms had applied to do business in the North. Only two, however, had decided to set up there. Reasons for corporate reluctance to go to the North ranged from the low skill levels of North Korean workers to bureaucratic hassles to difficulties obtaining credit from banks in South Korea for such high-risk ventures in view of the South's own economic problems. South Korean officials promised not to "put up any barriers or obstacles for those wanting to make business contacts and cooperation in the North," but the policy had its limits.

In the case of Tongil, the government was troubled not only by the political record of its leaders but also by its financial problems, worse than those of most other chaebol. The Reverend Moon opened a "world culture and sports festival" in Seoul on February 4, 1999, asking participants, "What happens to us when we die?" while his temporal business empire in Korea faced the question of how to repay debts now approaching $2 billion. The "spirit world" after death transcends "time and space," said Moon, talking in Korean after a six-course banquet as interpreters translated into seven languages for an audience that included former third-world heads of state and ambassadors. "There are no factories there to produce food. There are no automobile factories. There is nothing like that."[62]

In the real world, the weakness of the crumbling Tongil group emerged as Moon toured South Korea drumming up flagging support for his church and his companies. The debts of the group's 16 companies now exceeded two trillion won, $1.7 billion. Its four biggest companies, manufacturing products ranging from ginseng tea to tank guns, were bankrupt, awaiting reorganization under court supervision.[63] The flagship Tongil Heavy Industries, founded in 1962, had suffered the most. Its plant, which produced motor vehicle transmissions, had had to lay off 40 percent of its 2,000 workers but survived on military sales. Church donations "have helped but are very much limited," said Seo Pyong Kyu, a manager in the group's financial planning department. "We make rifles and cannon for tanks. The contract is going on."[64]

Tongil was no match for Hyundai. While the Little Angels were still in Pyongyang, Chung Ju Yung said that he expected North Korea to approve his request to return to the North leading 1,000 head of cattle through Panmunjom. Chung made much of the fact that he had stolen a cow from his father for the cash he needed to flee from home and seek his fortune in Seoul. Now he wanted to repay the debt with interest—and open the North to investment by Hyundai.[65] He would focus on a plan that he had first broached during a trip to the North in 1989 for building a resort near Kumkang. Hyundai also was interested in enterprises ranging from textiles to electronics, for which the North had a large pool of inexpensive labor.

Chung's cattle drive to the North required approval to transport them by the seldom-used land route through Panmunjom. Chung and his family would go with them. Use of the Panmunjom crossing, heavily guarded by troops on both sides, would set a precedent for future commerce. Almost all travelers went to North Korea by air, flying to Beijing and then to Pyongyang, while goods moved by ship. Hyundai officials said it would be impractical if not impossible to send so many cows to the North by any way but land. Chung also said he would donate 50,000 tons of corn to North Korea. The first shipment of 10,000 tons left Pusan on May 7, 1998, bound for Wonsan on the east coast north of Kumkang.[66]

Chung Ju Yung, three of his brothers, and two of his sons, Chung Mong Ku and Chung Mong Hun, cochairmen of the group, passed through the border in limousines behind 100 truckloads of 10 cows apiece on the morning of June 16, 1998. Another 501 cows were to go through Panmunjom to seal the deal. The 100 flatbed trucks on which the cows were riding were to remain in the North as part of the package. Younger brother Chung Se Yung, who had served for many years as chairman of Hyundai Motor, smiled when asked if the Hyundai trucks might be the precursors to shipments of Hyundai cars. SY, still honorary chairman of Hyundai

Motor, his son still HMC chairman, smiled again when asked whether Hyundai might build motor vehicle plants in the North. "So many people are wanting to go to North Korea," said SY, who had left for the South in 1945 to join his older brother. "We'll talk over and see what they want."[67]

Chung Mong Hun, group cochairman and chairman of Hyundai Engineering and Construction, facing hard times as contracts dwindled, envisioned all types of projects in a land left far behind in the rush for development. "We hope first we can develop Kumkang," he said. The Kumkang project and several others had fallen apart after Chung Ju Yung's 1989 trip partly because of disagreement over financing of a project to exploit timber in the Russian Far East using cheap North Korean labor. "We hope this trip will be different," said a Hyundai official as the Chung family rested briefly on the South Korean side of the line at Panmunjom. "Much has changed."[68]

From a personal viewpoint, the biggest difference between 1989 and 1998 was that the elder Chung was bringing the most influential members of his family with him. On the way to and from Asan, midway between Wonsan and Mount Kumkang, Chung and his family inspected the port facilities at Wonsan, which Chung had agreed in 1989 to turn into a world-class harbor. Chung Mong Ku had prepared for Hyundai's role by arranging for the purchase of 50 freight cars from North Korea by Hyundai Precision, which he served as chairman. The cars, built with Hyundai Precision technology, were shipped to the South from Wonsan.[69]

The spirit of cooperation encountered a severe test, however, just as the Chung family was about to return from their eight-day trip. A 70-ton North Korean Yugo-class midget submarine was snared in the nets of a South Korean fisherman on June 22, 1998, 11.5 miles east of the South Korean port of Sokcho, 20 miles south of the DMZ. The spot was about 30 miles north of the beach where the North Korean Shark-class submarine had run aground in September 1996.

Initially, a crew member of the South Korean boat, Han Ki Chul, told South Korean reporters, "I saw one or two people working on the submarine," North Koreans trying to free its propeller from the net, at 4:33 P.M. on June 22. "The submarine was afloat," he was sure. "It was not sinking." The submarine arrived under tow early on June 23 at a point about three miles east of the South Korean navy base of Kisamun, near Sokcho, but the water was too shallow to pull it the rest of the way until high tide. After the sailors aboard ignored demands broadcast over loudspeakers for their surrender, the South Koreans were afraid to board the vessel for fear it was rigged with explosives. About five or six North Korean sailors were believed to be still on board as a South Korean corvette towed the vessel toward Sokcho. The corvette diverted south to another South Korean navy base at Donghae after the submarine got bogged in weeds and mud outside Sokcho. On June 24, the submarine sank entirely, slipping 90 feet below the surface. Raised by four huge airbags, the sub was pulled by tugboats to the breakwater of Donghae harbor.[70]

The episode vastly complicated efforts at bringing representatives from North and South together as part of DJ's sunshine policy. South Korean officials planned to raise the topic at talks on June 23 between generals from the United Nations Command, including the United States and South Korea, and North Korea at Panmunjom. The talks would be the first on the general-officer level at Panmunjom in seven years. "We don't understand why the North Koreans are doing this when talks are being arranged," said a commentary on South Korea's Munwha Broadcasting."[71] Adding to the puzzle was that Chung Ju Yung was to return through Panmunjom with his brothers and sons at the same time as the talks between generals.

"This incident will have a serious impact on North-South policy," said Korea Broadcasting System, reminding viewers that talks between the United States, China, and the two Koreas were delayed for three months after the 1996 submarine incident.[72] North Korea several hours before the incident threatened to resume its nuclear program if the United States did not withdraw sanctions. American intransigence "will inevitably encourage us to take the road of our own choice," warned the North before the meeting of generals in Panmunjom.[73]

A demolition team broke through the top hatch of the sub on the night of June 25th but waited until daylight to go into the main pressurized chamber. Lee Kwang Suu, the only known survivor of the North Korean submarine that had gone aground in September 1996, warned of bombs possibly attached to the second hatch and advised the team to bore holes rather than pull the hatch open. South Korean navy divers expected to find "a few dead bodies." Divers, using welding torches, broke through one hatch and discovered an empty plastic cider bottle from South Korea along with flippers and other infiltration equipment, suggesting the vessel had picked up North Korean commandos from an incursion into the South.[74]

The bodies of nine North Korean sailors and agents, clad in thermal underwear and thick jackets, were discovered on June 26 after the underwater demolition team drilled into the main chamber. The chief of operations of South Korea's joint chiefs of staff, Lieutenant General Chung

Yong Jin, described "the most plausible scenario" of how the nine had died: "Four trained agents killed themselves after mowing down five crew members who had resisted an order to commit suicide to avoid capture." He drew the scenario from the grisly scene; the bodies of the four agents were found with bullet wounds in their heads, while the five sailors, apparently after a struggle, had been killed by shots to various parts of their bodies. Nearby were two AK-47 rifles used for the killing as well as two machine guns, two hand grenades, two Czech pistols, and an antitank rocket-propelled-grenade tube.[75]

The agents in charge of the submarine had evidently decided on the murder-suicide plan after South Korean sailors removed the submarine's antennae, closing communications with their commanders in North Korea. South Korean military officers said the agents had been picked up the night of June 21 from missions in the South. They found three sets of American-made infiltration gear and an empty pear juice container made by a South Korean manufacturer. "They were probably trying to contact spies here," KBS reported. "Their purpose is to instigate turbulence in South Korean society regardless of the South's policy toward the North."[76] Military officers speculated that the submarine had dropped off agents before picking up the ones who were killed but did not order a massive manhunt similar to the 53-day search after the 1996 episode.

The defense ministry demanded that North Korea admit "this act of aggression," explain why it did it, and again guarantee "never again."[77] Kim Dae Jung, however, stuck to his moderate position at a reception marking the forty-eighth anniversary of the date on which North Korean troops had attacked the South, beginning the Korean War. Acknowledging that "tensions are continuing," he said his government would hew to "a flexible North Korea policy."[78] North Korea was not expected to issue an apology. "North Korea stated this submarine was on an exercise," said Oh San Yul, research fellow at the Korea Institute for National Unification, an arm of the ministry of unification. "It is very difficult to prove their real intention, and it is more important that we get rid of these tensions between North and South."[79]

Hyundai, most of all, was undeterred. As a first step toward manufacturing in North Korea, Chung Mong Hun announced on July 2, 1998, that Hyundai and the North Koreans would form a joint venture to make such basic products as textiles, shoes, and toys in a vast new free export zone. He forecast exports climbing to $4.1 billion a year, about 10 percent of it paying the wages of thousands of North Korean laborers. The bold plan represented perhaps the most concrete sign so far that Kim Dae Jung intended to pursue his sunshine policy despite the capture of the midget submarine. "We just hope both North and South can attain international competitiveness as a result of the venture," said Chung. "It's time to cooperate in this difficult economic situation."[80] The same day, officers from the UN Command agreed with a North Korean officer at Panmunjom to return the bodies of the crew members and DJ endorsed Hyundai's efforts in the North, remarking, "The North is showing positive attitudes toward economic cooperation and tourism."[81]

Hyundai planned to lease two ferryboats in time to open tours to Mount Kumkang by late September 1998. Tourists would get to spend four or five days in the Kumkang region. "If the project is successful, we will open up several other areas," said Chung Ju Yung, citing Mount Paektu, the Korean peninsula's highest peak, on the Chinese border, as next choice. "From this tourism, North Korea will earn money and learn about economic matters," said Chung, linking tourism to the plan for building an export zone.[82] Authorities saw tourism to the North as a first step toward family visits. The government also would approve shipment of the remaining 501 cows after postponing it during talks over return of the bodies.

The North, however, seemed determined to frustrate the cause. Ten days later, on July 12, the government ordered a manhunt by land, sea, and air for possible infiltrators and the vessel that carried them after the body of a heavily armed man identified as a North Korean commando washed ashore near Donghae. Witnesses on the beachfront said they had seen "three people acting strangely" near where a shopkeeper found the body. The man, in his 30s, did not appear to have been wounded or injured. Blood around his mouth had not coagulated, and his body had not shown signs of rigor mortis, suggesting that he had died between 11 P.M. on July 11 and 2 A.M. on July 12. Air was still leaking from an oxygen tank found nearby. Korean television stations showed the body on the beach, in front of a stone embankment, clad in a black diving suit with fins and mask, carrying a Czech-made rifle as well as several knives. He also had chocolate and a powder substitute for natural foods. A submersible, 1.47 meters long and 3.31 meters wide, large enough to carry three to five infiltrators, was found 100 meters offshore. A South Korean minesweeper scoured the waters in search of the submarine that might have launched the submersible.

The discovery rattled government leaders, who believed they had dealt calmly with the midget submarine incident north of the same port. "This is proof that North Korea's violent strategy toward the South has not changed," said Munwha Broadcasting. "They want all the economic aid they can get, but they do not abandon their dual plans to attack us."[83]

Defense officials expressed alarm at the porous nature of defenses along the northeastern coastline, studded with military installations.[84] Three other commandos were believed to have gone ashore, disappearing into nearby mountains.

General John H. Tilelli, Jr., commander of U.S. forces in Korea, pledged on July 14 to increase military cooperation with the South as thousands of South Korean troops searched the slopes. He and the South Korean commander agreed on strengthening the "cooperation system for coastal submarine watch," according to the combined forces command. The statement said that "we responded immediately to North Korea's provocation," but the South Koreans wanted the United States to assist with planes and ships equipped for antisubmarine warfare.[85]

The wording of the agreement suggested lack of coordination between U.S. and South Korean forces. "In all future operations against limited provocation," it said, "both sides will consult closely regarding timely and appropriate support by U.S. forces." The subtext was consternation among South Koreans about the ease with which North Korea had infiltrated the area. Adding to the anxiety was the realization that the northeastern coastal region of the South lay across a mountain range from the nearest U.S. troops, concentrated on the "invasion route" leading to Seoul, about 140 miles to the west. The nearest American unit, a helicopter brigade, was headquartered at Camp Eagle, outside Wonju, between Seoul and the east coast.

Northeastern South Korea "is vulnerable," said David Steinberg, a scholar with long experience in Korea. "It's very easy to infiltrate. If you come in at the right tide, there's no detection installations visible on the beaches."[86] Once past the beach, commandos could hide out in near-impenetrable mountains while gathering material on South Korean naval and air bases as well as radar and missile facilities. Those same mountains formed the southern tier of a chain that extended to the Kumkang region.

Did North Korean strategists dream of attacking South Korea's northeastern region, much of which jutted north of the 38th parallel, which had divided the peninsula before the outbreak of the Korean War? Under the 1953 armistice, North Korea had ceded that region while the United States ceded areas south of the 38th parallel in the west, including the Koryo dynasty capital of Kaesong. The infiltration of commandos across the beaches into northeastern South Korea endangered not only that corner of the country but also DJ's sunshine policy.

Would DJ pursue this policy or evolve as a hard-liner in the mold of every other South Korean president? "The government is trying to promote economic exchange and trying to give the benefit of the doubt in dealing with North Korea," observed Han Sung Joo, who had served as foreign minister under Kim Young Sam. "These are reasonable and necessary policies, but to put all these things under the umbrella of the sunshine policy and advertise it that way was not such a good idea."[87] DJ needed to satisfy the conservatives, the vast majority of the electorate, and to guarantee the defense of the northeast. He got a chance to balance the pros and cons of reconciliation and toughness when his national security council convened on July 15, 1998, for its first meeting since his inauguration. A sign of the alarm caused by the infiltration was that the constitution stipulated that only "grave external threats or provocations" could justify such a meeting.

Under the circumstances, South Korea needed all the friends it could get. The expulsion of a South Korean diplomat from Moscow on July 6 shocked the government at a critical stage of relations between South Korea and Russia. "We are torpedoed by their Federal Security Service," said a foreign ministry official awaiting the arrival on July 7 of Cho Sung Woo, sent home after Russian security officials charged him with paying bribes for secrets. "We may need some cooling-off period." The cash-strapped Korean government had been attempting to get Russia to pay back $1.5 billion in overdue loans while also hoping the Russians would put off negotiating with Pyongyang for a new agreement in place of the old military assistance agreement that had been abrogated two years earlier.[88]

The real fear was that the episode might derail a prolonged effort to guarantee Russia's tacit support against North Korea, Russia's protégé in the Korean War and staunch ally until the demise of the old Soviet Union. The arrest might also indicate a desire in some quarters in Moscow to improve relations with North Korea. Pyongyang's Korea Central News Agency described Cho on July 6 as "a career agent of a special agency of the South," and the South Korean media reported that he was an agent of the National Security Planning Agency.[89]

South Korean officials denied that Cho's arrest in Moscow had anything to do with negotiations on the loans, most of which were extended around the time Moscow was opening diplomatic relations with Seoul in 1990. They expressed their dismay, however, over an unfortunate coincidence in timing. Russia's deputy prime minister, Oleg Sysuyev, had canceled a trip to Seoul to discuss the debt payments and other problems shortly before Cho was picked up on July 4 in the act of handing over cash to Valentin Moiseyev, deputy director of the First Asian Department. The Federal Security Service, successor to the KGB of the Soviet era, charged that Cho had been "systematically supplying" South Korean intelligence agencies with political and economic material.[90]

The arrest suggested that the Federal Security Service was engaged in a power struggle with the Russian foreign ministry and wanted to show it had the last word. Evidence for this view was that Cho was questioned for several hours before the South Korean embassy was notified. He claimed that he was merely handing over payment for a lecture that Moiseyev had given. Han Sung Joo as foreign minister had accompanied Kim Young Sam to Moscow in 1994 when Russia's President Boris Yeltsin said Russia would no longer provide arms for North Korea. "It was a big triumph," said Han, but Russia since then had been disappointed by the low level of South Korean investment in Russia as well as pressure for loan repayment. "The Russians have not been happy with the amount of attention they were getting from Seoul," said Han.[91]

In the battle between national intelligence agencies, South Korea on July 8 expelled a Russian embassy counselor who it said was an agent for Russia's Federal Security Service. The foreign ministry, echoing the Russian charges against Cho, charged that Oleg Abramkin had engaged in "activities which do not suit the status of a diplomat" since arriving nearly four years earlier. South Korean officials implied that the Russians had violated a tacit understanding under which a Korean intelligence agent could also operate from the Korean embassy in Moscow. "The foreign ministries of the two countries were waging a proxy war for intelligence agencies of the two countries," said Yonhap.[92]

The foreign ministry in Seoul worried that conservatives had gained the upper hand in Moscow, hoping to turn Russia away from South Korea and back toward North Korea. Foreign ministry officials noted that Moscow continued to supply spare parts to the North Korean armed forces. At the same time, Moscow had been negotiating to ship arms to South Korea in an effort to pay back some of the debt. "Why not sell arms to both?" Gennady Isaev, political counselor of the Russian embassy, asked me after Abramkin's expulsion. "The quality of Russian arms is one of the best in the world. We are free countries. There are no sanctions for us, so what's the big deal?"[93] In their anxiety to ensure good relations with Russia as a counterbalance to the North, South Korea rescinded the expulsion of the Russian diplomat. Park Chung Soo was dismissed as foreign minister—a casualty of an all-consuming effort to guarantee both North Korean isolation and compliance within the 1994 Geneva framework.

The framework was soon to face its greatest test. On August 31, 1998, in a show of strength designed to impress its starving people as much as the rest of the world, North Korea test-fired a long-range ballistic missile 1,380 kilometers into waters between Vladivostok and northern Japan.

Military authorities in Seoul, in Washington, and in Tokyo confirmed that the Taepodong 1 missile, capable of carrying a nuclear warhead 2,000 kilometers, knifed into the sea at 12:12 P.M. local time after having been fired from the village of Taepodong on the northeast coast. It was the first time North Korea had fired one of its missiles, all designed with technology supplied by Russian advisers in the 1970s and 1980s, in more than five years. North Korea had fired a much smaller Rodong 1 missile about 550 kilometers on May 29, 1993, and fired Scud missiles in 1984 and 1986. The North had designed the Taepodong, with twice the range of the Rodong, in the 1980s. Satellite imagery had long since shown such a missile under construction near Taepodong.

Russian and Japanese warships converged on where the missile landed, about 300 kilometers southeast of Vladivostok, 80 kilometers east of the main Japanese island of Honshu. The Russian news agency Tass said that a U.S. aircraft, "an American Orion spy plane," had been able "to spot the rocket's launch and fall."[94] The Pentagon said it viewed the missile-firing as "a serious development" but did not say how it learned about it. The firing sent shock waves through Seoul, pursuing reconciliation with North Korea, and jeopardized the $5 billion project by the United States, South Korea, Japan, and the European Community for building twin light-water nuclear reactors in the North.

The missile firing also came as a shock to Israel, hit by North Korean missiles fired by Iraq during the Gulf War and alarmed by the continued export of North Korean Scud and Rodong missiles to Middle Eastern countries. Israeli Defense Minister Yitzhak Mordechai, in Beijing, cited "two dangers"—that North Korea had "this capability" and had ties "with countries like Iran and Syria" as well as Iraq. He asked China's President Jiang Zemin "that China refrain as much as possible from giving technological assistance to countries that are still hostile to us."[95] North Korea had exported 250 Scud and Scud-based Rodong missiles to Pakistan and Middle Eastern countries, including Iran, Iraq, Syria, Libya, and the United Arab Emirates, between 1992 and 1997, earning about $580 million. More recently, the North had been exporting "components and technology," refinements on technology originally provided by the Russians during the Soviet era. The North was likely to be ready to export the liquid-fueled Taepodong 1 by 2000 at a cost of $6 million each and was working on the Taepodong 2, with twice the range.[96]

Seoul, however, minimized the implications, interpreting the firing as a politically inspired device to build up Kim Jong Il. The North fired the missile as delegates to the Supreme People's Assembly were gathering in

Pyongyang for its first session in eight years. The 700-member Assembly was convening for the purpose, it was widely believed, of electing Kim as president on September 5, 1998. "One interpretation is they are preparing for the official inauguration of Kim and fired the missile as a symbol of their achievement," said a senior aide to Kim Dae Jung.[97] Japanese newspapers had reported two weeks earlier that North Korea was planning to test-fire a missile before Kim's election.[98] "It's a very good method to assure the people of North Korea that it is a very powerful country even though they have severe economic problems and do not have enough to eat," said Chung Young Tae at the Korea Institute for National Unification."[99]

South Korean officials also believed the firing was intended to strengthen North Korea's position in talks between U.S. and North Korean diplomats in New York. Charles Kartman, leader of the U.S. team, had planned on the day of the incident to ask the North to permit U.S. inspectors to look at an underground facility where the North was suspected of planning to build nuclear warheads in violation of the 1994 Geneva agreement. U.S. satellite photographs showed hundreds of workers building the facility. "It's part of the pre-negotiation process," said Lee Ho, a director at the ministry of unification, responsible for dealings with the North.[100] "Firing the missile is a negotiation card for talks," said Huh Moon Young, leader of the national policy team at the Korea Institute for National Unification. "That means they can develop their capability to strike Japan, and someday they can strike the United States."[101] In any case, the North was angry about delays in shipment of the heavy oil the United States was supposed to send under the agreed framework.

By extraordinary coincidence, the Korean Peninsula Energy Development Organization, KEDO, had issued a draft resolution on August 31, the same day the missile was fired, reconfirming the commitment of all the governments. Shortly after the news of the missile launch was confirmed, however, both Japan and South Korea were reconsidering their roles. Outraged Japanese officials said they had "suspended signing" the resolution. Japan had only recently restarted talks on normalizing relations with North Korea, whose missile capability was viewed as an immediate threat to Japanese security. Japanese Prime Minister Keizo Obuchi said he was considering "measures" but did not elaborate. South Korea warned the resolution was in danger but was distinctly less upset. DJ, practicing forbearance, preferred to await an explanation from the North.

The test-firing might have reflected differences between North Korean military and civilian officials. "The military has not necessarily bought into peace talks," said a western diplomat. "Perhaps Kim Jong Il is saying okay to both sides, to the diplomats to talk, to the military to test missiles."[102] In its first public hint of its role in the incident, North Korea berated Japan on September 2 "for making a fuss." It was, Pyongyang's Korea Central News Agency advised, "imprudent for Japan to say this or that, unaware of what the Democratic People's Republic of Korea did, a missile test or anything else." The commentary observed that "many countries around Japan possess or have deployed missiles" but did not single out the United States or South Korea. The same day, September 2, U.S. and North Korean negotiators kept their date in New York after the North Korean team had failed to show up the day before.[103] South Korea's response remained low-key. The government on September 2 permitted a freighter laden with a donation of 6,000 tons of corn from Chung Ju Yung to go to North Korea's west coast port of Nampo.[104]

On September 4, North Korea came up with an explanation. It was a satellite, not a missile, that had flown over Japan, said Radio Pyongyang, and it was orbiting the globe broadcasting revolutionary hymns of praise for Great Leader Kim Jong Il and (the late) Great Leader Kim Il Sung. "Our scientists and technicians succeeded in launching the first artificial satellite on a multistage rocket and getting it successfully into the correct orbit," said the broadcast at 5 P.M. local time. The satellite, it said, took off at an angle of 80 degrees and went into orbit 4 minutes and 53 seconds later, at 11 minutes and 53 seconds after noon.[105] The broadcast suggested listeners tune in to "the immortal revolutionary hymns 'Song of General Kim Il Sung' and 'Song of Kim Jong Il,'" wafting over the airwaves at 27 megahertz. The satellite was also broadcasting in Morse code the term *juche*, the slogan in the North for self-reliance, it said.

The broadcast began as a joyous reminder of the founding of the Democratic People's Republic on September 5, 1948. "All North Korean people are celebrating the 50th anniversary of the state," said the announcer. The broadcast did not mention what North Korea watchers assumed would happen—the formal naming of Kim Jong Il by the 687-member Supreme People's Assembly as head of state. It did, however, provide an advance buildup by remarking that "the great success" of the launch "makes our socialist state great under the leadership of the Great Leader Kim Jong Il." Besides emitting revolutionary music, said the broadcast, the satellite had a practical purpose "for the peaceful use of the cosmos and for scientific research."[106]

No one ever picked up the broadcast. The best guess was that North Korea had launched a three-stage missile bearing a satellite but the satellite had failed to launch properly from the missile. "It's conceivable they could put a satellite up there, but I have serious reservations," said Kenneth Quinones, former North Korean expert at the U.S. State Department. "If there is a North Korean satellite up there, we would have heard about it by now." Quinones, who had spent several months in the North before and after the United States negotiated the 1994 Geneva agreement, noted that Japan had satellite-tracking facilities on Hokkaido. The Japanese had originally built them, he said, to monitor the launch of Russian satellites under Soviet rule.[107]

One of North Korea's basic purposes, whatever it fired, was to bargain for hundreds of millions of dollars from the United States. Members of an American congressional delegation were told during a visit to Pyongyang in August that the North would stop exporting missiles in exchange for $500 million a year. Quinones said the North had been hinting at a deal for months. "The fact that we are seeing dollar figures cited means the North Koreans are defining a negotiating position," said Quinones. "They're trying to drive the hardest bargain. Their aim is not to start a war but to maximize gains for minimal concessions." One problem, he said, was that the market for North Korean Scud and Rodong missiles in the Middle East had dried up since the end of the Gulf War in early 1992. "The missile market got saturated," he said. "That's why they're willing to negotiate or trade now."[108]

Possibly the worst fallout from the missile firing was that it might have triggered enough insecurity in Japan to bring about a significant escalation in Japan's military posture. The Japanese, after years of debate, were considering two multibillion-dollar options—a variation on "theater missile defense," TMD, propounded in Washington, and an "intelligence satellite" by which they could analyze photographs and data without having to rely on the Americans. The Japanese might cooperate in setting up sites for weapons capable of shooting down missiles such as the Taepodong 1 but did not want to be accused of expanding militarily in violation of their own "peace constitution," drafted during the American occupation under General Douglas McArthur. The constitution limited Japanese "Self-Defense Forces" to Japanese soil. Hence the Japanese preferred to talk about "ballistic missile defense," BMD, not TMD.[109]

The problem was that TMD, an outgrowth of the Strategic Defense Initiative proposed by President Reagan in 1983, had been in the talking stage between Washington and Tokyo for five years and would not become operational for five more years at a cost of $20 billion. Meanwhile, in response to a groundswell of demand for a vastly improved advance warning system, the Japanese might go ahead with an intelligence satellite. Underlying Japan's decision to plunge into a project that could cost $2 billion was chagrin over the ease with which North Korea had violated Japanese territory—or at least its airspace. North Korea's claim that it had launched a satellite, not a missile, hardly dimmed the fury, especially since Japan and the United States had spotted no trace of it. Most disappointing: the failure of the United States to provide advance high-tech intelligence. "The United States wants to monopolize satellite information so it will be easy to manipulate Japan," said Toshimitsu Shigemura, Northeast Asian expert at *Mainichi Shimbun,* a Japanese daily.[110] American officials responded that the United States shared intelligence in accordance with the U.S.-Japan security treaty, under which 47,000 U.S. troops were based in Japan, 60 percent on Okinawa, backing up 37,000 U.S. troops in South Korea.

Japan could escalate sharply first by counting the intelligence satellite as a nonmilitary expense and then by deciding that the threat of attack by medium-range missiles from North Korea justified a special budget for missile defense. The result could be a shift in the regional military balance that would raise tensions not only between Japan and North Korea but also between Japan and China, much of which the Japanese had overrun in the 1930s after having taken over the Korean peninsula in 1910. The impact could be to jeopardize the equilibrium that had contributed to peace since the Korean War. If the Japanese worked with the United States to build a missile defense system, Chinese and Koreans, both North and South, might ask, "How do you separate defense from an offensive system?" China would worry about Taiwan getting its own missile defense system while South Korea feared a renaissance of Japanese militarism even though both Japan and South Korea were allies of the United States.

The exigencies of North Korean politics deepened the sense of armed confrontation. Kim Jong Il was "head of the party, army, and state," said *Rodong Sinmun,* the newspaper of the Workers' Party—the first time that a North Korean organ had described him in those terms.[111] Under the constitution promulgated by the Supreme People's Assembly on September 5, 1998, 50 years after the installation of the first North Korean government, the National Defense Commission, of which Kim was chairman, would oversee social and political as well as military affairs. More than ever, the military under Kim Jong Il, already general secretary of the party and commander in chief of the armed forces, was in the ascendancy. Kim did not, however, become president. That title, said the new constitution, belonged to Kim Il Sung as "eternal president."

# Chapter 11

# Of Missiles and Nukes

S outh Korea again looked for sunshine in the clouds. The day after the North Korean constitution was published, on September 6, 1998, the government authorized Hyundai to begin carrying boatloads of South Korean tourists on five-day trips to view Mount Kumkang. Hyundai's total investment in the North would come to a minimum of $1 billion. Unification Minister Kang In Duk said that he perceived "some changes in the economic sector" in the new constitution and predicted the North would break the communist mold and "employ some basic market system."[1] One sign was the farmers' markets that had opened in the North as basic distribution broke down. With the appointment of a cabinet of technocrats, might the North convert to a basic market system?

Far from retaliating with sanctions for the missile and submarine incidents, South Korea was undeterred in its sunshine policy. Kang In Duk reaffirmed the South's commitment to 70 percent of the cost of building the nuclear power plant in the North and urged Japan to pledge its commitment too. Kang's remarks deepened the sense of contrast, if not conflict, between Seoul and Tokyo over the firing of the missile. Could it be that South Korea did not mind if North Korea upset Japan, the historic foe of all Korea? While the Japanese imposed sanctions as punishment, South Korea was more anxious than ever to pursue economic ties as the best antidote to North Korean threats. That way the North might come to terms with its own economic crisis—and open up to full-scale trade and investment.

The economic desperation of the North was clearly a major factor in official thinking. While North Korea in 1997 earned a surplus of $78 million from two-way trade of $308 million, the South in the first seven months of 1998 had a surplus of $20 million on trade of $100 million. Most of the trade consisted of textiles shipped to the North, where small factories made them into finished items for shipment to the South. The need to make up for loss in revenue from trade might drive the North to open up to manufacturing ventures.

How logical, however, was the North? "There has been no reform at all," said Cho Dong Ho, research fellow at the Korea Development Institute. "Free markets are prevailing in rural areas, but the volume of trade in those markets is limited. They barter for rice and daily necessities because the official distribution system has collapsed." The new constitution provided for the northeastern free trade zone as well as an independent accounting system for state enterprises, but the zone had existed since 1991 and independent accounting had been in effect for 15 years. Even so, business throughout the country, including the zone, was declining. The KDI estimated that industry outside the zone was going at 30 percent of capacity while the average household relied on barter trade for 70 percent of consumption.[2]

Removal of signs saying "Free Trade Zone" at the gates to the northeastern zone signaled a reluctance to accept significant investment. North Korean officials viewed the concept of such a zone as "too far from their society," said a European businessman who had visited the 746-square-kilometer zone by the Chinese and Russian borders. While declaring their openness to foreign investment, he said, "they like to bite the hand that feeds them." Approximately "95 percent of the projects they talk about are not happening." Kim Jong Il was said to have ordered removal of the signs, facing the North Korean countryside beyond the zone, so citizens would never see them—and get strange ideas about capitalism.[3]

Making matters worse, the economic crisis that now afflicted the rest of Asia from Thailand to Japan and Korea had forced investors either to slow down projects or to cancel them entirely. Since the zone opened in 1991, hundreds of firms had signed agreements to set up factories or offices, but only about six of them were doing real business there by the fall of 1998. Actual investment totaled only about $65 million, according to the United Nations Development Program in Beijing, even though North Korea had said that 111 foreign contractors had promised to pour in about $750 million. "The zone is not ready for investment so far," said Cho Eun Ho, monitoring North Korean economic problems for the Korea Trade-Investment Promotion Agency in Seoul. "Companies will not invest according to their contracts because the situation is not clear." Cho questioned the United Nations' figures for investment, maintaining that UN officials "want the zone to succeed." By his calculation, investment in the zone since 1991 had reached about $40 million.[4]

South Korean companies, despite their own economic difficulties, were interested in exploring the zone's possibilities, but the North refused admittance to most of them. One reason was the North Koreans did not want South Koreans to get a firsthand view of an infrastructure that lagged far behind plan. Failure to develop roads and water, sewage, and other facilities in the zone helped explain why so few companies were there. The most successful was Roxley Pacific, set up by two Thai companies, Roxley Public and Charoong Thai, to operate telecommunications in the zone and establish service elsewhere in North Korea. Together they ran a joint venture called Northeast Asia Telephone and Telecom, which had installed 5,000 telephone lines inside the zone. Callers could dial long distance at about $5 or $6 a minute, the highest prices in Asia. Roxley had scaled down or canceled other projects in the zone while waiting for the North Koreans to pave the roads and persuade more investors to come in.[5]

Another organization firmly committed to the zone was the Emperor group of Hong Kong, which had invested $20 million in a hotel and casino near Rajin. That was the first stage of a complex that might someday cost $180 million. One problem, said Iris Ying, senior vice president for the project, was that North Korean bureaucrats were not receptive to suggestions, much less complaints. "Most of the time they won't listen to you," she said. She did, however, get a positive response to one complaint. "I asked them why the toll on the 60-kilometer stretch of unpaved road from the Chinese border to Rajin was so high," she said. "The road is terrible. They said, 'All countries charge tolls,' but next time I made the trip the toll was much lower."[6]

The problems created by North Korea were compounded by America's own domestic crisis. The U.S. ambassador to Korea, Stephen Bosworth, left for Washington on September 17, 1998, on an exercise in reverse diplomacy, to try to convince recalcitrant members of Congress to appropriate funds for shipping heavy oil to North Korea. The question was whether Washington, in the midst of the furor over President Clinton's affair with Monica Lewinsky, could deal effectively not only with a heightened North Korean threat but also with the economic crisis that had devastated the region. "Koreans have been asking me that question," said Robert Gallucci, who had led the American team in negotiations on the Geneva agreement. "This issue has never been free of politics. We've always had a hard time on the Hill"—that is, in persuading Congress to appropriate funds for heavy oil for North Korea.[7] "I would hope people in

Congress would say, 'This is a bipartisan issue,'" said Donald Gregg, former ambassador to Korea.[8]

South Korean officials responded with alarm on September 18 to a vote in the House of Representatives for a foreign aid bill that did not include funding for both the IMF and the heavy oil. From the immediate viewpoint, officials appeared most concerned about the implications of failure of the Congress to approve $50 million for the oil. If the United States did not find the funds, North Korea could delay the canning of heavy spent fuel from the facility at Yongbyon, where it had been developing nuclear weapons before the signing of the Geneva agreement. South Koreans said there was little chance that construction could begin soon on building a light-water nuclear reactor in the North.

When the two Koreas, China, and the United States resumed talks in Geneva on October 21, 1998, for the first time in seven months, one thing was certain: the best one could hope for was that this round, like the two before it, would conclude in agreement to meet again. The talks would be the first since Kim Jong Il was named, in effect, head of state. "From his policy and behavior so far, we cannot be optimistic," said Park Kun Woo, a former South Korean ambassador to Washington, who would chair this round.[9] In fact, it would be a minor triumph of diplomacy if the delegations, besides deciding to get together again, agreed on the time of their next meeting before going home after four more days of getting nowhere. "The last two four-party talks did not set a date," said Kim Gook Jin, professor at the Institute of Foreign Affairs and National Security, affiliated with the South Korean foreign ministry. "Maybe they can talk about when to proceed with the talks." Beyond that, he said, "You cannot expect progress."[10]

The question was how long the four-party talks, first held in Geneva in December 1997 after two and a half years of fitful talks about talks, could go on before the process collapsed altogether. "The last round in March broke down over whether the withdrawal of U.S. forces from South Korea should be on the agenda," said John Barry Kotch, political historian at Hanyang University. "A repeat performance would cripple the four-party process at a time when new uncertainties have sprung up."[11]

Once suspicious that the United States was trying to sell them out by negotiating with the North, South Koreans now viewed talking as preferable to not talking even if the talking got nowhere. "I don't think the talks will break down," said Lee Soh Hang, at the Institute of Foreign Affairs and National Security. "They will meet again and talk and talk. It will be just a continuation of the crisis." As long as the talks dragged on, "We will not have any new problem" that might upset the equilibrium on the peninsula.[12]

Bosworth returned from Washington with a sense of triumph after Congress approved $35 million for the oil—with conditions. Clinton had to certify to Congress that North Korea was not using an underground facility discovered from satellite photographs to build nuclear warheads. Clinton's certification also had to assure Congress of progress in talks with the North. "It was essential," said Bosworth, that the United States, Korea, and Japan "continue to meet our commitments to supply North Korea the heavy fuel oil and light-water reactors which are the quid pro quo of the nuclear freeze." The ambassador did not mention the congressional conditions, except to say the United States had had "one round of talks with North Korea" on the missile issue "and another round of talks is planned."[13]

Kim Dae Jung's sunshine policy and the North's need for money ensured the success of one project that now seemed separate from diplomatic or military confrontation. The venture into North-South tourism by the Hyundai group, with the full backing of the governments of North and South Korea, came to symbolize the opening of the country to a range of opportunities. Hyundai now planned to open the service in November, packing a maximum of 1,250 tourists for each cruise aboard a chartered vessel from the east coast port of Donghae on a 12-hour journey of 105 miles. The tourists would never meet or even see any North Koreans other than their tour guides and security guards. Just to make sure, they had to spend all five nights of the trip aboard the boat. Still, South Korean officials as well as private business people saw the trip as leading to much more in the long run. For starters, Hyundai was building the pier where the ship would dock in the North, along with an auditorium, and envisioned opening a hotel at the base of Mount Kumkang in a few years.

The 28,000-ton *Hyundai Kumkang* left November 14, 1998, with 442 passengers and 415 crew members. Among the passengers was Chung Se Yung, founder Chung Ju Yung's younger brother. Chung Se Yung said before boarding that he hoped the trip would "lay the foundation of reunification." North Korea's Korean Asia-Pacific Peace Committee credited "the idea of the Workers' Party of Korea of the great unity of the entire nation and compatriotic measures of the government" for making the tours possible. The ship arrived after 9 hours and 50 minutes at 4 A.M. on November 15, but it was not until 8 A.M. that a North Korean pilot boat led it into port. "The ship had to wait for four hours until the North Korean pilot boat was ready," South Korea's Munwha Broadcasting reported. "Only at 11:10 A.M. could the passengers step on North Korean soil."[14]

Since there was no direct communication between the two Koreas, news of the arrival of the first cruise ship from South to North Korea was relayed by a North Korean travel agency in Hong Kong called Korpen, for "Korean peninsula." All four South Korean TV networks carried the voice of a woman who worked for the agency, talking in Korean, assuring listeners, "Everything is fine, the ship has arrived." In case there was any doubt, she added, "The weather is fine, there are no waves." Once cleared through immigration, the passengers boarded buses for a 10-mile ride and a 2-mile hike, then went on a 14-mile bus ride and a walk of more than a mile.[15]

The tourists, mostly Hyundai executives and managers testing the waters before regular tourist service began on November 18, returned to Donghae just as a top-level U.S. team was arriving in Pyongyang on November 16 to press for inspection of the suspected underground nuclear site. U.S. negotiator Charles Kartman assured South Korean officials in a stopover on the 14th that Washington was prepared to get tough. Kartman, returning through Seoul on November 19, said that he had flatly rejected a North Korean demand for a sizable fee for the right to inspect the facility.[16]

For the first time, Kartman pinpointed the site, in Kumchangri, about 25 miles northwest of North Korea's nuclear facilities at Yongbyon. He said he had warned North Koreans that refusal to open the site for inspection might jeopardize the Geneva agreement. "We asked the North Korean side to remove our suspicions," said Kartman. "My presentation contained a very clear element about the danger the failure to resolve those suspicions could pose to the viability" of the agreement. He said it was "imperative these suspicions be resolved" but admitted "there is still a rather wide gap between our positions" and "we are still not satisfied." Kartman refused to reveal how much money the North Koreans had asked for inspection of the site, saying rejection of the demand made the issue "sort of irrelevant." Clinton administration officials said the North had asked for $300 million—demanded, said Kartman, as "compensation for the insult" of Washington's suggestion that it was violating the Geneva agreement.[17]

Kartman's briefings to South Korean officials were intended to allay fears that the United States might be reluctant to compel the North to comply with "the agreed framework" for discarding its nuclear program. He indicated the frustration of the mission in his sardonic response to a question from the Reuters bureau chief about whether the United States contemplated a "surgical strike" against Kumchangri. "If I have to go back to Pyongyang, I may need surgery," he replied.[18]

Kartman remained in Seoul until the weekend in order to brief skeptical South Korean officials as well as aides of Clinton, who arrived on

November 20 from Tokyo for a summit on November 21, 1998, with Kim Dae Jung. The issue of close cooperation between the United States and South Korea would dominate the summit amid mounting concern about how to get the North to open Kumchangri for inspection. DJ worried the United States might provoke a crisis similar to the nuclear crisis of the summer of 1994. He planned to suggest Washington climb down from its tough stance and adopt a more conciliatory policy. "There is no evidence," he said, that North Korea was violating the Geneva agreement.[19]

Taking their cue from DJ, South Korean officials disagreed with some of Kartman's remarks. "The United States tells the South they found 'compelling evidence' of nuclear activities," said Kim Gook Jin, at the Institute of Foreign Affairs and National Security, "but the Korean side says the evidence is not yet conclusive." He said that conservative opinion in the U.S. Congress had helped to bring about "a change of opinion in Washington regarding the nuclear issue."[20] Hong Yong Pyo, a research fellow with the Korea Institute for National Unification, said there was "a hard-line mood in the United States Congress." Kim "will try to persuade President Clinton of the engagement policy," he said, while Clinton "wants to urge more strong observation of the nuclear site."[21]

There was no doubt the United States and South Korea were at odds over Kumchangri. Kartman, on November 21, as he was leaving Seoul just before the summit, issued an extraordinary "clarification" to allay Kim's concerns. He quoted himself as having said earlier that there was "compelling evidence" that the site was "intended to be used for nuclear-related activities." In his clarification, however, Kartman stated, as DJ had pointedly done the day before, that "we lack conclusive evidence that the intended purpose" of the site was nuclear-related. If the site was for nuclear activities, he added, analysts did not know "what type of nuclear facility it might be." Thus, he concluded, "full access to this site" was required.[22]

Clinton and Kim appeared almost entirely in agreement as they stood side by side in the Blue House several hours later discussing policy vis-à-vis the North. Clinton said the United States had made clear that the North "must satisfy our concerns and that further provocation will threaten the progress we have made." He was enthusiastic about the tours that Hyundai was arranging to Kumkang. In the local media "the picture was the tourist ship" bound for North Korea, not the underground site, said Clinton. "To us, this was amazing"—and a clear sign of North Korea's own best interests.[23]

In a visit to Osan Air Base the next day, however, Clinton pledged a tough policy in a rousing finale to a weekend dedicated in part to bucking up a government that seemed inclined to go easy against the North. Addressing several thousand U.S. troops at the base 25 miles south of Seoul, he declared that North Korea had to stop developing nuclear weapons as well as a wide range of non-nuclear weapons. Clinton's remarks, given the setting and the timing, appeared as one of the strongest administration challenges so far to Pyongyang. "North Korea must maintain its freeze on and move ahead to dismantle its nuclear weapons program," said Clinton, standing in front of two U.S. Air Force fighter planes. The danger signs, ranging from the submarine incursions to the missile firing to the suspect site, "have intensified," he told the troops. "So we must remain vigilant, and thanks to you, we are."[24]

The morale boost provided by Clinton's visit was forgotten in an unseemly debate between U.S. and South Korean forces over who should command a battalion of American and South Korean troops at the critical Panmunjom crossing. The debate was triggered by the death in February 1998 of a South Korean army lieutenant that American and Korean investigators had listed as a suicide. The United States and South Korea agreed on December 13 to form a joint team to investigate the killing, while the U.S. commander in Korea, General John H. Tilelli, Jr., rebuffed suggestions that South Korea completely take over the battalion of 150 troops. Tilelli, as commander in chief of the UN Command, which included South Korean as well as American troops, said that he was "committed to the combined structure of the joint security battalion" at Panmunjom.[25]

The decisive nature of the statement reflected U.S. concern about a rift between American and South Korean forces stemming from the death. The South Korean army reopened the investigation in response to claims that a South Korean soldier might have killed the lieutenant to cover up illegal contacts with North Korean troops. The South Koreans, investigating the case as a possible murder, arrested one of their soldiers after learning that South Korean soldiers had had dozens of contacts with North Korean soldiers a few feet away. The South Koreans admitted the North Koreans had plied them with gifts, including watches, cigarettes, and whiskey, apparently to extract intelligence information.

Tilelli defended the role of U.S. and South Korean forces on Asia's most sensitive military fault line. The joint battalion, commanded by an American lieutenant colonel, covered the southern side of a 124-acre Joint Security Area, set up under the Korean War armistice. The battalion represented the "solidarity and commitment" of both countries to the armistice, said Tilelli. "If in fact there have been acts of indiscipline," said the command, Tilelli was confident "good leadership" would overcome them.[26]

As the investigation unfolded, however, Kenneth Quinones at the Asia Foundation's Seoul office revealed that U.S. and North Korean soldiers had once partied regularly in what the Americans called "the party room" straddling the line between North and South Korea in Panmunjom. "The United Nations Command was hosting weekly beer and pizza parties for some years," said Quinones. "They have a game room where both sides got together. At first it was just U.S. and North Korean soldiers smiling through their teeth at one another to open up informal channels."

Quinones said that a North Korean colonel had escorted him in July 1996 to the building at Panmunjom where American and North Korean soldiers held their weekly parties. "The American side knew we were there," said Quinones, then in the North to negotiate repatriation of remains of American soldiers killed in the Korean War. "We had to call across to the American side so they could let us in." Inside the building next to the one where U.S. and North Korean military officers held periodic talks, Quinones said he saw "color TV, air-conditioning, nice furniture, and an unlimited supply of beer and liquor" supplied by the Americans. "I heard from the North Koreans about how much fun it was," he said, and "confirmed all this by talking to an American officer" at Panmunjom.[27]

The get-togethers, oiled by beer and liquor from the U.S. military post exchange system, were authorized by the UN Command. The Americans believed they were obtaining information from the North Koreans, but "you could never separate the wheat from the chaff" and they got "substantial misinformation." Quinones was "sure" that all the North Koreans said at the party nights was "well choreographed" by their superior officers. The gatherings in the party room were "bizarre at a time when we have so much tension," he said, and may have inspired the illegal contacts between South and North Korean soldiers. Such contacts "are not sanctioned under the [South Korean] national security law," said Quinones, but he believed that South Korean authorities knew their own soldiers were meeting with the North Koreans. "It's impossible for a South Korean to do anything up there without their knowledge. Everything is covered by videotapes and recording devices." After the contacts were revealed, he understood, "the party room was shut down."[28]

The Americans were embarrassed by my story on Quinones's remarks. Air Force Major General Michael V. Hayden, deputy chief of staff for the UN Command, responsible for negotiations with North Korean military officers, escorted me through the buildings where the contacts were made, introducing me to subordinate officers on the scene. One of them, Lieutenant Colonel Stephen Tharp, said, "Some of the North Korean officers are big-time alcoholics and some of them are teetotalers." Usually they drank beer and whisky offered by the Americans, but "when they want to drink hard they bring snake wine"—wine in a bottle with a dead snake for flavoring.[29] "We try to keep this a very social relationship," said Navy Commander Lennart Wendel, who manned a desk in a building just behind the North-South line.[30] Wendel arranged routine meetings with North Koreans through a hand-cranked telephone supplied by the North Koreans, who apparently wanted their own phones at both ends of the line so they would not have to ask the Americans to go to their side for repairs. If the phone broke down at either end, officers shouted over loudspeakers.

Hayden talked about the contacts in the building where informal talks were held. Wine and liquor glasses lined the shelves of a cabinet. A refrigerator was tucked in a corner. "I haven't checked out what's in there," said Hayden, ordering Wendel to open it. The fridge was stuffed with six-packs of American beer. "In meetings we exchange formal statements," said Tharp. "We come in here, we sound each other out for ideas."[31] Hayden was emphatic about the need to keep up such contacts. "These are not social get-togethers," he said. "They were tightly tied to the purposes of Panmunjom. This is a very narrowly focused venue for clear objectives. Despite 45 years of tension, the fact that we still have the dialogue channel is something we've worked hard at. It achieves what we set out to achieve."[32]

Formal talks between generals on both sides were canceled by the North in 1991 when the UN Command assigned a South Korean general instead of an American one to represent the command. They did not resume until June 1998, when North Korea agreed to send a general to negotiate the return of the bodies of the nine North Koreans who died in the midget submarine incident of June 22–26. The difference between those and the earlier talks was that they were called "GO" for "general officer" rather than "MAC" for Military Armistice Commission negotiations.[33]

Hayden led the negotiations in June and July 1998, sitting opposite North Korean Lieutenant General Li Chan Bok, who had served in Panmunjom since the 1960s. General Li's recent "election" to the Supreme People's Assembly underlined the significance of the talks. Hayden faced Li three times, twice to arrange for the return of the bodies and again to talk about the discovery of the body of the commando found on the beach three weeks later. Hayden said that Li later requested an "informal meeting without press, without a formal agenda." What for? "We talked about the future of the talks."[34]

Hayden saw growing potential in a channel that at times had seemed completely closed. "My role in all of this is to pass to the KPA [Korean People's Army] that we want to meet with them." The talks were another sign of a broadening dialogue in which South and North Korean diplomats had met in four-party talks with Americans and Chinese at Geneva and American and North Korean diplomats had met in New York and Washington.[35] Hayden had been a member of the U.S. delegation to the four-party talks in October 1998 and was going to Geneva again for the next round in January 1999. His experience at Panmunjom might be especially useful, since a North Korean would be chairing the talks for the first time. At Geneva, "They're talking about replacing the armistice with a peace treaty," said political historian John Barry Kotch. "At Panmunjom, they talk about maintaining the armistice. Both are vital to keeping the peace."[36]

Amid negotiations, however, the South could never let its guard down. On December 18, 1998, the first anniversary of DJ's election, South Korean aircraft and naval vessels pursued a North Korean "spy boat" for 100 kilometers off the southern coast before sinking it in a furious gun battle. The body of one frogman with a grenade in his wetsuit was discovered floating on the surface, but it took another month before a special naval mine vehicle located the sunken boat beneath 150 meters of water and divers found the bodies of several more infiltrators, along with their weapons.

Was a "second Korean War" a possibility? The American defense secretary, William Cohen, visiting Seoul on January 15, 1999, warned of an end to dialogue with North Korea "if it undermines" the agreement reached at Geneva more than four years earlier to stop attempting to produce nuclear warheads. Cohen, on the eve of talks between U.S. and North Korean negotiators in Geneva, stated bluntly that the North "will dash any hope of realizing the benefits of dialogue" if it failed to back down from its hard-line opposition to opening up an underground site to inspection.[37] Cohen's remarks appeared as a challenge to North Korean negotiators not only in the bilateral talks but in the four-party talks opening on January 18.

The tough U.S. position limned the differences that still threatened peace on the Korean peninsula as the North bargained for aid to rescue its people. Cohen and the South Korean defense minister, Chun Yong Taek, called on the North to clarify "the nature of this construction" at the underground site and warned against efforts at building any type of weapon of mass destruction. A joint communiqué said that Cohen had not only "reaffirmed the U.S. commitment to render prompt and effective assistance" to South Korea in case of war but also had promised to "provide a nuclear umbrella" covering the South. Cohen did not allude publicly to defending the South with nuclear weapons but promised, "We will maintain a strong deterrent as we pursue dialogue with North Korea."[38]

The secretary's remarks drew a prompt rejoinder from China, North Korea's main source of diplomatic and economic support. The official New China News Agency, Xinhua, blasted U.S. demands for inspection of the site at Kumchangri as "unjustified and ridiculous." The commentary also appeared sympathetic to North Korea's efforts at producing missiles similar to the one fired on August 31, 1998, across northern Japan, saying that "the United States seems to be the only country in the world allowed to launch rockets and missiles."[39]

The North clearly hoped to use the issue as a lever for persuading the United States to lift economic sanctions that had barred trade and investment in the North since the Korean War. "North Korea would allow inspection of Kumchangri as the price of relieving economic sanctions," said Park Young Kyu, a senior fellow at the Korea Institute for National Unification.[40] He doubted, however, if the North would stop producing missiles, one of its main export items. Han Sung Joo, South Korea's foreign minister at the time of the signing of the Geneva agreement, believed both North Korea and the United States "see that it is in their interests to keep the Geneva framework." In the bargaining game, said Han, "the North Koreans don't see they have reached the brink yet" and would move "only when they have to, depending on the price of the payoff."[41]

The fear in Seoul was that the mood in Washington was hardening. William Perry, former U.S. defense secretary, coordinating U.S. policy toward North Korea, flew into a storm of controversy over Kim Dae Jung's sunshine policy when he arrived on March 8, 1999, on the second leg of a mission to determine how tough a line to adopt toward the North. His arrival coincided with publication of a book that he had coauthored with his top aide, Ashton Carter, that appeared to dispute Korea's basic policy. Perry and Carter concluded in an epilogue that "it might not be safe to follow Kim Dae Jung's advice and wait for change to come to North Korea." They argued that North Korea's program "would rob us of the time needed for Kim Dae Jung's engagement policy to work."[42]

South Korean officials made no secret of their concern that Perry would derail a soft-line policy they had carefully nurtured, encouraging commercial contacts with the North as well as the first tourist visits there since before the Korean War. One sign of official unhappiness was a remark attributed to the U.S. ambassador, Stephen Bosworth, reported to have indicated his own worries about DJ's policies in a private conversation. A source in the ruling party quoted Bosworth as saying that "any complaisant pursuit of the engagement policy on North Korea might be quote dangerous," according to the English-language *Korea Herald.* Bosworth denied the quote, telling me it was "the opposite" of what he believed.[43]

Perry was caught between the extremes of sympathy for North Korea offered by China and Japan's hard line toward the North. In consultations in Beijing, he sought to win assurances that China would try to persuade the North to open up Kumchangri for inspection. Adding urgency to Perry's mission was the failure of talks the week before in New York between American and North Korean diplomats. The United States had indicated its willingness to increase food aid to the North but insisted on multiple visits to the site.

Perry hoped to obtain support for U.S. demands in Tokyo, still furious over the missile incident. The American director of central intelligence, George Tenet, fueled fears in Seoul and Tokyo by warning before a congressional committee in February that North Korea might be able to fire an intercontinental nuclear warhead in the near future. South Korea's defense minister, Chun Yong Taek, meanwhile, revived historic Korean fears of Japan, warning of "escalation of the situation" if Japan "takes unilateral action without consultation with Korea."[44]

Immediately after stepping off the plane at Seoul's Kimpo Airport on the evening of March 8, Perry sought to dispel South Korean fears that he would insist on a hard line. He had made no "final recommendations," he said, reading a prepared statement to the local press, and he viewed Kim Dae Jung's "engagement policy" as a "very important factor on which to build." He qualified that remark, however, by observing that "North Korea has created real problems and serious problems with their nuclear and missile programs." Still, the United States and South Korea "must stand together" while seeking "possibilities of engagement in other areas."[45]

Perry concluded his arrival statement by defending his book, which he said "was written just as this policy review began." While the book "raises issues relative to our policy review," he said, "it does not reach any conclusions." Rather, he said, "it frames problems," leaving conclusions to be reached "in coordination" with South Korea. Perry now wanted Korean leaders to define just how far they wanted to pursue reconciliation. There had to be a "red line," said the Americans, beyond which action would replace words—storm clouds obscuring the sunshine. In the final stop of his mission in Japan, Perry faced a somewhat different problem—how to convince the Japanese that North Korea's nuclear ambitions presented a more serious danger than North Korean missiles. One concern of American as well as South Korean officials was that Japan would withdraw entirely from KEDO, the Korean Peninsula Energy Development Organization, if the North again fired a long-range missile.

Fears of an imminent crisis were resolved the next week when U.S. and North Korean negotiators, on March 16 in New York, agreed on U.S. inspections of the Kumchangri site in exchange for food aid for the starving country. "We really hope that suspicions about the site will be removed," said Song Min Soon, a security expert on DJ's staff. "The trend is now moving toward negotiated settlement rather than crisis."[46] That view summarized the outlook of a government that appeared to have gained new confidence in Kim's policy of reconciliation. South Korean officials, once sensitive to any sign of direct talks between North Korea and the United States, not only welcomed the agreement but hoped the United States and the North would form diplomatic relations.

"Normalization of ties with the United States would lead to further opening of North Korea with the global community," said South Korea's foreign minister, Hong Soon Young. The ultimate result, he predicted, would be "normalization of relations on the Korean peninsula"—that is, broad agreement between North and South Korea to deal directly with each other.[47] U.S. and North Korean negotiators had come to terms in New York after the United States spurned North Korea's demand for $300 million for one inspection of the site and persuaded the North to stop insisting on a specific commitment on food. Debate over that detail, said a foreign ministry official, was "the last stumbling block."[48]

The terms called for two inspections, one in May 1999, the next a year later. Crucial, however, was the understanding that America would provide more than 500,000 tons of food, the amount it had supplied in 1998 through the World Food Program. The WFP provided a convenient channel for Washington to live up to its refusal to tie inspections to any specific amount, whether in money or food. State Department spokesman James Rubin said the United States had reaffirmed that "our decisions on humanitarian assistance are based on need." A U.S. pledge to support a program for potato production in the North improved the atmosphere.[49]

South Korean officials did not seem worried that the agreement on Kumchangri provided for only two inspections. "Once a year is enough," said Song Min Soon at the Blue House. Nor did he think the North would build nuclear facilities elsewhere. "I am not so sure they can make other sites," said Song. "It is too early to say they have more things in their pockets." The North seemed likely to keep up an appearance of confrontation while avoiding hostilities. "War is impossible," said Huh Moon Young at the Korea Institute for National Unification, "but there will always be the threat of military action in accordance with the tactic of brinkmanship."[50]

Coordination of a joint policy with Japan was sure to dominate talks on March 20, 1999, between President Kim and Prime Minister Obuchi. Kim hoped to persuade Obuchi to offer full support for reconciliation with the North despite Japanese outrage over the missile. South Korean officials were confident that talks between the United States and North Korea on the missile issue would relieve Japan of some of its concerns. Then, they believed, Japan would provide its share of funding for KEDO.

Both Japanese and Korean officials used such terms as "cooperation" and a "common strategy" toward the North to characterize the dominant theme of the summit to be held in Seoul on March 20. They agreed the task would be easier now that the North had promised to permit the United States to inspect the underground site. The Japanese were determined, however, not to gloss over the missile launch in the artificial atmosphere of a meeting seen as another major step toward removing the stigma of Japanese colonialism from the Korean consciousness. Obuchi, in Tokyo, underlined this aim by saying that Japan would welcome the chance to conduct its own inspection of the site. That way, he said, his government would "find it easier to solicit cooperation from the Japanese people" on relations with North Korea. He noted, moreover, that his parliament still had to approve Japan's contribution of $1 billion to KEDO.[51]

The Seoul summit itself was not a great success. While "cooperation and coordination" among Japan, South Korea, and the United States were "essential," said Obuchi, "we do not need to adopt exactly identical policies at exactly the same pace." Thus he sought to close the gap between Japanese outrage over the missile and Kim's repeated pleas for dialogue. If such efforts as inspection of Kumchangri "can build mutual confidence between us and North Korea," Obuchi conceded, "it will be possible to replace the current structure of confrontation on the Korean peninsula with a structure of peace."[52]

Obuchi faced a critical test when he spoke on the afternoon of March 20 at Korea University against a background of militant student protest. His remarks proved how unconvinced he was of the need for a softer stance. Three hours after agreeing with DJ on the importance of "engagement," Obuchi made clear that the North had to be the first to "respond constructively" to "concerns and anxieties" aroused by the missile firing.

The situation on the Korean peninsula was so "extremely bleak," he said, as to make it "difficult to build amicable bilateral relations" with North Korea.[53] As Obuchi outlined his unrelenting stance, several hundred students chanted anti-Japanese epithets a few hundred yards away. Thousands of policemen were waiting nearby in case the students posed a serious threat. Downtown, in Pagoda Park, a small crowd, many of them elderly people with memories of the colonial era, gathered to denounce his visit and march to the Japanese embassy.

The United States also wanted assurances that the North would not fire another missile—and would stop building missiles for export. Robert Einhorn, deputy assistant secretary of state for nonproliferation, after two days of talks in Pyongyang on March 29 and 30, 1999, admitted the North had not budged from declaring its right to test missiles. The asking price for shutting down the program, including four factories and ten launch sites, was now said to be about $1 billion.[54]

Perry himself hoped to go to Pyongyang to discuss both the missile issue and Kumchangri. "The consequences of military confrontation are so serious that we should exhaust all diplomatic measures," said Perry. "None of us wants a war." While emphasizing a conciliatory approach, he defended U.S. military policy in the region, notably deployment of about 100,000 U.S. troops. If American troops "were to leave," he said, "the departure would involve an arms race" in which "each nation would have less security at greater cost." The United States in 1994 "came close to armed conflict with North Korea" before reaching agreement at Geneva the same year, he reminded the conference.[55] Similarly, the Kumchangri agreement had defused a budding crisis.

Others questioned, however, whether the American administration as well as members of Congress would have the patience. "The United States is burnt out and suspicious of North Korea," said L. Gordon Flake, director of the Mansfield Center for Pacific Affairs in Washington, citing "increasing worries" that Pyongyang no longer wanted to try to normalize commercial and diplomatic relations. There was "increasing concern about the role of the military" in the North, said Flake. "The primary power structure goes through the National Defense Commission," chaired by Kim Jong Il, rather than the Workers' Party, of which he was general secretary.[56]

Still, there seemed to be an easing of tensions. The State Department on May 17, 1999, announced it was sending 400,000 tons of food aid to Pyongyang. The announcement, on May 18, Korea time, coincided with the arrival in Pyongyang that day of a 14-person U.S. team on a one-week inspection mission of the cave at Kumchangri.

The mission was likely to fail on one level but succeed on another. No one gave the team a chance at finding evidence that the North Koreans had dug an enormous cave in Kumchangri for the purpose of building nuclear warheads. "The cave is vacant," said Chun Hyon Joon, senior researcher at the Korea Institute for National Unification. "There's nothing there. It's only a bargaining chip." The mission, however, would delay if not deter the North's plans for developing nuclear warheads in other caves that American satellites were believed to have discovered. "The mission is useful because it helps to contain North Korea's proliferation program," said Yoon Duk Min, a North Korean expert at the Institute of Foreign Affairs and National Security. "It has stopped a certain pattern of behavior." So doing, the inspection would set precedents on what equipment to carry and what they could do with it. "Agreement on details is more important than anything else," said Choi Won Ki at the Unification Research Institute of *Joongang Ilbo,* a leading newspaper. "Maybe there will be more inspections of other sites."[57]

On May 25, the day after the team's departure, armed with letters for Kim Jong Il from Presidents Clinton and Kim Dae Jung, Perry flew to Pyongyang to present a comprehensive peace package. DJ's senior secretary for security, Lim Dong Won, saw Perry beforehand in Tokyo to fine-tune the message of reconciliation the South wanted him to deliver, including a proposal from Kim Dae Jung for an "inter-Korean summit" with Kim Jong Il. DJ set "no time limit" on a response from the North. Rather he said he would keep trying "with the expectation that North Korea will come forward when it is convinced of our true intentions."[58] On the day Lim was in Tokyo, DJ named him unification minister in place of Kang In Duk, whose presence was viewed as a barrier to negotiations in view of his background with the Korean Central Intelligence Agency.

Kim Dae Jung was also anxious to improve relations on another front—with North Korea's Korean War ally, Russia. "The first thing President Kim wants to do is finish his summit diplomacy with the four surrounding powers," said Park Young Ho, a senior researcher at the Korea Institute for National Unification. "He has held summits with the leaders of the United States, China, and Japan but has yet to meet Mr. Yeltsin. The second is to resolve the issue of the Russian debt."[59] DJ took off for Moscow on May 27, 1999, for a summit at which he would discuss both his sunshine policy and $1.7 billion that Moscow owed on $1.47 billion in loans, plus interest, extended from 1990 to 1992 just as Seoul was drawing the Kremlin from Pyongyang.

In payment for the loans, the Russians offered three 2,200-ton Kilo-class submarines at 400 billion won—$330 million—apiece. In turn, "we hope to get support from the Russian side for our policy with North Korea," said Lee Sang Chul, a secretary for international security on Kim's staff.[60] The Russians for their part proposed either expanding the four-party talks, giving Russia a seat at the table, or opening an entirely new "multiple regional dialogue" including all powers in the area. "We are ready to take part in the four-plus or inevitably a multiple dialogue," said Moscow's ambassador to Seoul, Yevgeny Afanasiev, calling for a "very urgent multiple-dialogue structure surrounding the Korean peninsula." Multiple-party talks, he argued, would have the advantage of encompassing economic as well as security issues.[61]

One factor that might influence South Korea was the desire to make up for the tiff in July 1998 in which Moscow had expelled a South Korean diplomat for spying and South Korea had retaliated by expelling a Russian diplomat. The proposal for buying Russian submarines opened a bitter debate among defense and military officials—and big business interests as well. "Russian submarines have many problems," said Brigadier General Cha Yung Koo, citing the need for "logistical support" for refitting and repairs.[62] The Korean navy far preferred German, French, or Swedish models for speed and maneuverability. Hyundai Heavy Industries and Daewoo Heavy Industries both pressured for contracts to build the submarines with European technology. Daewoo had already built nine 1,200-ton submarines for the navy with German technology.

Policies of "engagement" and "reconciliation," however, made war seem like a distant prospect. The inspection team returned from Kumchangri having found "an extensive empty tunnel complex"—and nothing else.[63] Perry, in Seoul after his visit to Pyongyang, foresaw "major expansion" in relations with the North, though he failed to meet Kim Jong Il. The U.S. offer included diplomatic recognition and easing of sanctions if the North stopped selling missiles abroad, again forswore building nuclear warheads, and opened up to normal relations with the South and the world.[64]

The offer seemed to justify all North Korea's military and diplomatic posturing, but Pyongyang wanted more. In a bid to increase its bargaining power, the North on June 7, 1999, opened a challenge to South Korean navy vessels, sending fishing vessels south of a "Northern Limit Line" in the Yellow Sea. DJ, anxious to keep up his sunshine policy, called an emergency meeting of top aides and advisers after six North Korean patrol boats on June 10 accompanied 20 to 30 fishing vessels on their largest sortie yet. North Korean authorities denied the existence of the line, established unilaterally by the UN Command after the Korean War, insisting on their right to enter what the command called a "buffer zone," 2 to 14 kilometers wide, south of the line. South Korea prepared for a "worst-case scenario" in which the North fired first. "If North Korea uses gunfire, we will respond," said Brigadier General Cha Yung Ku, spokesman for the command. "Our first mission is to keep the peace." South Korean commanders had orders to follow "our basic guideline" of holding fire unless "absolutely needed for self-defense."[65]

Military experts saw the forays across the line as reflecting disagreement between those in Pyongyang who favored a strong policy against the South and others who might be open to DJ's repeated efforts at rapprochement. "Maybe the hawks up there are not happy with the idea of dialogue between North and South Korea," said General Cha. "They have never accepted officially the Northern Limit Line. They may want a showdown." He also suspected a diversionary tactic. "Maybe they want us to pay attention to that area so they can prepare another intrusion somewhere else."[66]

There were other reasons for the North to have sent its boats into the buffer zone. "North Korea intends to show there remains a terrible military situation on the Korean peninsula," said Park Young Ho. "It is a negotiating tactic before talks between North and South Korean diplomats in Beijing on June 21"—even though tensions had eased since the South had promised to ship 100,000 tons of fertilizer to the North as a prelude to the talks.[67] Another compelling reason was that the North needed the crab found in the waters in the area in the early summer. "It's about food," said John Barry Kotch at Hanyang University. "The North is starving and may see no other way to feed their people. Unfortunately, they don't have an agreement on fishing with the South and have to feed themselves any way they can."[68]

The North Korean fleet remained between one kilometer and five kilometers south of the Northern Limit Line for 14 hours on June 10 while South Korean boats weaved among them, ordering them by loudspeakers to go back. South Korean boats rammed North Korean boats on June 11 to get them to turn around. On June 13, North Korea showed its resolve not to back down by repeatedly sending groups of three to six patrol boats south of the line as a reinforced South Korean flotilla, including destroyers and frigates, pursued them back and forth across the buffer zone. Despite the unlikelihood of reaching any understanding, the North agreed to come to a talk between generals at Panmunjom on June 15 after having twice refused requests for negotiations.

The prospect of talks came as a relief to Kim Dae Jung, at pains to defend his sunshine policy in the face of escalating tensions. DJ, visiting the island of Cheju, reiterated his desire to pursue his policy but said his government was determined "to defend our sovereignty and territory." One ominous sign, however, was a claim by South Korean fishermen that North Korean troops were aboard the fishing vessels. The defense ministry "asked the Americans to consider" sending their own warships to the area, said General Cha. "Maybe it's useful to show them some force." The risk of sudden escalation was clear. "If they open fire, you must be prepared to return fire immediately," Admiral Lee Soo Yong, navy chief of staff, told his men.[69]

At 9:25 A.M. on June 15, 1999, 35 minutes before North Korean and American generals were to face off in Panmunjom, South Korean naval ships sank a North Korean gunboat in a furious ten-minute gun battle. South Korean officials claimed, however, the North Koreans had timed the incident to precede the talks. The evidence was that officers from the UN Command had no idea the incident had occurred when North Korean officers, shortly after the talks began, said that South Korean forces had opened fire and North Korean soldiers "are dying now."[70]

South Korean officials said three North Korean torpedo boats fired first, provoking eight South Korean vessels to return fire with 35-milimeter guns, but they admitted the South Korean boats had rammed the North Korean boats to get them to turn around before the North Koreans opened fire. The South Koreans fired about 100 rounds, setting ablaze a 40-ton North Korean torpedo boat. No survivors were seen as the ship sank with at least a dozen men aboard. Two South Korean boats were slightly damaged, and seven South Korean sailors were wounded before the other two North Korean boats fled. One of them, a relatively large vessel weighing 420 tons, was heavily damaged. All told, 40 North Korean sailors may have died, and perhaps 80 more were wounded, but there was no way to know the exact numbers.[71]

The incident, the first naval battle in the Yellow Sea since the Korean War, shocked both government officials and average Koreans as they gathered around television screens. Kim Dae Jung and his aides "are very embarrassed because he has been pushing his sunshine policy," said Korea Broadcasting System.[72] Several hours later, top U.S. and South Korean commanders, denouncing "a manifest act of provocation" on the part of North Korea, issued "a strong warning" against "further provocation." General Tilelli, meeting General Kim Jin Ho, chairman of the South Korean joint chiefs of staff, promised "prompt US support to reinforce combined military posture." Such escalation, however, would test the strength of U.S. forces depleted by the military build-up in the Balkans. A squadron of U.S. Air Force F15E fighters had to be dispatched to South Korea in May as partial replacement for American planes aboard the carrier USS Kitty Hawk, on its way to support the bombing of Serbia.[73]

Kim Dae Jung, informed of the incident as he was about to receive Singapore Prime Minister Goh Chok Tong, urged Koreans to remain "cool-minded" but "determined to protect our territory." While the North accused the South of "reckless provocations" designed to push North and South Korea "to the brink of war," DJ vowed to keep up his efforts to engage the North in dialogue. One hopeful sign was that the North had guaranteed the safety of South Korean tourists visiting Kumkang. DJ's foreign security chief, Hwang Suu Sok, said a ship containing 100,000 tons of fertilizer for North Korea had returned to Inchon but would deliver the fertilizer when assured of safe passage.[74] U.S. officials were just as anxious to play down the incident, refraining from putting the 37,000 U.S. troops on alert. Instead, they were placed on "watchcon," a category described as "awareness" of the danger.[75]

"It does look as if someone in Pyongyang is trying to sabotage the sunshine policy," said Aidan Foster-Carter, North Korean specialist at Leeds University, England. "There may be people up there who don't want an outbreak of peace."[76] More likely, however, some analysts believed the North was pursuing a fight-talk policy in anticipation of the talks opening on June 21 in Beijing. "They need some leverage with South Korea," said Yoon Duk Min at the Institute of Foreign Affairs and National Security. "They usually use a kind of brinksmanship. They don't want to break out of the whole peace framework. The sunshine policy is very good for them."[77]

The next day, June 16, at the opening of the one hundred-ninth session of the International Olympic Committee in Seoul, DJ made one of his strongest arguments so far in defense of his "policy of warm partnership" with North Korea as the way to bring about "peace on the Korean peninsula." While he addressed a glittering audience of IOC members in the Seoul Arts Center, American warplanes crisscrossed the skies over the Yellow Sea and South Korean navy boats waited to ward off any fresh North Korean challenge. Ignoring the standoff, DJ cited both the opening of Kumkang to South Korean tourists and resumption of talks between senior North and South Korean diplomats in Beijing as proof of his point. DJ made an easy transition in his remarks from praise of the ideals of the Olympic movement. While "negative factors have not been eliminated completely," he said North and South were "beginning to cooperate on exchanges in many areas including economic and cultural fields."[78]

North Korea's leading newspaper, *Rodong Sinmun,* editorialized that North Korea's armed forces "must be strengthened in every way," but there were no signs of any change in their posture.[79] "I think that's enough for them," said Choi Jin Wook, research fellow at the Korea Institute for National Unification. "Too much aggression is not a good thing for them. If they overdo it, it will not help." The "major purpose" of the North, he believed, was to get into talks with the United States for a Korean War peace treaty in place of the 46-year-old armistice.[80]

With no signs of renewed fighting, tensions eased among South Koreans who had watched the showdown on television for much of the day. "I was afraid at first," said businessman Kim Yong Han, 46, "but now I'm all right." He said the clash was "rather unexpected, but I don't think it will lead to war." The spectacle forced many to consider consequences that seemed unimaginable. "I was very scared," said Lee In Hae, 19, a student. "Nobody is prepared" for expanded conflict. "Young people don't have a sense of war," said Kim Soo Yeun, also 19. "To us, war is just an abstract idea." Lee Jong Kwon, 49, believed the North might have learned a lesson from the sinking of the boat. "North Korea must know it cannot win a war," he said, "so probably the incident won't lead to a second Korean War."[81]

Even so, the guided missile cruiser USS Vincennes steamed into South Korean waters on June 17 in a dramatic show of support for a South Korean flotilla on guard against North Korean forays. The Vincennes and another guided missile cruiser, the USS Mobile Bay, due in the area on June 18, provided a powerful signal of U.S. determination to force the North Koreans to stay on their side of the line. The cruisers were coordinating closely with four U.S. Navy EA6B Prowler reconnaissance planes, dispatched to relay information on North Korean naval and ground movements. Two submarines, the Buffalo and the Kamehameha, also arrived in Korean waters, calling at the small U.S. navy base at Chinhae on Korea's southern coast. In another display of solidarity with the South Koreans, the aircraft carrier USS Constellation was to pass through South Korean waters as part of a battle group en route from the U.S. navy base at San Diego to the Middle East.[82]

The show of American power at sea was designed to plug a gaping hole in defenses created by the deployment of U.S. forces to the Persian Gulf and the Adriatic Sea during the Balkan War. South Korean military leaders feared U.S. forces in the region were dangerously under strength since the departure of the Kitty Hawk, by now in the Persian Gulf. "The U.S. ships will have an impact on the North Korean attitude toward the South," said Park Young Ho at the Korea Institute for National Unification. "North Korean leaders will perceive the U.S. military posture as a sign to stop the pressure. They will refrain from escalating."[83] A sign of the North Korean response was that North Korean military vessels had not crossed the Northern Limit Line since the gunfight on June 15 after having entered the buffer zone for eight days before then.

The North, however, maintained a base for Rodong 1 missiles, with a range of 300 kilometers, north of the western end of the DMZ. North Korea also stationed Frog rockets, with a range of 70 kilometers, along the DMZ within easy range of Seoul. Adding to concerns, NHK, the Japan Broadcasting System, reported on June 17 that North Korea had been conducting experimental propulsion tests of its Taepodong 2 missile since April.[84] The Taepodong 2, successor to the Taepodong 1, had a range of about 6,000 kilometers—enough to reach the American mainland. The next few days would be critical in discerning North Korea's basic intentions.

The talks in Beijing on June 21 were sure to turn into an exchange of charges and countercharges. While Kim Dae Jung steadfastly upheld his sunshine policy, a spokesman for the North Korean navy described him as a "puppet leader." The North "will answer retaliatory blow of the enemy with retaliatory blow, an all-out war with an all-out war," Pyongyang's Korea Central News Agency said on June 20 as the diplomats arrived in Beijing.[85] South Korean officials for their part spoke proudly of the success of the South Korean navy. "Our readiness has been swift and firm," said Foreign Minister Hong Soon Young, calling the use of force a "demonstration of our readiness to retaliate against any attempt by North Korea" to intrude upon South Korean territory.[86]

Gearing themselves for a defense of the Yellow Sea incident, South Korean diplomats still hoped the talks would get around to their original purpose—negotiations on aid to the North and reunions among several million families divided by the Korean War. The real position of the North might not emerge until two days later, June 23, when the American negotiator, Charles Kartman, also in Beijing, was to meet Kim Gye Gwan, North Korean vice foreign minister, to talk about the U.S. report on Kumchangri.

Just as the North-South talks were to have begun, however, the diplomatic maneuvering got considerably more complicated. On the afternoon of June 21, a 36-year-old South Korean woman, Min Young Mi, was detained by North Korean immigration officials amid a tour of Kumkang after telling a North Korean guide that North Korean defectors were living well in the South—and suggesting the guide go South to see for herself. While the woman was being detained, a boat with 560 tourists aboard remained at Changjun, the port serving Mount Kumkang. Another boat, bound for North Korea from the South Korean port of Donghae, was ordered back with 524 tourists aboard.[87] The detention dramatized the sensitivities of the standoff between North and South Korea, made all the more acute since the sinking of the North Korean boat.

A Hyundai official telephoned Hyundai Merchant Marine officials from the boat at Changjun, saying there had been "a misunderstanding" but that North Korean immigration officials needed authorization from higher officials before they could release Min Young Mi. Meanwhile, he said, the North Koreans were fining her $100 in American currency and seizing the paper authorizing her to go on the tour. South Korea's national security council called an emergency meeting. "If there is any insecurity about our citizens, the entire program will be canceled," said Hwang Won Tak, chief of the council. Hyundai announced shortly after midnight that the tourist boat on which the woman had sailed to Changjun would leave without her. The state-owned Korea Broadcasting System said the government had ordered Hyundai to "protest strongly."[88]

Both sides gave conflicting signals. DJ sought to head off criticism of his sunshine policy, saying "We will not make one-way concessions" to the North, and "should be on alert all the time" lest the North "take revenge" for the Yellow Sea incident. A unification ministry official said the South had made good on its pledge to give the North 100,000 tons of fertilizer as promised, finishing the shipment the day the woman was seized, but would not deliver another 100,000 tons promised after the talks if the North refused to talk.[89] North Korea, delaying the opening of the talks with the South, accepted a request from the UN Command for a meeting between American and North Korean generals at Panmunjom on June 22.

The Kumkang incident, along with the Yellow Sea skirmish, did not just cast a shadow over sunshine. It warned of a much larger event waiting to happen—an outbreak of real hostilities. The problem with sunshine for the North was the fear in Pyongyang of an opening that might expose the weakness, corruption, and failure of its ruling elite. The problem for the South was that the happy talk promoting sunshine might blind people to the danger of a war for which no one was prepared. In Seoul, the rich were getting richer and good times were back with a rebounding stock market. Shooting at sea, talks in Beijing, suspension of tours were far removed from hustling markets and go-go boardrooms.

The United States was even less prepared. President Clinton and his cabinet had little to say during all these events about trouble in Korea. In two hours of talks with North Korea's crusty General Li Chan Bok at Panmunjom, a newly arrived American major general, Michael Dunn, successor to General Hayden, got nowhere suggesting ways for everyone to stay on the proper side of the imaginary line in the Yellow Sea. Those talks, in the austere one-room negotiating hootch on the DMZ, paralleled the meeting in Beijing between senior South and North Korean diplomats, first shaking hands in a blaze of strobe lights, then sitting down on either side of a flower-strewn table. Both talks devolved into North Korean demands for an "apology" for the Yellow Sea incident. The diplomats in Beijing never got to family visits, which the North would not have endorsed for fear such exchanges would reveal too much about how North Koreans were living—and dying.

Conservatives in the South Korean National Assembly and the American Congress had long opposed aid for North Korea. Were these hardliners ready, however, for the ultimate incident—the dare and double-dare that might trigger the guns massed on each side of the DMZ? As American warships hove into Korean waters, the question was more than hypothetical. Suddenly the stakes were raised on the Korean peninsula. The risk was not that of North Korean nukes or missiles but of an oldfashioned shoot-'em-up along Asia's most heavily armed frontier. The bait of massive aid had compelled the North nearly five years earlier to stop building a nuclear warhead, but few figured on conventional war, up close, personal and potentially as deadly. Kartman, meeting Kim Gye Gwan on June 23, now had a real threat to paper over. Slick diplomacy would be needed to keep all sides, South and North Korea, the United States, even China, North Korea's last ally, from stumbling blindfolded into more than another exercise in diplomatic dueling.

The detention of Min Young Mi, like the shootout in the Yellow Sea, was an incident waiting to happen. With thousands of South Koreans on the trails of Kumkang, someone, some time, had to say something the North could seize upon. Not even the hermetically sealed walking route, shielded by barbed wire from any ordinary North Koreans curious about these well-fed apparitions from the South, could guarantee immunity. The woman's release on June 25, with a face-saving statement from the North that she could have been tried as a criminal, was less than reassuring. The woman was turned over to South Korean tour managers in Changjun after North Korean officials decided not to sacrifice a lucrative contract with the Hyundai group for the sake of holding her as a South Korean spy.

North Korea's Korea Central News Agency said Min was freed in deference both to "desires of South Koreans to visit Mount Kumkang as early as possible" and to its "relations with the Hyundai group." For her crime, said the statement, issued in the name of the Korean Asia-Pacific Peace Committee, she "should have been judged" according to North Korean law.[90] Thus North Korean authorities acknowledged the importance of a contract in which Hyundai had guaranteed payment of $942 million and also had promised to invest another $300 million in an enormous hotel and recreation complex. The entire program—and plans for Hyundai to build an industrial park across the peninsula on North Korea's southwest coast for $20 billion—appeared in jeopardy as Hyundai executives, in talks with the North Koreans in Beijing, strove to explain away any "misunderstanding."[91]

The tours were the most visible sign of rapprochement between the two Koreas, but Hyundai officials revealed that North Korean guides had been eager to pounce on tourists for trivial offenses. Until Min's arrest, the most serious offenses were for threatening North Korea's national security by photographing the harbor at Changjun. North Koreans got especially angry with a tourist who shouted, "Daehan Minguk Mansei," meaning "Long Live [the Republic of] Korea"—South Korea—at the end of a walking tour. Three times, said a Hyundai official, North Korean guards had confiscated camcorders from tourists—once after a tourist refused to sign a "confession" admitting his guilt.[92]

Guides routinely imposed fines, payable in U.S. dollars, for these and lesser offenses, such as spitting beside one of the tourist trails and posing for photographs on stones inscribed with the names of Kim Il Sung or Kim Jong Il. One man was fined for wearing a tee-shirt with an American flag stamped on it; another for washing his socks in a stream, and two people were fined for tossing food from their bus to North Korean children. Hyundai officials tried to be understanding. "When you visit your neighbor, you should respect their customs and traditions," said Lee Yong Jun, sales manager for the tours. "Slowly, we are opening it up."[93]

The fears engendered by the shootout in the Yellow Sea, the detention of Min Young Mi—and then reports of North Korean plans to launch Taepodong 2—opened up the question of what the South could do to defend itself. South Korean and American officials were already engaged in sensitive talks over development of a new South Korean missile capable of reaching Pyongyang and beyond. Washington had imposed an understanding on Chun Doo Hwan after he seized power in late 1979 that South

Korean missiles would have a range of no more than 180 kilometers, 20 kilometers shy of Pyongyang. The understanding was part of a tradeoff in which the United States agreed to recognize Chun's takeover provided he bowed to U.S. wishes militarily and promised not to execute Kim Dae Jung. As a reward for good behavior, Ronald Reagan, then president, had Chun Doo Hwan over as the first foreign head of state to visit the White House in 1981. For years LG Precision produced the mainstay Hyunmoo missile, named after a mythical figure, part dragon, part phoenix, modeled after the Nike Hercules, but the South Koreans complained it was no match for North Korea's creations.[94]

U.S. officials sympathized but said Seoul had to conform with standards set by the Missile Technology Control Regime, signed by 32 countries. Seoul, not a signatory, pledged to abide by conditions of the regime, which limited a missile's range to 300 kilometers, considerably further than Pyongyang, but Washington suspected the South Koreans were plotting to develop a much more powerful device. South Korea's Agency for Defense Development deepened suspicions in April 1999 when it test fired a model that flew 50 kilometers off the South Korean west coast. The missile had a far longer range, it was said, but was tested for only a short distance in order to allay American suspicions. Seoul denied the charge. It was all "a misunderstanding," said Brigadier General You Jin Kyu, director of the defense ministry's arms control bureau.[95]

In a spirit of defiance South Koreans said they might develop new missiles on their own—with technology from European countries. "The U.S. people don't even want Korea to study long-range missiles," said Sohn Young Hwan, a missile expert with the Korea Institute of Defense Analyses. "We think that kind of limitation is not proper." He was not certain if the next missile would be "solely U.S.," noting that "for technology we have cooperation with other foreign countries." The issue was delicate. "The United States doesn't want to antagonize North Korea," said Kim Chang Suu at the same institute. "The United States wants us to report all information on test-firing."[96]

Against this background, South Koreans pressed for a new understanding on missiles during Kim Dae Jung's next summit with President Clinton on July 2, 1999, in Washington. "We have sort of a balance of terror on the Korean peninsula," said a senior official on DJ's staff. "We want a reasonable level of capabilities." Although DJ failed to win a commitment in his meeting with Clinton, Defense Secretary Cohen assured South Koreans during a stop in Seoul on July 29 that Washington favored an increase in the number and range of their missiles. South Korean defense officials said the United States agreed to supply 100 air-to-surface "Popeye" missiles over a three-year period from 2000 to 2003. "It is very significant for South Korea," said Park Young Ho at the Korea Institute for National Unification. "It gives some psychological relaxation."[97] The Popeyes, produced by Lockheed Martin at a cost of $800,000 apiece, were capable of wiping out installations from 110 kilometers—perfect for attacking North Korean positions on shore in case of a confrontation similar to the one in the Yellow Sea.

Regarding South Korea's request for revision of the 1979 agreement with Chun Doo Hwan, Cohen was vague but encouraging. Washington supported Seoul's desire to join the Missile Technology Control Regime, he said, and would work with the South Koreans "to accommodate their needs as far as their missile capabilities." Simultaneously, Robert Einhorn, an assistant secretary of state, discussed with South Korean diplomats the possibilities of Seoul's joining the regime. Thus Korean officials could say what the Americans were still not ready to enunciate—that Washington had agreed "in principle" that the South should develop missiles with a 300-kilometer range.[98]

North Korea's apparent determination to test-fire Taepodong 2 goaded Tokyo as well into an urgent view of the need for building up its defenses—perhaps on a much greater scale than envisioned for the South Koreans. In Tokyo on July 28, the day before his meetings in Seoul, Cohen struck a bargain of long-range import for the region. Cohen called it part of "theater missile defense;" Hosei Norota, minister in charge of the Japan Defense Agency, referred to "ballistic missile defense." Either way, the deal called for the United States and Japan to jointly develop the means for spotting and shooting down enemy missiles before they hit Japanese territory.[99]

The Japanese if anything were more concerned about Taepodong 2 than were the South Koreans. The obvious reason was that Taepodong 1 had flown over Japanese territory, and Taepodong 2 would probably represent an even greater infringement on Japanese sovereignty. Beyond the territorial issue, however, many South Koreans privately did not object if the North Koreans wanted to test fire a missile over Korea's historic enemy, Japan. There was an underlying Korean pride in the success of the North Koreans in developing such a missile, defying some of the world's most powerful countries. Idealistic students—not just radicals—went so far as to praise the North Koreans for their achievement.

Nonetheless, concern over North Korea prompted Japan and South Korea, at the instigation of the United States, to work together more closely than ever before—not just on commercial and cultural exchanges but militarily. Thus, on August 5, Japanese and South Korean navy ships maneuvered in the choppy waters of the Tsushima Straits, at the closest point between the two countries while patrol planes and helicopters circled above in a search-and-rescue exercise that was small in scope but large in implications. The exercise was the first military operation ever conducted jointly by Japanese and Koreans since the Japanese ruled the entire Korean peninsula as a colony.

Equally significant was the timing of the operation, conducted on the first day of another round of "four-party" talks in Geneva among diplomats from North and South Korea, the United States, and China. While the diplomats were discussing how to work out a peace treaty in place of the armistice that ended the Korean War, North Korea attacked the exercise as not only "reckless" but part of a Japanese effort to foment another conflict on the peninsula. "Japan is about to make a direct involvement in a new Korean War," said *Rodong Sinmun*. The exercise, it said, was "aimed at making a surprise attack."[100]

The exercise was a warning of the possibilities for much larger cooperation between Japan and South Korea if North Korea persisted in plans for test firing Taepodong 2. "The political symbolism is important," said Tomohisa Sakanaka, president of the Research Institute for Peace and Security, a think tank with close connections to the Japan Defense Agency. Sakanaka said South Korea had been reluctant to join the exercise, viewing Japan's Self-Defence Forces as "a symbol of Japanese imperialism."[101] Keenly aware of such sentiment, the Japanese minimized the significance of the exercise. The setting, they noted, was as far away from North Korea as possible—in the straits between the southern Japanese island of Kyushu and South Korea's Cheju Island, off the southern tip of the Korean peninsula.

The Japan Defense Agency said only 630 members of its Maritime Self-Defence Force and almost as many from the South Korean navy were involved in what was described as a simple simulated rescue of sailors and passengers from a civilian vessel. Three Japanese destroyers, one South Korean destroyer, and one South Korean frigate crisscrossed the area for most of the day after having spent two days in their own waters planning the operation and sending out patrol planes. While flares were dropped and helicopters swooped low, no weapons were fired, and on-scene media coverage of what would have provided easy television footage was barred lest it offend Pyongyang.

Japanese saw South Korean participation as reflecting rising concern in

Seoul about Taepodong 2 despite DJ's policy. "The attitude of the Republic of Korea has toughened," said Sadaaki Numata, foreign ministry spokesman, countering a widespread view in Japan that South Korea was not strong enough in condemning the North for launching the first Taepodong.[102] Public-opinion polls revealed the sense of vulnerability among Japanese, many of whom looked on North Korea, like China, as a long-range foe. More than 70 percent of those polled by *Yomiuri Shimbun,* Japan's largest-selling newspaper, said they were worried about a war that would threaten Japanese security. The newspaper said 57 percent feared Japan would be "directly attacked by military force."[103]

The North Koreans played upon Japanese concerns for all the impact they could achieve. Evoking memories of Japanese colonialism and World War II, North Korea on August 10 warned of "merciless retaliation" if the Japanese attempted "reckless provocation" on the Korean peninsula. Pyongyang issued the statement amid mounting concerns in Tokyo that North Korean leaders were developing a rationale for ignoring the pleas of Japan, the United States, and South Korea not to launch Taepodong 2. The statement, timed to precede the fifty-fourth anniversary of the Japanese surrender and liberation of the Korean peninsula on August 15, 1945, said North Korea would exact "a high price for the blood shed by the nation and give vent to its century-old wrath."[104]

North Korea's denunciation of Japan was one of several signals of a gathering confrontation of forces as leaders in Pyongyang tried to demonstrate their resolve before their own people as well as the region. The North Korean diatribe came just a day after the breakup of the sixth round of four-party talks in Geneva on August 9. North Korean negotiator Kim Gye Gwan insisted the talks could not resume until the United States had agreed to discuss withdrawing its troops from South Korea. He also said the talks had to include negotiation of a peace treaty between the United States and North Korea—a demand routinely rejected by Washington as an effort to exclude South Korea. In a calculated decision not to be deterred by rhetoric, U.S. and South Korean leaders in Seoul decided to go ahead with a computer-generated war game involving 70,000 troops, including 14,000 American and 56,000 South Koreans. Both the South Korean defense ministry and the U.S. command in Seoul described the exercise as "routine." The American commander, General Tilelli, did not comment on North Korea's denunciations but twitted Pyongyang by remarking that North Korea, by launching a missile, would be making the "wrong choice."[105]

In another sign of the tensions surrounding Taepodong 2, three U.S.

Navy vessels left Japan to monitor North Korean activities. The 2,262-ton USS Invincible, designed to track missiles and watch for other signs of military activity, and a nuclear-powered submarine, the 6,080-ton USS Los Angeles, both left Sasebo on August 9. The 17,000-ton missile-tracking vessel USS Observation Island steamed out of Sasebo Navy Base near Nagasaki on August 10. The Observation Island, which had monitored the launch of Taepodong 1, and the Invincible were officially on a mission to conduct "hydrographic studies."[106]

The level of American, Japanese, and South Korean alarm increased as Taepodong 2 was rolled onto the launch pad, under the watchful eyes of spy satellites. For North Korea, however, the payoff would not be armed conflict but just what Perry had offered, the lifting of sanctions. South Korea's unification minister, Lim Dong Won, back from talks with Perry and others in Washington, said on August 29 that the United States was ready to "relax economic punishment" and expand relations with the North if the North abandoned plans to test-fire Taepodong 2. Lim hoped the North and the United States would come to terms in another round of talks between Kartman and Kim Gye Gwan—this time in Berlin.[107]

The timing was crucial. The talks were to begin on September 7, two days before the fifty-first anniversary of the founding of the Democratic People's Republic of Korea on September 9, 1948, and go on for five days. Many observers had predicted the North, if it did not fire Taepodong 2 on August 31, the first anniversary of the firing of Taepodong 1, would surely do so by September 9. Lim was "encouraged" by the fact that there were still "no signs" that the North would really launch Taepodong 2. Pressure from China would also be influential. Li Ruihuan, chairman of the National Committee of the Chinese People's Political Consultative Conference, was quoted on August 29 by China's official Xianhua news agency as describing both South and North Korea as "good friends of China." Thus Beijing pointedly viewed the two Koreas equally even though China remained North Korea's only real ally. China, Li said, was "concerned" about continued conflict between North and South.[108]

North Korea's isolation was evident in a visit to Seoul in early September by a Russian military delegation. In meetings with Kim Dae Jung and Defense Minister Cho Sung Tae, Russia's defense minister, Igor Sergeyev, discussed not only arms sales, broached during DJ's trip to Moscow in May, but also training of troops. The Koreans, as part of the bargain, extracted from Sergeyev a promise of support in dissuading the North from firing the missile. In the meeting of ministers on September 2, said the South Korean defense ministry, "Defense authorities of the two countries agreed to closely cooperate in curbing North Korea's missile test-firing"—and "to launch joint naval exercises" beginning in 2000.[109] Could there be a better way to show off the submarines that the South, not the North, had the money to buy? Moscow, with an eye toward balancing forces on the peninsula, still exercised leverage on its one-time Korean War ally through its embassy in Pyongyang.

North Korea's missile diplomacy worked to perfection. At the end of a week's hard bargaining in Berlin, the North agreed on September 12, 1999, not to launch Taepodong 2. Kim Dae Jung, Clinton, and Obuchi got the word in Auckland, New Zealand, where they were attending the annual session of APEC, the Asia-Pacific Economic Cooperation forum of leaders of Pacific Rim economies. "We agreed to make further efforts to solve apprehensions both sides face," the North Korean vice foreign minister, Kim Gye Gwan, remarked enigmatically in Berlin.[110] The quid pro quo, not immediately stated, was that Washington would lift sanctions, opening the North to normal trade and investment by American interests. As in the critical period before the conclusion of the Geneva framework agreement nearly five years earlier, negotiations had averted a showdown that nobody wanted, for which neither side was prepared.

Several days later, on September 17, the details emerged. The United States would indeed remove all restrictions on just about every kind of normal peacetime transaction. "For more than 40 years, the threat of another war on the Korean peninsula has hung over our heads like a dark cloud," said Perry as the White House announced the lifting of sanctions. "Today, that cloud is beginning to drift away." He was "hopeful," he said, "that we are finally moving on the path to normalization."[111] There would be a testing period. North Korea was still classified as a "terrorist" state. Americans could not sell weapons or components or technology that might be used for military purposes. While American citizens could send money to relatives in North Korea, they could not engage in financial transactions with the North Korean government, and there would be no U.S. support for international loans to the North.[112]

The agreement reached in Berlin, however, would not suddenly lower the level of confrontation. Rather, the North saw the missile threat as a device to turn on and off during more talks for ever more concessions. A spokesman for the foreign ministry in Pyongyang indicated this strategy on September 24, one week after the White House announcement. The Korea Central News Agency quoted him as saying the North would "not launch a missile while the talks are underway with a view to creating an atmosphere more favorable for the talks." It was up to the United States, he said, to "respond with good faith and strive to remove the U.S. suspicions and apprehensions in the interests of the two sides."[113]

The success of the agreement depended on how the North used the lever that it possessed in the form of Taepodong 2 in projects to build and export other missiles, and in the extent to which it opened up to aid, trade, and investment. It was easy to accuse North Korea of "blackmail," to say and write, as did some politicians and journalists, that the North was guilty of "extortion" and the United States of "appeasement." It was equally easy for proponents of the agreement to hail it as a "landmark," a turn down what Secretary of State Madeleine Albright called "a new and more hopeful road." There was also another disquieting possibility—that the road would go nowhere, that the North would create periodic unexpected crises, that many thousands more would starve. The North could no longer blame all its troubles on American sanctions, but it could refuse to cooperate as long as the United States called it a "terrorist" state and did not lift the remaining restrictions.

The most disheartening aspect of the agreement was not the threat the North still posed but the realization that Washington, Tokyo, and Seoul were keeping the North Korean regime on a system of life support that had no foreseeable ending. The opening of an American liaison office in Pyongyang, perhaps even an embassy, would add legitimacy if not prestige to a government that remained one of the world's most oppressive. Perry denied that the lifting of sanctions was a reward, but the sanctions would surely have remained in effect if North Korea had not so effectively rattled its missile. Perry's prediction that the North would someday comply with the Missile Technology Control Regime appeared far too optimistic. The North's promises were all negative. The North would *not* fire Taepodong 2, just as it would *not* build nuclear warheads. Beyond saying what it would *not* do, the North did *not* offer anything.

A meeting in Pyongyang on October 1 between Kim Jong Il and the Hyundai group's octogenarian founder, Chung Ju Yung, and Chung's fifth son, Chung Mong Hun, group chairman, added to the puzzle. Chung Ju Yung, returning to Seoul, said that Kim Jong Il had accepted his proposal to build an industrial park on 66 million square meters on North Korea's southwest coast not far from the scene of the Yellow Sea incident. The park, said Chung, would be large enough for 800 to 900 factories, including a motor vehicle assembly plant and facilities for producing electronic equipment. Earlier in their visit, Chung Ju Yung and Chung Mong Hun watched basketball teams sponsored by Hyundai play "friendly" games with North Korean opponents—the first North-South athletic events to which the South sent teams sponsored by private enterprise.[114]

The announcement, however, contained no hint of change in the North. There was no sign the complex would do anything other than attract money and expertise. Who would work at the complex? Would North Koreans see its products outside the gates? Was there any guarantee that Hyundai, already burdened by debt, would profit at all when it was already losing heavily on its Kumkang operation? Would the North seal off the area, as it had the Kumkang tour route, from North Koreans who might emulate the capitalists from the South? The failure of the industrial zone in the northeast corner of North Korea had set an ominous precedent. If there was no progress there, what reason was there to believe that the complex envisioned by Hyundai could go beyond the talking stage?

North Korea's armed forces, with Kim Jong Il as chairman of the defense commission, remained in the ascendancy. They formed the pinnacle of a hierarchy confirmed in its reliance upon foreign investment and diplomacy. Under the circumstances, what incentive was there for the North to change, to consider serious peace talks with South Korea, to open up normal contact in the form of cross-border commerce, visits, mail and telephone calls? The leaders of this edifice could prosper, at the expense of their own people, by remaining remote and belligerent, as had Korea's dynastic emperors in bygone centuries, forcing their former enemies to bow before them, begging for the chance to help them still more. The Korean peninsula remained in crisis, wary and uncertain except for the certainty of more crises, in a manner, a time that no one dared predict.

Perry's "review," when it finally came out on October 12, was disconcertingly circumspect. The overriding consideration was to stay out of war. "The United States and its allies would swiftly and surely win," said the review, "but the destruction of life and property would far surpass anything in recent American experience." Thus the United States had to "pursue its objectives with respect to nuclear weapons and ballistic missiles in the DPRK without taking actions that would weaken deterrence or increase the probability of DPRK miscalculation." In the process, the United States "should be prepared to establish more normal diplomatic relations with the DPRK and join in the ROK's policy of engagement and peaceful coexistence." Self-consciously, as if to anticipate all manner of criticism, the review touched all the bases, ranging from the interests of surrounding countries to the "humanitarian tragedy" of the North.[115]

Always, the point was to sidestep confrontation either by aggressive military threats or demands for "hastening the advent of democracy and market reform that will better the lot of the North's people and provide the basis for the DPRK's integration into the international community in a peaceful fashion." The North, said the review, "would strongly resist such reform, viewing it as indistinguishable from a policy of undermining." North Korea, the review noted, would then "proceed with nuclear weapons and ballistic missile programs." The result could be a "destructive war" that "would not win the support of U.S. allies in the region" and "might harm the people of North Korea more than its government."[116]

Was a showdown, however, unavoidable? A report issued by a group of Republican members of Congress on November 4 said the threat posed by North Korea since the signing of the Geneva agreement five years earlier had "advanced considerably" while the United States, lavishing the North with a grand total of $270 million in aid in 1999, had replaced Russia as North Korea's main source of aid. "N. Korea's Best Pal: The U.S.?" asked the headline atop the front page of the Investor's Business Daily, a paper not noted for its interest in foreign affairs. "Some Fear Policy of Aid, 'Engagement,' Rewards Bad Behavior."[117] U.S. policy no doubt was an experiment in appeasement. Unless the United States were willing to go to war, however, there might be no choice but to negotiate in hopes that practical considerations would prevail. The alternative might be to elevate the Korean crisis to a regional conflagration that Americans wanted at all costs to avoid amid unprecedented good times at home.

For all their outrage over Taepodong, the Japanese soon followed the American example, lifting sanctions on December 14, 1999, in the long-range hope of opening up the North as they had so many other markets around the world. The next day, in Seoul, officials for KEDO and the Korea Electric Power Co. signed the contract under which KEPCO would actually build the reactors. Amid a possible thaw in tensions on the peninsula, the signing was a watershed. To North Korean complaints that the project was far behind schedule, Kartman had a ready reply. "They complained, 'Where's the beef,'" he said. "Now you can see there's a lot of beef out there."[118]

# Chapter 12

# Crisis of Identity*

A debate in the fall of 1998 over why North Korea invaded the South in 1950 touched the deepest sensitivities of a nation puzzled by how much freedom to allow commentary on the left and the right. The protagonists were a well-known professor, Choi Jang Jip, who served as a close adviser to President Kim Dae Jung, and *Chosun Ilbo,* a leading newspaper noted for attacks on officials viewed as leftist. A court ruled on November 11 that *Chosun Ilbo*'s monthly magazine had libeled him in an article attacking him for his analysis of the launching of the Korean War in June 1950.[1]

The critical question was why Choi, in a book published eight years earlier, stated that "the war in the initial period was fundamentally a national liberation war." Acceptance of that view, said Korean conservatives, would mean that North Korea, backed by China and the Soviet Union, was justified in its attack on the South, defended by U.S. forces. Choi, on the faculty of Korea University and chairman of the Presidential Commission on Policy Planning, charged that the magazine had crafted the article to make him appear sympathetic with the North Korean regime. He noted that he did not question whether Kim Il Sung, the North Korean leader who passed on power to his son, Kim Jong Il, before dying on July 8, 1994, had started the war.

As for conservative complaints that he had described Kim Il Sung's decision to go to war as "historic," Choi countered that the Korean word that he used was misinterpreted. "I did not use *yoksajok* ("historic" or "historical") to positively portray Kim Il Sung's decision," said Choi, "but to explain its historical significance—that it had since continued to influence Korean society." The debate aroused concern at the top levels of government, since Choi had a hand in writing Kim Dae Jung's speeches, and was partly responsible for his sunshine policy. Unlike previous governments, however, this one was anxious to defend rather than to censure. A spokesman for Kim said the government had "concluded there are no problems" with Choi's writing.[2]

The debate raised an intriguing question about freedom in Korea. Here was a country where people expressing Choi's views a few years earlier would have gone to jail. Now the conservatives were under fire for exercising their right to criticize. Running like a subtext through the controversy were memories of the leftist revolts that had periodically shaken South Korean society since the Japanese occupation. Campus turmoil had a special place in social and political evolution if not revolution. Student revolt helped bring down the government of Rhee Syngman in 1960; student revolt shook the country for months after the assassination of Park Chung Hee by his intelligence chief in 1979. Revolt on the campuses helped bring about the end of military dictatorship and the introduction of democracy in June 1987.

More than a decade later, the issues—and attitudes—were different. In mid-1998, on the third floor of the student union at Korea University, Lee Han Byul, vice chief of a student union affiliated with Hanchongryun, the Korean Federation of University Students, spoke earnestly about the problems. "Statistics show that only 20 percent of college graduates these days find jobs," he said, talking in a milieu of sloganeering banners, piles of leaflets, and computers where students were eagerly writing more diatribes against the government. "We try to deal with that problem," said Lee. "Those who have power say simply, 'Study hard,' but the problem comes from much deeper sources. We will keep fighting."[3]

Three floors below, in the student center coffee shop, students talked derisively of the relatively few activists on a campus once rocked by demonstrations and riots in which Molotov cocktails were routinely piled up, ready for instant use by agitators, in the same student center building. "Nowadays there are no demonstrations," said Park Sung Hwan, who had decided to continue his studies in English literature in graduate school after failing to find a job before getting his bachelor's degree in 1997. "Students are too busy looking for jobs. Most students are just not interested."[4] Beside him, Cho Jae Hyun, a sophomore, said he was considering one way out—a leave of absence from college to do his military service. "Students are not dropping out," said Cho, denying claims that many students had to quit school when their parents ran out of funds.

"Instead, most students go to the army early. By the time they get back from the army, things will be fine."[5]

Such approaches to "the IMF crisis" typified the mood on Korean campuses. While the headlines were full of reports of walkouts and strikes by workers fearful of losing their jobs, only a relatively small minority of students were joining them on the picket lines. The focus on campuses in the IMF era was on scraping together enough funds to get through the next year or two of school and on competing for the few jobs offered by companies that once hired scores of students each year with promises of lifetime employment.

The shifting mood extended to the upper levels of faculties. Suddenly an often-criticized leader from the era of the Korean War emerged as a respected figure among scholars looking for the origins of the "Korean miracle" in the current period of economic turmoil. The legacy of Rhee Syngman assumed fresh significance as Koreans celebrated the fiftieth anniversary on August 15, 1998, of Rhee's inauguration as first president of the newly formed Republic of Korea, South Korea. "He was the architect of modern Korea," said Lew Young Ick, director of the Institute for Modern Korean Studies at Yonsei University. "He was the best-prepared man to lead Korea through the war and into the new era of peace."[6]

The respect accorded Rhee, who was forced into exile in the student-led revolution of April 1960 and died five years later in Hawaii, was just one aspect of an outpouring of patriotism surrounding the anniversary. It was as if the celebrating might help ease the pain of the IMF crisis. "In the last half century, Koreans have recovered their pride," said Kim Ki Hwan, ambassador-at-large and a key figure in arranging the IMF negotiations. "This economic crisis causes difficulty, but I don't think it will last that long."[7] Officials anxiously saw the anniversary as a reminder of the optimism engendered by Rhee Syngman's assumption of power half a century earlier. "In 1948 we established a new government with a bright future," said Choi Kyu Hak, a director at the Korean Overseas Cultural and Information Service. Half a century later, ran the litany, we can do it all over again, without the tragedy of a second Korean War.

Korean flags sprouting from taxicabs, apartment buildings, shops, and factories conveyed the message. It was a show of nationalism that transcended the usual patriotism displayed on August 15, not only the date of Rhee's inaugural but also "independence day" commemorating Japan's World War II surrender, the end of 35 years of Japanese rule. Newspapers before the anniversary serialized histories of South Korea's great leap from abject poverty and suffering to relative prosperity, and television documentaries portrayed the success of a country that had appeared by the end of the Korean War to be incapable of surviving on its own.

The government whipped up the spirit with a commercial featuring champion golfer Pak Se Ri taking off her shoes before stepping in the water and chipping the ball from the rough, all to background voices singing what had once been an anthem of antigovernment radicals. "Although we have a long way to go," ran the lyrics in rough translation, "in the end we can triumph." On the streets of the capital, Koreans mingled pride with concern as they looked back on the ups and downs of a turbulent history marked not only by war but also by military dictatorship, coups, antigovernment demonstrations, labor unrest—and the miracles of economic success and democratic forms of governance.

For some Koreans, the memories went deeper still. Han Hae Soo remembered the Japanese who ruled his village as if he had seen them just yesterday. "The Japanese didn't treat Koreans like human beings," said Han, now 70 and working as a security guard in a busy shopping district. "They treated dogs better than they treated us. They made us children kneel down on the road in apology if we got in their way." Now Han believed it was "absolutely time for the Japanese to apologize to the Koreans" even though such an apology "will not be easy to get and anyway will not be sincere."[8]

The issue of "the apology" turned into the primary concern of Kim Dae Jung as he flew to Tokyo on October 7, 1998, for a four-day visit highlighted by his first summit on October 8 with the Japanese prime minister, Keizo Obuchi. The two leaders were to sign a joint statement that included Japan's first apology specifically to Koreans for Japan's colonial occupation. For Korea, wresting what the government called a "full apology" from Japan was vital to opening a wide range of relations with a people still viewed with the deepest suspicion, if not hatred. Japanese cars, cameras, and television sets were still banned from import into Korea, and television, radio stations, and theaters could not present Japanese dramas, shows, movies, or play Japanese songs. "Of course, the Japanese have to make the apology first," said Shin Hyung Sook, a librarian. "We still will not trust, and we are not sure if it's a good thing ever to let them bring in their culture."[9]

For many young people, the past was all but forgotten. In a warren of electronics shops in central Seoul, the young customers outside a tiny store with no name and no license to do business testified to its local fame as a purveyor of illicit Japanese songs on compact disks and diskettes. "Many people buy Japanese music here," said Park Chul Ho, lounging on his bicycle outside the shop. "We want to hear Japanese rock and ballads. We love to listen to pop music from Japan."[10] Although Japanese music, movies, magazines, comic books, and popular novels were all banned, they had long since flooded the black markets of Seoul and other cities and towns throughout the country.

Well before Obuchi expressed his "remorseful repentance" to Kim Dae Jung in Tokyo, DJ had said that Korea would open up to forbidden Japanese products ranging from music to cars. With the apology, the government was now working on the "timing" of a process that would legitimize the trend in cultural as well as consumer products. Park Moon Suk, director-general of the cultural policy bureau in the ministry of culture, suggested the confusion among bureaucrats as they sorted out what Japanese songs, shows, and books were fit for Korean ears and eyes. "Japanese culture has two sides," said Park. "One side is bad and harmful to teenagers. The other side has good aspects." As an example of the bad, he cited Japanese *manga,* mass-selling comic books that he described as "harmful and dirty." On the good was "old traditional culture." Too much of either would pose the danger of "Japanese cultural imperialism" in which "Japanese culture intrudes on other countries."[11]

The forces of nationalism and conservatism, however, might be no match for the desire of young Koreans to absorb influences from Japan, just as they absorbed American songs and movies, often merging them into their own styles and forms. Hwang Hye Joung, a secretary in downtown Seoul, laughed derisively at the Japanese apology but welcomed the prospect of being able to buy Japanese CDs and videos openly at major stores rather than in obscure back-alley shops. "They have apologized many times," she said. "It's not a practical thing, and it's not enough." Even so, she said, "Japanese culture is unique and modern, it's very different from ours, and we would like in the future to see it clearly."[12]

The specter of a renaissance of Japanese domination faded from the national consciousness as Koreans contemplated the import of Japanese goods after the lifting of anti-Japanese trade barriers. "Japanese cars are well known," said restaurant manager Kim Hyun Jong. "If the price is suitable, why not buy them? We should open up our markets."[13] As Kim Dae Jung returned on October 10 from his four-day visit to Japan, Korean officials said they were now ready to end historic bans on almost all Japanese consumer products, notably motor vehicles. Only six Japanese cars were imported in 1997, all of them Toyotas assembled not in Japan but in the United States and listed as American imports.

After years of waiting for Korea to open its markets to Japanese motor vehicles and big-screen television sets, Japanese and Korean officials and business leaders were beginning to wonder what all the fuss was about. One thing they did not expect when the barriers were removed at the end of June 1999 was a major influx of Japanese consumer items, notably motor vehicles. "Japanese exports to Korea will not increase automatically," said Yoshioki Nakamura, general manager of the Seoul branch of Mitsubishi Corporation, Japan's largest trading company. "Japanese automobiles cannot get in here."[14]

Japanese and Koreans cited Korea's long-term economic problems as the major barrier to a significant increase in Japanese imports despite the summits of their leaders—Kim Dae Jung's visit to Japan in October 1998 and Obuchi's return visit six months later. While Obuchi offered less than whole-hearted endorsement of DJ's sunshine policy, he smoothed over Korean sensitivities by extending another $1 billion loan to the South, in addition to Japanese loans of $10 billion since the onset of the economic crisis. Despite its own economic troubles, Japanese loans to South Korea exceeded those of any other country.

Korean officials denied that lingering Korean suspicion of Japanese intentions was maintaining unwritten barriers that might discourage entry of Japanese consumer products as much as formal regulations. "We have no problem between Japan and Korea," said Choi Chul Woo, deputy director of the import division of the commerce ministry. The consumer market at the end of June 1999 opened up to 15 items previously barred from import, including 7 types of motor vehicles as well as engine parts and automotive machine centers, but Choi observed confidently, "Our consumers have no money to buy these items."

As Korea lifted barriers, Koreans worried about a balance of trade that was overwhelmingly in Japan's favor even though total trade between the two countries fell by 33 percent in 1998 from 1997. Korean exports to Japan totaled $11.87 billion in 1998, a drop of 19.7 percent, while Japanese imports fell by 41.5 percent to $16.28 billion, still heavily in Japan's favor. The worst problem was not the deficit with Japan but the fact that Korean industries did not have money to buy machinery they were accustomed to getting from abroad. The result was a vicious circle in which exports of finished Korean products were decreasing and manufacturers had still less for capital equipment.

"The trade volume is contracting due to the economic crisis," said Shin Hyoung Oen, deputy director of the division in charge of trade with Asian countries at the ministry of foreign affairs and trade. Capital goods were 90 percent of Korea's imports from Japan, while Korean exports to Japan ranged from semiconductors and steel to clothing, petrochemicals, audio equipment, and dairy products. Shin saw a direct relationship between declining capital imports and the efforts of the chaebol to restructure. "They do not have much money for machinery from Japan," he said, predicting that "the decrease of machinery imports will further adversely affect our economy."[15]

There was a bright side, however, in the form of a possible increase in Japanese investment in an economy in need of foreign funding. A bilateral investment treaty between Japan and Korea would be an important dividend of the economic crisis for both countries. "These days we like to talk more about investment than trade," said a Japanese diplomat. "We want to see investment promotion take effect." The Japanese hoped to negotiate removal of barriers that had held direct investment from Japan in 1998 to just $500 million of a total of $8.8 billion invested by foreign countries. Among them were restrictions on remittances of funds from Korea to Japan, high taxes, and rules that discouraged or prevented Japanese banks from setting up branches in Korea. So difficult were the problems that total Japanese direct investment in South Korea was only $6.3 billion, 1 percent of all Japanese investment abroad.[16]

The prospect of an influx of Japanese funds and goods recalled an era when Korea was still independent but at the mercy of Japanese merchants. The ironies of Korean history echoed across the generations in a Korean musical that blended the influences of Broadway, European opera, and Korea's own musical tradition in a saga of nationalist frustration, foreign skullduggery, and personal tragedy of imperial dimensions. The musical, *The Last Empress,* was about the legendary Queen Min, who had risen from common origins to marry King Kojong in the late nineteenth century and then strove mightily to protect him, their son, the throne, and the country from the conniving Japanese. She paid with her life as the victim of a meticulously arranged Japanese assassination plot, a tragic precursor to half a century of Japanese domination.

"This production is a lesson from history," said Yun Ho Jin, who directed *The Last Empress* for its initial opening on December 30, 1995, the one hundredth anniversary of Queen Min's assassination, and for a revival in 1998. "The form is western-style, the sets are western, but the songs carry a lot of Korean rhythm and emotion."[17] Kim Hee Gab, who had written songs for Korea's popular singers, had composed a score that mingled operatic arias of love and longing with lighthearted numbers evoking the streets of the capital and a palace that was isolated from its own people as well as the world at large. Yun was responsible for bringing Kim together with Lee Moon Roel, poet and novelist, who fashioned the words, all in Korean, for nearly three hours of almost continuous music interspersed with very little dialogue.

"I want to emphasize a little more Queen Min's emotional side," said Yun, but it was hard to imagine a more emotional creature than the torn character of Queen Min, an egotistical, power-minded woman whose dilemma epitomized that of Korea a century ago—and possibly today as well. "Your Highness is like Elizabeth the First of England," trilled her flattering French teacher, but in her emotional and patriotic quandary she was more reminiscent of Lady Macbeth. "It's a look at ourselves, how nothing has changed," said Kolleen Park, the Korean-American musical director. "The audience is looking out and saying, 'It's the same situation now.'"[18] English subtitles flashed above the stage for the benefit of the few foreigners in a typical audience. "I wish I could see beyond his mask," sang Queen Min of the wily Japanese ambassador. "Behind his words lies poison and behind his sly smile hides a sharp knife."

That was a lesson the Koreans had never forgotten about the Japanese, not to mention a host of other foreigners whom they still saw as threatening a peninsular culture caught in the vortex of great powers near and far. One ominously hulking reminder of Japanese culture was the great granite-and-marble structure built by the Japanese as a colonial capitol building, then turned into a national museum in the 1980s.

On June 1, 1996, the building, conceived by the Japanese at the outset of their rule in 1910, designed by a German architect, begun in 1915, and completed in 1926, was on its way to the ash heap of history. Outside, scaffolding hid the dome above a large banner announcing, in Hangul, "Demolition of the old capital." Banners on either side of the massive front doors demanded, "Boost Our National Spirit to Achieve Reunification and Globalization," and "Outcry for Independence, the Basis of New Korea's Power." Korea's largest builder, Hyundai Engineering and Construction, whose green diamond also adorned the banner, was tearing down the structure, giant block by giant block, an expensive operation that lasted for two years. It was all part of a cultural reawakening, a campaign to reassert national identity in a time of crowning success. The new national museum, on the site of another affront to nationalist sensitivities, the golf course on the sweeping grounds of the American military headquarters, would not open until 2003, by which time Koreans hoped the Americans would have retreated to a base outside the city.

No one doubted that the Japanese had chosen the site and size of the building, in the Kyongbok Palace complex, as an insult to Koreans. The enormous structure was designed to block the view from Kwangwhamun, the gate in front, to the ceremonial hall where Korean kings had received visitors and gazed down the long avenue through the capital. "Why did the Japanese build it here?" asked a guard rhetorically. "They wanted to hide our emperor. The Japanese tried to cut the straight line of the Korean spirit." The line extended through the palace grounds and the gate, on down Seoul's most impressive avenue, past the statue of Admiral Yi Sun Shin, the naval commander whose ironbound "turtle ships" had turned back a Japanese invasion in 1592, then on to Namdaemun, the historic South Gate to the city.

The other way, beyond the back wall of the palace grounds, was Chong Wa Dae, the Blue House, built by President Park Chung Hee, also with help from Hyundai, soon after he seized power in May 1961. The blue-tiled roof of the presidential mansion was all that distinguished it from the ceremonial hall of emperors. The style was much the same—inspired by Beijing's Forbidden City, from which Korean kings had derived protection and inspiration while paying homage for centuries until Korea's humiliation by the Japanese.

Inside the Blue House complex, Park Jin, presidential assistant, talked to me one day in 1996 about the new nationalism. Phrases like "glorious revolution" and "reform policy" and "dynamic democracy" tripped lightly off the tongue of an erudite man who had picked up a bit of an English accent during studies for a doctorate at Oxford and three years teaching political science at Newcastle. "We see Korea as a dynamic democracy which would transform itself," he said. "One century ago we were swallowed by other countries"—mostly Japan. With the destruction of the capitol, the hated monument of the Japanese, in the Blue House view, Korea could emerge from the cocoon in which it was enveloped through centuries of dynastic, feudal, self-defeating rule. Next stops on the deconstruction train were Seoul City Hall, an ugly mass of concrete on the square opposite the Toksu Palace, and Seoul Station, a lesser version of Tokyo Station. "We don't like to satisfy ourselves with just being an assertive nationalist nation," said Park, but he saw both the city hall and Seoul Station as affronts that had to go.

The economic crisis of the late 1990s undoubtedly sped up the process of globalization, transparency, and openness to foreigners. In central Seoul, across a broad avenue from Pagoda Park, thousands of students immersed themselves in the pursuit of a different kind of foreign discipline. From 7 A.M. to well into the evening, in small classrooms in a score of office buildings, they struggled to learn English. It was a battle they believed they had to wage in order to compete for jobs or promotions in the jobs they already held—though dwindling funds in a period of economic turmoil had forced many to abandon the quest.

"Koreans know we very seriously need English-speaking capability," said Chung Young Sam, president of YBM Sisa-Yong-o-sa, a language institute (*hakwon*) whose 600 teachers were busy drumming English conversation into 20,000 students, mostly young people in college or first jobs, in 10 schools scattered around metropolitan Seoul. The students, like most of the others in about 130 registered *hakwon* and 10 times as many nonregistered ones, attended one 50-minute class a day, 5 days a week, 20 times a month. The goal of many was to learn the language well enough to pass a test administered by the Educational Testing Service in Princeton, New Jersey, called the TOEIC—Test of English for International Communications. Armed with a good score on the TOEIC, job applicants and those already working for companies had a distinct advantage over less tutored colleagues whose knowledge of English was generally a few phrases from high school.

Where language, ideology, nationalism, and economy were not issues, however, Koreans were still often divided among themselves. Inside a basement tearoom in central Seoul, secluded from the wreckage of nights of mayhem outside the Chogye Temple, Korea's leading Buddhist place of worship, the Venerable Mu Hyu boasted of his ability to survive against the "gangsters" who he said were trying to regain control of the order for personal profit and power.[19] He would need all his strength to defend the 1,000-year-old Chogye Temple and the headquarters beside it of the Chogye order, Korea's largest Buddhist sect. Ranged against him and 200 other monks in the barricaded headquarters were upward of 500 other monks and laypeople who said they were in charge of the order and its eight million adherents.

Periodically, zealots had been tossing Molotov cocktails along with sticks and rocks around the temple, festooned with signs urging support for the monks, who had seized it on November 11, 1998, in the name of a Committee for Purification and Reform. The physical appearance of some of the monks and their followers on both sides suggested the real cause of the struggle for control of the order, which included about 2,000 temples around the country. The victor got the purse strings of a treasury overflowing with contributions. "There are millions involved in donations and a certain percentage that everyone pays for entering a national park with a temple," said Frank Tedesco, an American Buddhist living in Seoul. "They've gotten rich off real estate. There are companies that are building new temples."[20]

Other Koreans looked outward, attempting to resolve their role on a wider stage. Peter Hyun, eyeing a world that had opened to him as a child during Japanese colonial rule, wondered where on earth he belonged. "I'm a gypsy, I'm stateless," said Hyun, balancing a cup of tea during one of his infrequent returns to Korea. "I know a country, but I'm not a part of it."[21]

Author, editor, memoirist, Hyun wasn't exactly searching for his roots. Rather, at 70, he was trying in 1998 to make sense of a life that had begun in the industrial city of Hamhung, on the east coast of what is now North Korea, and deposited him in New York and France. "I want to go home to North Korea," he said, but mostly he would like to revisit the turbulent era in which he was born and raised, then flung into what was now South Korea on his way to alien western cultures. He evoked those days in a memoir published in Korean, serialized in the local press, and turned into a Korean television special. Now he was writing "a fictional account of my memoirs" in hopes the story, as a novel, would interest foreigners to whom factual accounts of Korean hardship and suffering remained elusive.[22]

Turning the story of his life into English should not be an impossible task for one who was inspired to write after reading Walt Whitman's *Leaves of Grass* while interpreting for American GI's shortly before the outbreak of the Korean War. "It was an exciting period for Korean artists and intellectuals," said Hyun, sipping his tea, looking toward Namsan, the small mountain that rises in the center of Seoul. "Many of them were radicals. The American military was not very happy with them. Some escaped to the North." Young Peter had other ideas. He had just lived through World War II in Hamhung under Japanese rule. "The Japanese were very cruel," he said. "We were not allowed to speak Korean. When I was 14, I wrote a short story called 'The First Kiss' in Korean. I was suspended from school."[23]

Now Hyun saw himself as a link between disparate cultures, the Korean one, often closed to foreigners, and the American and European ones, seemingly more open but difficult for Koreans to penetrate socially. "Because I happen to live in both worlds, I would like to play that role as a bridge," he said. "Koreans resist. They're kind of like frogs in a pond. Those guys who've spent time in grad school in the U.S. or London come back and are Koreans again." He was more optimistic about Koreans raised abroad. "They are globalized Koreans," he said. "I hate to say 'westernized,' but they see both sides of the coin, whereas Koreans here as a whole are narrow-minded. If an American writes a critical article, they aren't very pleased."[24]

The personal identity problem of Koreans was exacerbated by the economic crisis. I saw a young mother in tears as she and her landlady walked into the Seoul Children's Counseling Center, both of them tugging at her six-year-old daughter. "I was trying to keep my child, but I have to work," she said between sobs to a counselor. "It's too hard to be a single mother these days. I need to have two jobs to survive. There was no one to care for my child." With that, she hugged and kissed her wailing daughter, promised to visit her, then left to the landlady the details of abandoning her to the welfare system. "She is too exhausted," said the landlady as the child, still weeping, was led away to a dormitory already filled with several dozen other children. "She works night and day in restaurants. The girl's father has no job. He sends no money. There is nothing else to do."[25]

It was a familiar story. "We have more orphans than ever before," said Cho Kyu Hwan, director of an orphanage called Angel's Haven. "People lose jobs. They bring in their children. They say, 'Please take my child.' Some see their parents again. Some do not."[26] The plight of abandoned children in the IMF era would be sorrowful enough under any circumstances. In this society, deep-seated antiforeignism added immeasurably to the suffering of thousands of children who might otherwise find homes abroad. "We are constantly trying to reduce the numbers of children to be adopted overseas," said Lee Chang Jun, in charge of the problem of infant and nursery children in the child welfare division of the ministry of health and welfare. "Instead we are encouraging more and more adoptions within Korea. We believe we Koreans are the ones who will be responsible for Koreans."[27]

That outlook translated into a policy under which the number of Korean infants adopted by foreigners was declining at a prescribed rate of 3 percent a year until the IMF era. From a high of more than 8,000 in the late 1980s, 2,057 Korean children were sent abroad in 1997. The figure went up somewhat to 2,249 in 1998, but the government did not plan to raise it again even though Koreans in 1998 adopted only 1,426 children. More revealing, 846 of the babies sent overseas in 1998 were disabled while Koreans adopted only 6 with disabilities. Such statistics, though, hardly told the story. They included only the offspring of single mothers—married parents could not put a child up for adoption—and did not include thousands deposited in the welfare system. The number left in orphanages in 1998 went up 38 percent, from 6,734 in 1997 to 9,292. By now, 272 orphanages had nearly 20,000 children, none eligible for adoption. Orphanages held children in varying conditions, feeding and clothing them and sending them to local schools until they reached 18.[28]

The government preferred not to face the issue. "The problem is there are no more institutions for orphans in Korea," said a manager at the Eastern Child Welfare Society, one of four adoption agencies authorized to send babies for adoption abroad. "Until last year there were many rooms, with much space and many vacancies. Now they are full."[29]

If placing a baby in foreign hands was a fate to avoid, importing foreign films presented another threat to patrimony. Actor Ahn Sung Ki in 1992 played two of his greatest roles in Korean films—one as a North Korean soldier left in the South during the Korean War, the other as a South Korean intellectual who volunteered to go to Vietnam as a soldier in the South Korean army. In the year of economic turmoil, he was playing a new and different role—that of a crusader for a quota system that required Korean movie theaters to show Korean movies at least 106 days a year. "It is not only an issue of the Korean film industry but of all artists and scholars," said Ahn, talking in the office of the Motion Picture Directors' Association of Korea against a backdrop of black-and-white photographs of Korean directors. "This is a national issue. The pressure hurts the pride of our people."[30]

The pressure in this case was from the United States, calling for ending or at least sharply cutting the quota as it negotiated a bilateral investment treaty aimed at opening Korean markets that were previously closed or severely limited. The negotiations had galvanized the Korean film community into a campaign ranging from the streets of downtown Seoul to the steps of nearby Myongdong Cathedral to the American embassy. Korean moviemakers wore black clothing, mourning the coming death of their art, and carried banners and shouted slogans denouncing "the cultural imperialists" for "the murder of Korean films." In one outburst in December 1998, directors and producers led several hundred filmmakers in speeches, songs, and slogans for more than two hours in freezing temperatures. "Protect Korean films," the crowd shouted in response as technicians ran lighting and sound equipment from local film companies. "Let's fight against America."[31]

In the midst of the quota campaign, one director, Kang Je Gyu, showed it was possible to solve the great mystery of the local industry: how to make a movie that filled local theaters for weeks on end, reaped a small fortune, and trounced the competition from abroad. Kang overcame the barrier with an action drama called *Swiri* that had it all: fast-moving, vicious violence, a love story that had them dabbing at their eyes, and a political message that got at the tragedy of a divided nation.

By the time the film disappeared from local screens in June 1999 after a five-month run, more than 4.5 million people had seen it, setting a local record, and it had grossed about $20 million, another local record. Those numbers represented a return on an investment of about $2.5 million. "That's about 1 percent of the cost of a Hollywood film," said Kang, but it was the most money ever spent on a film in Korea, where movies are typically produced for less than $1 million.[32] So what did *Swiri* offer that compelled moviegoers to watch it rather than the nearest film from Hollywood or one of those Hong Kong action dramas that draw crowds throughout Asia? The distinctively Korean element was not the shoot-outs, the killings, the chases through the streets, or the heliborne pursuits but the drama of North versus South. The message was so political, so powerful as propaganda, that the armed forces were showing it to troops and national leaders endorsed it.

The story begins with horrifying shots of training terrorists in the North. One of them, a young woman with the taut facial muscles of the dedicated fanatic, bayonets prisoners bound to stakes, under the tutelage of a swarthy man in a leather jacket. The woman surfaces in the South as a terrorist. There she hooks up with a South Korean intelligence agent, living with him while taking orders from her old teacher, this time in charge of terrorists stalking the South. They communicate using a secret code in the form of a *swiri,* a freshwater fish unique to Korea, a symbol of the oneness of the divided Korean peninsula. A setup for a morality tale about the evils of communism? Perhaps, but the love triangle, superimposed on terrorism and counterterrorism, makes the story achingly emotional. "What's wonderful about this film," said actress Kim Yun Jin, the woman who terrorizes the South, "is that despite the conflict, the difference between North and South, love is the answer."[33]

Scenes of some of the film's carnage were set in the fashionable Kangnam district, south of the Han River that bisects Seoul. In Kangnam, however, the most visible sign of North Korean aggression as the South recovered from economic crisis in 1999 was a weapon that no one imagined through all the years of North-South confrontation—cold noodles, Pyongyang-style. The recipient of millions of tons of emergency food aid, the North in 1999 began shipping buckwheat, bean paste, and liquor to a restaurant off one of the wide boulevards running through Kangnam to satiate the appetites of South Koreans yearning for a taste of life above the DMZ. The buyer was an unabashedly capitalist entrepreneur whose restaurant took its ingredients, cuisine, and name from a restaurant, founded years before on orders from Kim Il Sung, in Pyongyang.

"We're getting ten tons of buckwheat a month for noodles," said owner

Kim Young Baek. "We don't think about politics."[34] Bags labeled "DPR Korea," shipped from Nampo to Inchon, littered the office of the Okryuk-wan Restaurant, named after the one in Pyongyang that once dished up 10,000 servings daily. Kim suspected the original Okryukwan, overlooking the Daedong River in Pyongyang, now served just occasional foreign visitors and VIPs while millions of North Koreans survived on tree bark and the husks of corn kernels. His own Okryukwan, meanwhile, could hardly cope with the crowds lined up at the door.

Inside, all 360 seats on three floors were filled with diners tucking in on the cold noodles for which Pyongyang remained famous in Seoul. A chef who held a Japanese passport and trained at the Okryukwan in Pyongyang and a defector who once worked at another restaurant there before escaping to Russia and fleeing to the South maintained the tradition. To guarantee authenticity, Kim Young Baek purchased hundreds of chopsticks, ceramic spoons, plates, and bowls from the Okryukwan Restaurant in Pyongyang, all displaying its label.

Kim, after importing metal products from the North for eight years, saw the restaurant as setting several precedents. "Since 1991, when there was the first economic cooperation between North and South, all the business was in the North," he said. "This is the first such business in the South." Another first: "Until now all economic transactions between North and South were based on crops and raw materials. Now Koreans can feel real cooperation, since this is a service sector." Pyongyang "is going to register the name of Okryukwan in the South, and the South Korean government has promised to accept it," he added. "That means if a South Korean company wants to sell anything in the North, North Korea will register its brand as well."[35]

The restaurant's popularity had surged so fast that Kim hoped to open another ten just like it. He had the permission of the South Korean government, eager to push reconciliation, as well as the cooperation of North Korean authorities, eager to export their products. The question, however, was how long supplies would last. "Because of the food shortage problem in the North, we anticipate that the supply will not be stable," said Kim, asking for three shipments of buckwheat a month as well as a ton a month of mung bean used for pancakes, also Pyongyang-style. "I'm not sure where they get the buckwheat and the mung bean," he said, but he believed the North Koreans purchased it on the free market that had grown as government agencies broke down. Kim talked every day to representatives of a trading company in North Korea, patching calls through Tokyo. A Korean-Japanese company handled paperwork for shipments that

Pyongyang still did not recognize as moving directly to the South.[36]

In the meantime, the Okryukwan sought to live up to the reputation of its namesake in Pyongyang. A painting of the real Okryukwan by a North Korean artist adorned the outside wall. There were no signs, however, of Kim Il Sung or Kim Jong Il. "We are completely nonpolitical," said Kim. The restaurant manager, Shin Dong Pyo, added an observation that said something about the North's military preparedness. "We are worried about our utensils," he said. "We need authentic brassware from the North, and the shipment is late. They are made by a factory that produces military products. They are quite slow."[37]

The popularity of the Okryukwan was one small sign of consumer spending in South Korea after more than a year and a half of IMF restraints. The Bank of Korea, after predicting in March 1999 that the economy would grow 3.1 percent for the year, was confident enough by May to proclaim the country as "escaping from the recession of 1998." Forecasts now showed growth of more than 5 percent for the year—a mirror image of the 5.8 percent decrease recorded in 1998. The desire to spend after 18 months of hoarding powered the increase as personal consumption soared 6.3 percent in the first quarter of 1999 compared with the corresponding quarter of 1998.[38]

"There's a pent-up demand," said Hyun Oh Seok, director-general of the finance ministry's bureau of economic planning. "Last year if you ordered a car, they delivered tomorrow. Now it takes three weeks."[39] The crowded shopping centers and streets of Seoul bore out the central bank's observation of "increased spending of households, especially on durable goods such as cars and personal computers." Said Kim Tae Ho, a salesman in a Hyundai Motor showroom in Kangnam: "Last year only those who needed cars bought them. Now that we're out of the IMF crisis, the big cars are selling better."[40]

The desire to spend spread to department stores, where sales managers and clerks reported customers returning in larger numbers and buying more. "Last month there was a big increase in sales," said Kim Hye Ok, a clerk at Hyundai Department Store in Seoul's upscale Apkujongdong district. "That was the year of the IMF crisis. This year there is not an IMF crisis anymore. The term 'IMF' is almost gone."[41] Sure enough, the signs on restaurants advertising "IMF menus" and stores announcing "IMF sales" were gradually disappearing. Sales of consumer items had done so well that some companies were taking on more workers again. Hyundai Motor was rehiring about 1,500 people let go in the summer of 1998. While many industries were downsizing or jettisoning unprofitable units, the government allocated more than $13 billion in 1999 to public works projects.[42]

Consumer confidence, though, was still not as high as during the years preceding Korea's plunge toward near bankruptcy in 1997. "Sales have been steadily increasing," said Chung Soo Young, selling men's suits in Hyundai Department Store. "The difference is, in good times before the IMF, people tended to buy two or three pieces at a time. Now things are better, but people are still complaining—and not buying that much."[43] Sales were strongest in the city's "night markets"—large buildings crammed with hundreds of shops selling consumer items all night long. "We can't tell if there's a real economic boost or people have just loosened up," said Whang Keong Bae, selling clothes in a night market near the city's Tongdaemun or East Gate. "Sales are up 10 to 20 percent. Many people who used to go to department stores are coming here."[44]

IMF Managing Director Michael Camdessus, visiting Seoul on May 20, 1999, as the central bank reported an array of promising statistics for the first quarter, was brimming with optimism. "Asia is now emerging from the crisis that engulfed the region," he said. "Most clearly in Korea, the Philippines, and increasingly in Thailand and Malaysia, we see an upturn." Korea, he said, might not have to draw the remaining $2 billion of $21 billion authorized by the IMF as its portion of the rescue package.[45]

The crisis, however, was not necessarily over. Finance Minister Lee Kyu Sung warned of the "downside risk" of economic problems in Japan and Europe slowing down exports and admitted the chaebol had been "slow to reform."[46] On May 24, Kang Bong Kyun, Kim Dae Jung's chief economics secretary, replaced Lee as finance minister. DJ also appointed ten other new ministers in his first cabinet shakeup. Chung Duck Koo, promoted from vice finance minister to minister of commerce, reminded finance officials as he left, "Remember you are not only competing with finance ministries around the world, you are being watched by Wall Street."[47]

DJ's greatest worries now were political, not economic. The reason for the reshuffle was to prepare for National Assembly elections in April 2000. "He's putting in all those guys who were close to him," said Park Nei Hei at Sogang University.[48] Most of the ministers who quit were members of the Assembly. Their mission was to expand the slim majority held by Kim's National Congress for New Politics and its coalition partner, the United Liberal Democrats, led by Kim Jong Pil. A strong opposition, led by DJ's 1997 campaign foe, Lee Hoi Chang, was waging a tough campaign. DJ's predecessor, Kim Young Sam, indicated the bitterness of the opposition, calling DJ a "dictator." DJ hinted he might ally with his old oppressor, Chun Doo Hwan. Full of hatred for YS for having had him tried and jailed, Chun attacked YS as a "barking tavern dog."[49]

More incredibly, DJ, visiting Taegu, the stronghold of the late Park Chung Hee as well as Chun and Roh and the rest of the "TK Mafia," suggested a monument immortalizing Park, his once mortal foe. It was all a cynical ploy for votes—and funds. How could DJ follow through on his policies, economic reform and sunshine, while courting his enemies—and facing the North? The politics of vengeance was back as Korea's leaders reverted to the custom of aggrandizing power and using reform as a campaign slogan. If the economy had survived the IMF era, the troubles of Daewoo again exposed the underlying weakness of not just the chaebol system but an entire business style. The priority was politics for a top-heavy ruling structure that might adjust somewhat but remained resistant to change, much less foreign intervention, until forced by crisis.

As he approached the second anniversary of the IMF agreement, Kim Dae Jung was confident enough to proclaim the country "completely out of crisis." On December 3, 1999, the anniversary date, a galaxy of speakers poured praise on DJ and his policies at a lavish government-staged conference. "It was the fastest recovery you could imagine," said Hubert Neiss of the IMF, donating his $15,000 speaker's fee to an orphanage.[50] Were more shocks possible? With the GNP up by 8 percent for the year, consultant Peter Bartholomew saw "a lot of premature elation and complacency." Choi Jang Jip, back at his post at Korea University, observed that ordinary people questioned the reality of reform. "Some are getting richer," he agreed, "but for most the economy is stagnant."[51] While the biggest chaebol grew bigger, the pattern was not likely to change as fast as the "economic indicators." For many at the dawn of the millennium, the renaissance of the miracle, like the second coming, would have to wait.

# Notes

## 1: Crossroads

1. The Fair Trade Commission provided all the statistics on assets, equity, and debt-equity ratios, 1997.
2. Fair Trade Commission.
3. Invitation, Samsung Motor, Seoul, February 1998.
4. Cho Yoon Je, interviewed by the author, Seoul, June 1997.
5. Kang Hee Bo, interviewed by the author, Seoul, June 1997.
6. Gong Byeong Ho, interviewed by the author, Seoul, June 1997.
7. Fair Trade Commission, 1997.
8. Ibid.
9. Ibid.
10. SaKong Il, Institute for Global Economics, interviewed by the author, Seoul, June 1997.
11. Kang Hee Bo, interviewed by the author, Seoul, June 1997.
12. Kim Dae Jung, "Conversation with Citizens," Korea Broadcasting System, Seoul, January 1998.
13. Ibid.

## 2: IMF vs. Korea, Inc.

1. Money changers, interviewed by the author, Seoul, December 1997.
2. Ibid.
3. Ibid.
4. Kim Jun Il, interviewed by the author, Seoul, November 1997.
5. Yonhap News Agency, November 14, 1997.
6. Alain Bellisard, interviewed by the author, Seoul, November 1997.
7. Cristoforo Rocco, interviewed by the author, Seoul, November 1997.

8. Jean Jacques Grauhar, interviewed by the author, Seoul, November 1997.
9. Yonhap News Agency, November 14, 1997.
10. Finance Ministry, Seoul, statement, disseminated in November 1997.
11. *Chosun Ilbo,* November 17, 1997.
12. Jean Jacques Grauhar, interviewed by the author, Seoul, November 1997.
13. Yonhap News Agency, November 19, 1997.
14. Kwang Sang Yang, interviewed by the author, Seoul, November 1997.
15. Daniel Harwood, interviewed by the author, Seoul, November 1997.
16. Jason Yu, interviewed by the author, Seoul, November 1997.
17. Yonhap News Agency, November 19, 1997.
18. Finance ministry, statement, November 19, 1997.
19. Lim Chang Yuel, remarks, Seoul Foreign Correspondents' Club, November 20, 1997.
20. Ibid.
21. Ibid.
22. Samsung group, statement, press conference, Seoul, November 26, 1997.
23. Ibid.
24. Cho Jang Won, interviewed by the author, Seoul, November 1997.
25. Federation of Korean Industries, statement, November 27, 1997.
26. Kim Woo Choong, statement, Seoul, November 27, 1997.
27. Korea Broadcasting System, November 30, 1997.
28. Ibid.
29. *Chosun Ilbo,* November 30, 1997.
30. Michel Camdessus, quoted by Reuters and the Associated Press, Kuala Lumpur, December 1, 1998.
31. The author was among the reporters in the crush outside the room in the Seoul Hilton.
32. Chung Eui Dong, interviewed by the author, December 2, 1997.
33. Michel Camdessus, quoted by Reuters and the Associated Press, Kuala Lumpur, December 2, 1997.
34. Ibid.
35. Yonhap News Agency, December 2, 1997.
36. Federation of Korean Industries, statement, December 3, 1997.
37. Hyundai director, interviewed by the author, Seoul, December 1997.
38. Park Ho Won, interviewed by the author, Seoul, December 1997.
39. Lee Jung Seung, interviewed by the author, Seoul, December 1997.
40. A Hyundai manager who attended the meeting provided this description.
41. Ibid.
42. Kim Young Sam, broadcast to the nation, December 4, 1997.
43. Chaebol and union leaders, interviewed by the author, Seoul, December 1997. They asked that their names not be used.
44. Michel Camdessus, press conference, Seoul, December 3, 1997.
45. Lim Chang Yuel, remarks, Seoul, December 3, 1997.

## 3: "DeeJay, DeeJay"

1. Ahn Hyo Hee, interviewed by the author, Pundang, December 1997.
2. Choi Yeun Hee, interviewed by the author, Pundang, December 1997.
3. Korea Broadcasting System, December 4, 1997.
4. Ibid.
5. Yonhap News Agency, December 6, 1997.
6. Ibid.
7. Ibid.
8. Kim Eun Sang, interviewed by the author, Seoul, December 1997.
9. Financial sources, interviewed by the author, Seoul, December 1997, did not want their names used.
10. Ibid.
11. Korea Broadcasting System, December 1997.
12. Cristoforo Rocco, interviewed by the author, Seoul, December 1997.
13. Yonhap News Agency, December 9, 1997.
14. Daniel Harwood, interviewed by the author, Seoul, December 1997.
15. Yonhap News Agency, December 10, 1997.
16. Ibid.
17. Korea Broadcasting System, December 10, 1997.
18. Ibid.
19. Korea Automobile Manufacturers' Association.
20. Korean Trade Association.
21. Lee Jeong Seung, interviewed by the author, Seoul, December 1997.
22. Korea Broadcasting System, December 10, 1997.
23. Park Hong Kyu, interviewed by the author, Seoul, December 1997.
24. Cho Jang Won, interviewed by the author, Seoul, December 1997.
25. Lee Jung Tae, interviewed by the author, Seoul, December 1997.
26. The author witnessed the rally, December 12, and called the Blue House for comment.
27. Yonhap News Agency, December 12, 1997.
28. Kim Dae Jung's comments, along with those of Shin Woo Jae, appeared in the Korean press on December 13, before the meeting.
29. Ibid., December 12, 1997.
30. Yonhap News Agency, December 15, 1997.
31. Chung Soon Jung, interviewed by the author, Kwangju, December 1997.
32. Yun Young Ju, interviewed by the author, Kwangju, December 1997.
33. Moody's Investors Service, report on Korea, December 22, 1997.
34. Standard & Poor's, report on Korea, December 22, 1997.
35. Shin Seung Yong, interviewed by the author, Seoul, December 1997.
36. Richard Samuelson, interviewed by the author, Seoul, December 1997.
37. Yonhap News Agency, December 23, 1997.
38. Finance ministry officials, interviewed by the author, December 23, 1997.

39. Korea Broadcasting System, December 23, 1997.
40. Senior U.S. diplomat, interviewed by the author, Seoul, December 1997. He did not want his name used.
41. Ibid.
42. Ibid.
43. The author visited Kwangju in December 1997 shortly after the election. Chang Sung Hee interpreted.
44. Kim Tuk Han, interviewed by the author, Kwangju, December 1997.
45. Park Jung Koo, interviewed by the author, Kwangju, December 1997.

## 4: Criminals on Parade

1. Donald Kirk, *Korean Dynasty: Hyundai and Chung Ju Yung* (Hong Kong: Asia 2000; Armonk, NY: M. E. Sharpe, 1994), 331-33.
2. Korean reporter, interviewed by the author, Seoul, July 1992.
3. Korean Broadcasting System, April 8, 1996.
4. Park Jin, interviewed by the author, April 1996.
5. Kim Kyung Woong, spokesman for South Korea's national unification ministry, provided this information.
6. Cho Jang Won, interviewed by the author, Seoul, April 1996.
7. Moon Young Ho, written response to the author's questions, April 1996.
8. Lee, interviewed by the author, Seoul, April 1996.
9. Lee Jeong Seung, interviewed by the author, Seoul, April 1996.
10. Lee Sung Bong, interviewed by the author, Seoul, April 1996.
11. Song Jin Myong, interviewed by the author, Seoul, April 1996.
12. Korea Broadcasting System, live report, August 26, 1996.
13. New Industry Management Academy, report on the chaebol, Seoul, 1996.
14. Park Il Kwon, interviewed by the author, Seoul, August 1996.
15. Nicholas Eberstadt, interviewed by the author, Washington, D.C., August 1996.
16. Nohn Kyung Hwe, interviewed by the author, Seoul, May 1997. The author witnessed the Hanyang riots.
17. The author interviewed the women in the cemetery, December 1997. Chang Sung Hee interpreted.
18. The author attended the inauguration.
19. Hahn Chai Bong, interviewed by the author, Seoul, February 1998.
20. Michael Breen, interviewed by the author, Seoul, February 1998.
21. Kang Kyung Shik, interviewed by the author, Kyonggi Province, July 1998.
22. Lee Han Dong, interviewed by the author, Seoul, July 1998.
23. Kang Kyung Shik, interviewed by the author, Kyonggi Province, July 1998.
24. Park Kyong San, interviewed by the author, Seoul, March 1998.

25. Yonhap News Agency, March 12, 1998.
26. Ibid.
27. Chun Jae Soon, interviewed by the author, Seoul, March 1999.
28. Nam Kyu Sun, interviewed by the author, Seoul, March 1999.
29. Yoon Yong Ki, interviewed by the author, Seoul, March 1999.
30. Ibid.
31. Oh Wan Ho, interviewed by the author, Seoul, August 1998.
32. Park No Hae, interviewed by the author, Seoul, August 1998.
33. Kim Seong Man, interviewed by the author, Seoul, August 1998.
34. Kim Nak Jong, interviewed by the author, Seoul, August 1998.
35. Yonhap News Agency, February 22, 1999.
36. Nam Kyu San, interviewed by the author, Seoul, August 1998.
37. Yonhap News Agency, January 22, 1999.
38. Bank of Korea, January 22, 1999.
39. Kim Tae Dong, interviewed by the author, Seoul, January 1999.
40. Korean military officer, interviewed by the author, Seoul, May 1999. He did not want his name used.
41. Lee Dong Bak, interviewed by the author, Seoul, May 1999.
42. Cha Young Ku, interviewed by the author, Seoul, May 1999.
43. Korea Broadcasting System, special report, May 1999.
44. Lee Bu Young, interviewed by the author, Seoul, July 1999.
45. Yonhap News Agency, July 23, 1999.
46. Peter Bartholomew, interviewed by the author, Seoul, July 1999.
47. Blue House, press office, July 1999.
48. Yonhap News Agency, August 20, 1999.

### 5: Policy and Politics

1. You Jong Keun, interviewed by the author, Seoul, January 1998.
2. Ibid.
3. Hyundai Electronics & Industries, announcement, January 1998.
4. Peter Bartholomew, interviewed by the author, Seoul, January 1998.
5. You Jong Keun, interviewed by the author, Seoul, January 1998.
6. GEC-Alsthom statement, Paris, January 16, 1998.
7. Federation of Korean Industries, January 13, 1998.
8. Michel Camdessus, press conference, January 13, 1998.
9. Ibid.
10. Hyundai group, announcement, January 13, 1998.
11. Korea Broadcasting System, January 13, 1998.
12. Park Se Yong, press conference, January 13, 1998.
13. LG announcement, January 20, 1998.
14. Samsung announcement, January 20, 1998.
15. Ibid.

16. David Young, interviewed by the author, Seoul, January 1998.
17. National Statistics Office, Seoul, January 21, 1998.
18. Hubert Neiss, interviewed by the author, Seoul, February 1998.
19. Korea Broadcasting System, February 1, 1998.
20. Yonhap News Agency, February 9, 1998.
21. Ibid.
22. You Jong Keun, interviewed by the author, Seoul, February 1998.
23. Kim Min Sok, interviewed by the author, Seoul, February, 1998.
24. You Jong Keun, interviewed by the author, Seoul, February 1998.
25. Peter Underwood, interviewed by the author, Seoul, February 1998.
26. You Jong Keun, interviewed by the author, Seoul, March 1998.
27. Kim Dae Jung, inaugural address, February 25. The author attended the inauguration.
28. *Maeil Kyungje,* February 26, 1998.
29. Korea Broadcasting System, live report, March 3, 1998.
30. Munwha Broadcasting System, March 3, 1998.
31. Lee Kyu Sung, interviewed by the author, Kwachon, May 1999.
32. Suh Jin Young, Chung Kap Yung, interviewed by the author, Seoul, March 1998.
33. Kim Min Sok, interviewed by the author, Seoul, March 1998.
34. Bae Ie Dong, interviewed by the author, Seoul, March 1998.
35. Daniel Harwood, interviewed by the author, Seoul, March 1998.
36. Lee Jee Woo, interviewed by the author, Seoul, March 1998.
37. Kang Hee Bo, interviewed by the author, February 1998.
38. Park Nei Hei, interviewed by the author, March 1998.
39. Fair Trade Commission, report, March 1998.
40. Finance ministry announcement, April 18, 1998.
41. Ibid.
42. Sri-Ram Aiyer, remarks, Seoul Foreign Correspondents' Club, April 21, 1999.
43. Bae Ie Dong, Lee Jae Woo, interviewed by the author, Seoul, April 1998.
44. The Fair Trade Commission provided the statistics.
45. Samsung headquarters, announcement, May 6, 1998.
46. Jang Ha Sung, interviewed by the author, Seoul, May 1998.
47. Hyundai group, announcement, May 7, 1998; Yonhap News Agency, May 7, 1998.
48. Korea Broadcasting System, June 3, 1998.
49. Kim Dae Jung, press conference, Seoul, June 5, 1998.
50. Choi Hye Bum, interviewed by the author, Seoul, April 1998.
51. Chung Eui Yong, interviewed by the author, Seoul, April 1998.
52. Semiconductor Industries Association, Seoul, 1998.
53. Ohn Un Ki, interviewed by the author, Seoul, June 1998.
54. Kim Chang Uh, interviewed by the author, Seoul, June 1998.

55. Park Young Man, interviewed by the author, Seoul, June 1998.
56. Del Ricks, interviewed by the author, Seoul, June 1998.
57. Yoshio Nakamura, interviewed by the author, Tokyo, June 1998.
58. Commercial Bank of Korea, "Results of Corporate Viability Assessment," Seoul, June 18, 1998.
59. David Kim, interviewed by the author, Seoul, June 1998.
60. Yonhap News Agency, June 20, 1998.
61. Edward Campbell-Harris, interviewed by the author, Seoul, June 1998.
62. Samsung official, interviewed by the author, Seoul, June 1998. He did not want his name used.
63. Park Dae Shik, interviewed by the author, Seoul, June 1998.
64. FKI officials and a foreign diplomat provided material on the meeting.
65. Planning and Budget Commission, July 3, 1998.
66. Lee Jong Sung, interviewed by the author, July 1998.
67. Planning and Budget Commission, July 3, 1998.
68. Korean Confederation of Trade Unions officials, interviewed by the author, Seoul, July 1998.
69. Kim Ho Sun, interviewed by the author, Seoul, July 1998.
70. Planning and Budget Commission, July 3, 1998.
71. Finance ministry, July 5, 1998.
72. Yonhap News Agency, July 5, 1998.
73. Leland Timblick, interviewed by the author, Seoul, July 1998.
74. Philip Uhm, interviewed by the author, Seoul, July 1998.
75. Finance ministry, 1998.
76. Bank of Korea, August 10, 1998.
77. Ibid., August 27, 1998.
78. Ibid.
79. Finance ministry, September 2, 1998.
80. You Jong Keun, speech at Seoul Foreign Correspondents' Club, September 2, 1998.
81. Financial Supervisory Commission, September 3, 1998.
82. Jwa Sung Hee, interviewed by the author, Seoul, September 1998.
83. Munwha Broadcasting System, September 3, 1998.
84. Korea Broadcasting System, September 3, 1998.
85. Kim Dae Jung and Finance Minister Lee, press conference, Blue House, September 28, 1998.
86. Jang Ha Sung, interviewed by the author, Seoul, September 1998.
87. Sri-Ram Aiyer, interviewed by the author, Seoul, September 1998.
88. Finance ministry press release, October 2, 1998.
89. Financial Supervisory Commission, statement, October 2, 1998.
90. Ibid.
91. Yonhap News Agency, Fair Trade Commission ruling, October 11, 1998.
92. Korea Development Institute, report on the economy, October 13, 1998.

## 6: Piggy-Banking

1. Won Bong Hee, interviewed by the author, Seoul, June 1997.
2. Kwak Manh Soon, interviewed by the author, Seoul, June 1997.
3. Won Bong Hee provided material on the report.
4. Korea Economic Research Institute provided the figures.
5. European banker, interviewed by the author, June 1997. He did not want his name used.
6. Ibid.
7. Won Bong Hee, interviewed by the author, Seoul, June 1997.
8. European banker, interviewed by the author, Seoul, June 1997.
9. Korea Broadcasting System, December 7, 1997.
10. Korea Broadcasting System, December 10, 1997.
11. Citibank executive, interviewed by the author, Seoul, December 1997. He did not want his name used.
12. Ibid.
13. Victor Kang, interviewed by the author, Seoul, December 1997.
14. Yonhap News Agency, December 16, 1997.
15. Michael Brown, speech, Federation of Korean Industries, December 15, 1997.
16. Ibid.
17. Ibid.
18. Lawrence Summers, press conference, Seoul, January 16, 1998.
19. Korea Broadcasting System, January 18, 1998.
20. Statement, Blue House, January 16, 1998.
21. Ibid.
22. Kim Dae Jung, remarks to foreign executives, Seoul, January 21, 1998.
23. Yonhap News Agency, January 29, 1998.
24. Ibid.
25. Ibid.
26. Bank of Korea, February 26, 1998.
27. Finance ministry, February 26, 1998.
28. Ibid.
29. *Maeil Kyungje,* February 27, 1998.
30. Kim Jun Kyung, interviewed by the author, Seoul, April 1998.
31. Jwa Sung Hee, interviewed by the author, Seoul, April 1998.
32. Kim Dae Jung, *Financial Times* forum, Seoul, April 23, 1998.
33. Lee Jung Yung, interviewed by the author, Seoul, April 1998.
34. Finance Minister Lee Kyu Sung, interviewed by the author, Seoul, April 1998.
35. Finance Minister Lee Kyu Sung, *Financial Times* Forum, Seoul, April 23, 1998.
36. Edward Campbell-Harris, interviewed by the author, Seoul, May 1998.

37. Richard Samuelson, interviewed by the author, Seoul, May 1998.
38. Financial Supervisory Commission, statement, June 3, 1998.
39. Del Ricks, interviewed by the author, Seoul, June 1998.
40. Shin Bok Young, remarks, Seoul Foreign Correspondents' Club, June 3, 1998.
41. Sohn Byung Soo, remarks, Seoul Foreign Correspondents' Club, June 3, 1998.
42. Stephen Marvin, interviewed by the author, Seoul, June 1998.
43. Ibid.
44. James Rooney, interviewed by the author, Seoul, June 1998.
45. Yonhap News Agency, June 10, 1998.
46. Ibid.
47. James Rooney, interviewed by the author, Seoul, June 1998.
48. Stephen Marvin, interviewed by the author, Seoul, June 1998.
49. Yonhap News Agency, June 28, 1998.
50. Korea Broadcasting System, June 28, 1998.
51. Moon Kyung Hee, interviewed by the author, Seoul, June 1998.
52. Korea Broadcasting System, June 29, 1998.
53. Yonhap News Agency, June 29, 1998.
54. Yoon Young Mo, interviewed by the author, Seoul, June 1998.
55. Financial Supervisory Commission, statement, October 16, 1998.
56. Yonhap News Agency, October 16, 1998.
57. Ibid., October 17, 1998.
58. Bank of Korea, October 17, 1998.
59. Jason Yu, interviewed by the author, Seoul, November 1998.
60. Stephen Marvin, interviewed by the author, Seoul, November 1998.
61. Peter Bartholomew, interviewed by the author, Seoul, November 1998.
62. Richard Samuelson, interviewed by the author, Seoul, November 1998.
63. Park Yung Chul, remarks, Seoul Foreign Correspondents' Club, December 4, 1998.
64. Lee Sung Gun, interviewed by the author, Seoul, December 1998.
65. Kim Young Mo, interviewed by the author, Seoul, December 1998.
66. Yonhap News Agency, December 4, 1998.
67. Fitch-IBCA, statement, London and New York, January 19, 1999.
68. Standard & Poor's, statement, New York, January 26, 1999.
69. Tom Byrne, interviewed by the author, Seoul, February 1999.
70. Byun Yang Ho, interviewed by the author, February 1999.
71. Philippe Delhaise, interviewed by the author, Seoul, February 1999.
72. Stephen Hess, interviewed by the author, Seoul, February 1999.
73. Tom Byrne, interviewed by the author, Seoul, February 1999.
74. Donald Johnston, remarks, Seoul Foreign Correspondents' Club, February 1999.
75. Park Nei Hei, interviewed by the author, February 1999.

76. James Rooney, interviewed by the author, Seoul, February 1999.
77. Jonathan Dutton, interviewed by the author, Seoul, February 1999.
78. Adrian Cowell, interviewed by the author, Seoul, February 1999.
79. Financial Supervisory Commission, announcement, February 22, 1999.
80. Greg Laroia, interviewed by the author, Seoul, February 1999.
81. Kim Joon Hyok, interviewed by the author, Seoul, February 1999.
82. Weijian Shan, press conference, Seoul, April 27, 1999.
83. Weijian Shan, interviewed by the author, Seoul, May 1999.
84. Koh Sung Soo, interviewed by the author, Seoul, May 1999.
85. Choi Won Ku, interviewed by the author, Seoul, May 1999.
86. Weijian Shan, interviewed by the author, Seoul, May 1999.
87. Lee Kyu Sung, interviewed by the author, Kwachon, May 1999.
88. Financial Supervisory Commission, announcement, February 19, 1999.
89. "A Test for Korea," the *New York Times,* August 7, 1999, p. A12.
90. Financial Supervisory Commission, Seoul, August 16, 1999.
91. Lee Hun Jai, Financial Supervisory Commission, Seoul, August 30, 1999.
92. Nahm Sang Duck, Financial Supervisory Commission, Seoul, August 31, 1999.
93. Todd Martin, interviewed by the author, Seoul, August 1999.
94. Standard & Poor's, statement, New York, August 31, 1999.
95. Bernhard Echweiler, interviewed by the author, Seoul, August 1999.
96. Bloomberg News, September 16, 1999.
97. Financial Supervisory Commission, Seoul, September 17, 1999.
98. Ibid., February 19, 1999.
99. Ibid., May 12, 1999.
100. Yonhap News Agency, September 16, 1999.
101. Financial Supervisory commission, October 1, 1999.
102. Lee Jong Koo, interviewed by the author, Seoul, May 1999.
103. Tim Ferdinand, interviewed by the author, Seoul, May 1999.
104. Ibid.
105. Son Won Kyung, interviewed by the author, Seoul, May 1999.
106. Moon Soon Min, interviewed by the author, Seoul, May 1999.
107. Lee Kyu Won, interviewed by the author, Seoul, May 1999.
108. Lee Kyu Sung, interviewed by the author, Kwachon, May 1999.
109. Hyundai Securities officials, interviewed by the author, Seoul, September 1999.
110. Yonhap News Agency, Seoul, September 10, 1999.
111. Ibid., September 19, 1999.
112. Ibid., September 10, 1999.
113. Financial Supervisory Commission, September 18, 1999.
114. Oh Ho Gen, interviewed by the author, confirmed the contents of the letter, September 28, 1999.

115. Ibid. He made these comments to explain the rationale for having written the letter.
116. Michael Schuman, "Seoul's Daewoo, Foreign Creditors Fail to Reach Pact," The Wall Street Journal, October 29, 1999, p. A17; Stephanie Strom, "Hopes for Rescue of Daewoo Again Dim," The New York Times, October 29, 1999, p. C4.
117. Financial Supervisory Commission, November 3, November 4, 1998; Jean Yoon, Reuters, Seoul, November 4, 1999.
118. Ibid., November 4, 1999.
119. Bloomberg News, Seoul, November 5, 1999.

## 7: Unrest in the Workplace

* Chang Sung Hee interpreted interviews with labor leaders and workers in Ulsan as well as Seoul.
1. Park Il Kwon, interviewed by the author, Seoul, June 1997.
2. Yonhap News Agency, December 22, 1997.
3. Yonhap News Agency, January 15, 1998.
4. Yoon Young Mo, interview by the author, Seoul, January 1998.
5. Ahn Dong Sul, interviewed by the author, Seoul, January 1998.
6. Federation of Korean Industries, January 15, 1998.
7. Yonhap News Agency, January 20, 1998.
8. Ibid.
9. Cho Hyung Ryu, interviewed by the author, Ulsan, January 1998.
10. Hyundai Motor Company, promotional video.
11. Lee Sang Yong, interviewed by the author, Ulsan, January 1998.
12. Kim Kwan Soo, interviewed by the author, Ulsan, January 1998.
13. Lee Byong Gil, interviewed by the author, Ulsan, January 1998.
14. Yoon Jae Kun, interviewed by the author, Ulsan, January 1998.
15. Yonhap News Agency, February 6, 1998.
16. Yoon Young Mo, interviewed by the author, Seoul, February 1998.
17. Ahn Dong Sul, interviewed by the author, February 1998.
18. Tripartite commission, agreement, February 6, 1998.
19. Yoon Young Mo provided information on the meeting.
20. Bae Ie Dong, interviewed by the author, Seoul, February 1998.
21. The pamphlets, in Korean, littered the streets during and after the demonstration, witnessed by the author.
22. Lee Kab Yong, union workers, interviewed by the author, Seoul, May 1998.
23. Shin Hyun Kyu, interviewed by the author, Seoul, May 1998.
24. Workers, interviewed by the author, Ulsan, May 1998.
25. Union and company officials, interviewed by the author, Ulsan, May 1998.
26. Park Sam Yuel, interviewed by the author, Ulsan, May 1998.

27. Chun Chang Soo, interviewed by the author, Ulsan, May 1998.
28. Chung Dal Ok, interviewed by the author, Ulsan, May 1998.
29. The author attended the rally.
30. Choo Won Suh, interviewed by the author, Seoul, July 1998.
31. Choi Jong Kuen, interviewed by the author, Seoul, July 1998.
32. Lee Kab Yong, interviewed by the author, Seoul, July 1998.
33. Kim Ho Sun, interviewed by the author, Seoul, July 1998.
34. Kong Sung Do, interviewed by the author, Seoul, July 1998.
35. Lee Byung Ho, interviewed by the author, Ulsan, August 1998.
36. Shim Sang Dahl, interviewed by the author, Seoul, August 1998.
37. Kim Jong Myung, interviewed by the author, Ulsan, August 1998.
38. Lee Byung Ho, interviewed by the author, Ulsan, August 1998.
39. Lee Young Ja, interviewed by the author, Ulsan, August 1998.
40. Union official, interviewed by the author, Ulsan, August 1998.
41. The author witnessed the confrontation, August 18, 1998.
42. Lee Ki Ho, press conference, Ulsan, August 18,1998.
43. Hyundai Motor, statistics, August 1998.
44. Chung Mong Gyu, comment to the author, August 18, 1998.
45. Munwha Broadcasting Company, August 18, 1998.
46. Shin Hyun Kyu, interviewed by the author, Seoul, August 1998.
47. Yoo Tae Ho, interviewed by the author, Seoul, August 1998.
48. Kim Kwang Shik, interviewed by the author, Ulsan, August 1998.
49. Korea Broadcasting system, August 24, 1998.
50. Ibid.
51. Shin Hyun Kyu, interviewed by the author, Seoul, August 1998.
52. Bae Ie Dong, interviewed by the author, Seoul, August 1998.
53. David Young, interviewed by the author, Seoul, August 1998.
54. Yoon Young Mo, interviewed by the author, Seoul, August 1998.
55. Choi Chang Jip, interviewed by the author, Seoul, September 1998.
56. Lee Ki Ho, interviewed by the author, Seoul, September 1998.
57. Tripartite commission, statement, January 19, 1999.
58. The National Statistics Office provided the figures, January 1999.
59. Lee Chung Yul, interviewed by the author, Seoul, February 1999.
60. Yonhap News Agency, February 24, 1999.
61. Park Nei Hei, interviewed by the author, Seoul, February 1999.
62. You Si Soo, interviewed by the author, Seoul, February 1999.
63. Workers at the center, interviewed by the author, Seoul, February 1999.
64. The author visited the strike headquarters outside the cathedral.
65. Lee Kab Yong, interviewed by the author, Seoul, April 1999.
66. Park Seok Woon, interviewed by the author, Seoul, April 1999.
67. Kim Chul Won, interviewed by the author, Seoul, April 1999.
68. Anthony Michell, remarks, American Chamber of Commerce, Seoul, February 1999.

69. Labor ministry, press office, September 20, 1999.
70. Ibid.
71. Ibid., December 1, 1999.
72. FKTU, International office, November 1999.

## 8: On/Off the Fast Track

1. The report and Chung Mong Gyu's retort were quoted in the Korean media on May 21, 1997.
2. Ibid.
3. David Young, interviewed by the author, Seoul, June 1997.
4. Korea Automobile Manufacturers' Association provided the statistics on production. Companies gave figures on profits.
5. *Chosun Ilbo,* December 7, 1997.
6. Korea Broadcasting System, December 7, 1997.
7. Hank Morris, interviewed by the author, Seoul, December 1997.
8. *Chosun Ilbo,* February 2, 1998.
9. Alan Perriton, interviewed by the author, Seoul, February 1998.
10. GM Korea, press release, February 2, 1998.
11. Ibid.
12. Lou Hughes, comment to the author, Seoul, April 24, 1998.
13. Chang Byung Ju, statement, Daewoo headquarters, November 20, 1998.
14. Koh Won Jong, "Alarm bells ringing for the Daewoo group," *Nomura Korea,* October 29, 1998, pp. 1-3.
15. Cho Jang Won, interviewed by the author, Seoul, December 1997.
16. Samsung Motor's public relations department, March 28, 1998.
17. Ibid., May 6, 1998.
18. Dan Dong Ho, interviewed by the author, Asan, December 1997.
19. Kia Motors, Asan plant, briefing, December 1997.
20. Um Sung Yong, interviewed by the author, Seoul, December 1997.
21. Jun Sang Jin, interviewed by the author, Seoul, December 1997.
22. Ibid.
23. Dan Dong Ho, interviewed by the author, Asan, December 1997.
24. Cho Jang Rae, interviewed by the author, Asan, December 1997.
25. Karl Moskowitz, interviewed by the author, Seoul, December 1997.
26. Jin Nyum, press conference, January 22, 1998.
27. Ibid.
28. Jun Sang Jin, interviewed by the author, Seoul, January 1998.
29. Korea Automobile Manufacturers' Association provided the statistics.
30. Yoo Seong Min, interviewed by the author, Seoul, March 1998.
31. Jun Sang Jin, interviewed by the author, Seoul, March 1998.
32. Shin Hyun Kyu, interviewed by the author, Seoul, March 1998, provided the statistics.

33. Shin Hyun Kyu provided the information. The author attended the reception.
34. You Keun Chan, Hyundai group public affairs office, interviewed by the author, Seoul, March 1998.
35. Chung Mong Gyu, press conference, Seoul, March 22, 1998.
36. Park Je Hyuk, press conference; Kia Motor, statement, Seoul, March 23, 1998.
37. Jun Sang Jin, interviewed by the author, Seoul, March 1998.
38. Robert Felton, interviewed by the author, Seoul, March 1998.
39. McKinsey Seoul Office, McKinsey Global Institute, "Automotive Industry: Executive Summary," in *Productivity-Led Growth in Korea* (Seoul and Washington: McKinsey & Co., March 1998).
40. James Tessada, interviewed by the author, Seoul, March 1998.
41. Jun Sang Jin, interviewed by the author, Seoul, April 1998.
42. Korea Broadcasting System, April 16, 1998.
43. Kia union leaflet, in Hangul, picked up by the author at a demonstration.
44. Kwon Yong Jun, interviewed by the author, Seoul, April 1998.
45. Yoo Jung Ryul, press conference, June 2, 1998. The author attended the press conference and witnessed the events outside Kia headquarters.
46. Ibid.
47. Lee Chong Dae, interviewed by the author, Seoul, June 2, 1998.
48. Kenneth Brown, interviewed by the author, Seoul, July 6, 1998.
49. Daewoo group, public relations department, July 6, 1998.
50. Korea Automobile Manufacturers' Association provided the figures.
51. Chang Sung Hyun, interviewed by the author, July 1998.
52. Yonhap News Agency, July 10, 1998.
53. *Nihon Keizai Shimbun,* Tokyo, July 10, 1998.
54. Lee Chang Won, Shin Hyun Kyu, and a Kia manager who did not want his name used all commented in response to the author's questions, July 1998.
55. Wayne Booker, interviewed by the author, Seoul, July 1998.
56. Ibid.
57. Korea Development Bank and Kia Motors announced the terms, July 15, 1998.
58. *Chosun Ilbo,* August 31, 1998.
59. *Korea Herald,* August 31, 1998.
60. Park Song Bae, interviewed by the author, Seoul, September 1998.
61. Peter Underwood, interviewed by the author, Seoul, September 1998.
62. Ibid.
63. John Spelich, statement to the media, Dearborn, Michigan, October 19, 1998.
64. Peter Underwood, interviewed by the author, Seoul, October 1998.
65. Park Chun Sop, interviewed by the author, Seoul, October 1998.
66. Yonhap News Agency, October 19, 1998.

67. A Hyundai Motor executive described the scene for the author, October 1998.

68. Hyundai Motor, press release, October 19, 1998.

69. Korea Automobile Manufacturers' Association provided the statistics.

70. *Chosun Ilbo,* December 24, 1998; *Nihon Keizai Shimbun,* September 22, 1999.

71. Spokesman, Daewoo Electronics, December 2 1998. He did not want his name used.

72. Financial Supervisory Commission, "Agreement for the Restructuring of the Top 5 Chaebol," December 7, 1998.

73. Kim Tae Gou talked to the author after a news conference, Seoul, December 8, 1998. Kim, formerly chairman of Daewoo Motor, took the title of president instead as part of "restructuring."

74. Lee Hun Jai, comments at a luncheon for foreign correspondents, Seoul, December 11, 1998.

75. Lee Jung Seung, interviewed by the author, Seoul, December 1998.

76. Financial Supervisory Commission, June 29, 1999.

77. Choi Jeung Jin, interviewed by the author, Seoul, June 1999.

78. Ibid., September 1999.

79. Financial Supervisory Commission, July 19, 1999.

80. Peter Underwood, interviewed by the author, Seoul, September 1999.

81. Financial Supervisory Commission, August 16, 1999.

82. Kim Woo Choong, *Every Street Is Paved With Gold: The Road to Real Success,* introduction by Louis Kraar, p. 245. (A note thanks Kraar "for his invaluable help in the expansion and adaptation of my original Korean book for this English-language edition.")

83. General Motors, press conference, Detroit, August 10, 1999.

84. Seong Jun Ke, interviewed by the author, Pusan, November, 1999.

85. *Chosun Ilbo,* March 2, 1999.

86. Hyundai Motor provided the statistics, March 1999.

87. Chung Se Yung, press conference, Seoul, March 5, 1998.

88. Hyundai Motor, Daewoo Motor, and Kia Motor provided the statistics, December 1999.

89. Bloomberg News, Seoul, September 16, 1999.

90. Alan Perriton, interviewed by the author, Seoul, December 1999.

91. Yonhap News Agency, December 19, 1999.

## 9: Selling Off/Selling Out

1. Tom Pinansky, interviewed by the author, Seoul, March 1998.

2. Robert Broadfoot, interviewed by the author, Seoul, March 1998.

3. Ahn Hun Mo, interviewed by the author, Seoul, March 1998.

4. Ssangyong headquarters announced the deal on March 11, 1999.

5. Yonhap News Agency, March 8, 1998.

6. Park Soon Baek, interviewed by the author, Seoul, June 1998.
7. Park Jie Won, press conference, June 24, 1998.
8. Korea Broadcasting System, Seoul, June 28, 1998.
9. *Chosun Ilbo,* June 28, 1998.
10. Lee Chan Jin, statement, June 26, 1998.
11. Kim Jung Soo, interviewed by the author, Seoul, June 1998.
12. Lee Chan Jin, Lee Min Hwa, press conference, Seoul, July 20, 1998.
13. Moon Jae Woo, interviewed by the author, Seoul, September 1998.
14. Hank Morris, interviewed by the author, Seoul, November 1998.
15. Sam Hageman, interviewed by the author, Seoul, November 1998.
16. Hong Too Pyo, president of the Korea National Tourism Organization, introduced the commercials to the foreign media on September 21, 1998, at the Seoul Foreign Correspondents' Club.
17. Richard Christenson, speech, Seoul, June 1997.
18. Michael Brown, "A Message from the President," *Journal,* American Chamber of Commerce in Korea, vol. 42, no. 9 (May-June 1997), p. 13.
19. John Alsbury, interviewed by the author, Seoul, May 1997.
20. Park Chan Sung, interviewed by the author, Seoul, May 1997.
21. Brian Chalmers, interviewed by the author, Seoul, May 1997.
22. Wayne Chumley, interviewed by the author, Seoul, May 1997.
23. Choi Byung Kwon, news conference, Seoul, March 11, 1998.
24. James Tessada, news conference, Seoul, March 11, 1998.
25. Yoon Dae Sung, news conference, Seoul, March 11, 1998.
26. Wayne Chumley, interviewed by the author, Seoul, September 1998.
27. Korea Automobile Importers and Distributors Association provided the statistics.
28. Kim Ho Shik, Yonhap News Agency, October 18, 1998.
29. The commercial section, U.S. embassy, Seoul, provided the statistics, November, 1998.
30. Bank of Korea, report, November 25, 1998.
31. Ibid.
32. Ibid.
33. Sohn Byung Doo, speech, Federation of Korean Industries, Seoul, December 4, 1998.
34. Kim Duc Choong, remarks, Seoul, December 3, 1998. Kim, president of Ajou University in Suwon, owned by Daewoo, may have been soothed somewhat when Kim Dae Jung, on May 24, 1999, named him education minister.
35. Kim Tae Dong, interviewed by the author, Seoul, December 1998.
36. Yonhap News Agency, December 7, 1998.
37. Financial Supervisory Commission, "Agreement for the Restructuring of the Top 5 Chaebol," December 7, 1998.
38. Ibid.

39. Yonhap News Agency, December 7, 1998.
40. Richard Samuelson, interviewed by the author, Seoul, December 7, 1998.
41. Daewoo group headquarters issued the statement on December 8 after Lee made his remark at a press conference for Korean reporters.
42. Lee Hun Jai, remarks, correspondents' luncheon, Seoul, December 11, 1998.
43. John Dodsworth, interview, Seoul, December 7, 1998.
44. Kim Dae Jung, luncheon for foreign business people, Blue House, Seoul, January 28, 1999.
45. James Rooney, remarks, Seoul, January 28, 1999.
46. Lawrence Summers, speech, American Chamber of Commerce in Korea, Seoul, February 25, 1999.
47. Stephen Marvin, "South Korean strategy: a tale of two rallies," Jardine Fleming Research, February 1999, Jardine Fleming Securities, Seoul, February 1999, p. 6.
48. National Statistics Office released the figures, February 25, 1998.
49. Michael Camdessus, remarks, New York, February 24, 1999.
50. Hilton L. Root, in collaboration with Mark Andrew Abdollahian, Greg Beier, and Jacek Kugler, "The New Korea: Crisis Brings Opportunity," Milken Institute, Santa Monica, February 1999, p. 1.
51. Lee Soo Hee, interviewed by the author, Seoul, March 1999.
52. David Young, interviewed by the author, Seoul, March 1999; Lawrence Summers, speech, American Chamber of Commerce in Korea, Seoul, February 25, 1999.
53. Kim Dae Jung, speech, Seoul, February 26, 1999.
54. The Bank of Korea and National Statistics Office provided the statistics, March 2, 1999.
55. American Chamber of Commerce in Korea, "Improving Korea's Business Climate: Recommendations from American Business" (Seoul, 1999).
56. *Maeil Kyongje,* March 15, 1999.
57. Jeffrey Jones, press conference, Seoul, March 25, 1999.
58. William Daley, remarks, Korean Chamber of Commerce and Industry, Seoul, March 26, 1999.
59. William Daley, press conference, Seoul, March 26, 1999.
60. Lee Soo Young, remarks, Korean Chamber of Commerce and Industry, March 26, 1999.
61. Bank of Korea, report, March 23, 1999.
62. Ibid.
63. Ibid.
64. Jang Ha Sung, remarks to reporters, Seoul, March 20, 1999.
65. Ibid.
66. Ibid.
67. Daewoo group, announcement, April 19, 1999.
68. Kim Tae Gou, FKI briefing, Seoul, April 23, 1999.

69. Hank Morris and Edward Campbell-Harris, interviewed by the author, April 1999.
70. Yonhap News Agency, April 21, 1999.
71. Hyundai group, announcement, April 23, 1999.
72. Ibid.
73. The government announced the decision but did not mention the quid pro quo of Hyundai and LG Semicon.
74. Lee Chong Suk and Choi Eui Jong spoke at the FKI briefing, Seoul, April 23, 1999.
75. Ibid.
76. LG group, announcement, Seoul, May 18, 1999.
77. Kang Yoo Shik, press conference, Seoul, May 18, 1999.
78. The Blue House released Kim Dae Jung's remarks to the cabinet, April 20, 1999.
79. Korean Air, statement, Seoul, March 10, 1999.
80. Shim Yi Taek, press conference, April 22, 1999. Cho Yang Ho was subsequently arrested and, on November 26, 1999, indicted on charges of tax evasion and embezzlement. His father, Cho Choong Hoon, also indicted, was not detained in deference to his age. A younger brother, Cho Su Ho, president of Hanjin Shipping, was indicted on lesser charges but not detained. A number of other Hanjin executives and government aviation officials were also arrested.
81. Korean Air, announcement, May 4, 1999.
82. Lee Yo Yul, interviewed by the author, Seoul, May 4, 1999.
83. Richard Samuelson, interviewed by the author, Seoul, May 6, 1999.
84. Stephen Marvin, interviewed by the author, Seoul, May 6, 1999.
85. Goldman Sachs, announcement, Hong Kong, May 6, 1999.
86. James Rooney, interviewed by the author, Seoul, May 6, 1999.
87. Park Jae Ho, interviewed by the author, Seoul, June 1997.
88. Bradley Geer, interviewed by the author, Seoul, June 1999.
89. Charles Carson, interviewed by the author, Seoul, June 1999.
90. Steven Lee, Harrison Jung, interviewed by the author, Seoul, June 1999.
91. Lee Keun Jang, interviewed by the author, Seoul, July 1999.
92. Jeffrey Jones, interviewed by the author, Seoul, July 1999.
93. Cho Won Dung, interviewed by the author, Seoul, July 1999.
94. Lee Keun Jang, interviewed by the author, Seoul, July 1999.
95. Chang Duck Koo, remarks, Seoul Foreign Correspondents' Club, July 23, 1999.
96. Lee Keun Jang, interviewed by the author, Seoul, July 1999.
97. Adrian Cowell, interviewed by the author, Seoul, July 1999.
98. Kim Woo Choong, press conference, Daewoo headquarters, Seoul, July 15, 1999.
99. A Daewoo group spokesman provided the figures, July 1999.

100. Kim Woo Choong, press conference, Daewoo headquarters, Seoul, July 25, 1999.
101. Yonhap News Agency, Seoul, July 25, 1999.
102. John Dodsworth, interviewed by the author, Seoul, July 1999.
103. Financial Supervisory Commission, July 27, 1999.
104. Korea Broadcasting System, August 11, 1999.
105. Edward Bang, interviewed by the author, Seoul, August 1999.
106. Peter Underwood, interviewed by the author, Seoul, August 1999.
107. Korea Broadcasting System, August 15, 1999.
108. Graham Courtney, interviewed by the author, Tokyo, August 1999.
109. Financial Supervisory Commission, Seoul, August 16, 1999.
110. Ibid.
111. Daewoo group, public relations department, November 2, 1999; Samuel Len, "Daewoo Founder and 12 Other Executives Offer to Resign," The New York Times, November 2, p. C4.
112. George Mansfield, press conference, Seoul, July 6, 1999. Doosan group headquarters responded in official statement the same day
113. Yonhap News Agency, Seoul, July 31, 1999
114. Financial Supervisory Commission, Seoul, July 7, 1999
115. Frazer Seitel, Tiger Management spokesman, interviewed by the author, Washington, September 1999.
116. Richard Samuelson, interviewed by the author, Seoul, July 1999.
117. Tiger Management, statement to U.S. Securities and Exchange Commission, August 11, 1999.
118. Ibid., August 23, 1999.
119. SK Telecom, communications office, Seoul, September 1999.

## 10: Framing a Peace

* Chang Sung Hee interpreted interviews with North Korean refugees and defectors, both in China and in Seoul, in late 1998 and early 1999.
1. Korean-Chinese businessman, who asked that his name not be used, interviewed by the author, Dandong, China, November 1998.
2. South Korean businessman, interviewed by the author, Dandong, November 1998. He did not want his name used.
3. Kim Byung Gon, interviewed by the author, aboard ferry from Inchon to Dandong, November 1998.
4. Waitress, interviewed by the author, Dandong, China, November 1997.
5. Chinese police commander, interviewed by the author, Dandong, China, November 1997. He did not want his name used.
6. Travelers obtained these statistics from Chinese officials. One source was Kenneth Quinones, director of the Asia Foundation in Seoul, who visited the region early in 1998.

7. North Korean refugees, interviewed by the author near the North Korean border, February 1999.
8. Park Chul, interviewed by the author, Dongsung village, China, February, 1999.
9. Ibid.
10. North Korean girl, interviewed by the author, Dongsung, China, February 1999. She did not want her name used.
11. Park Ji Hun, interviewed by the author, Seoul, February 1999.
12. National Intelligence Service spokesman, interviewed by the author, February 1999.
13. North Korean refugee, interviewed by the author, Seoul, February 1999. The woman, who had defected to South Korea, did not want her name used.
14. Kim Myung Goo, interviewed by the author, Yanji, China, February 1999.
15. The American embassy in Seoul reported evidence of the coup in 1998.
16. Kim Myung Goo, interviewed by the author, Yanji, China, February 1999.
17. Ibid.
18. Ibid.
19. Lee Han Seung, interviewed by the author, Yanji, China, February 1999.
20. Kim Chul Soo, interviewed by the author, Yanji, China, February 1999.
21. Kim Yun, interviewed by the author, Tumen, China, February 1999.
22. Choi Han, interviewed by the author, Tumen, China, February 1999.
23. Yoo Sun Eung, interviewed by the author, Tumen, China, February 1999.
24. Choi Han, interviewed by the author, Tumen, China, February 1999.
25. Han Sung Chul, interviewed by the author, Yanji, China, February 1999.
26. Nam Bok Ja, interviewed by the author, Yanji, China, February 1999.
27. Ibid.
28. Ibid.
29. Ibid.
30. Han Chang Kwon, interviewed by the author, Seoul, February 1999.
31. Ibid.
32. National Intelligence Service, statement, Seoul, February 20, 1999. The statement was issued in English without attribution; an NIS official confirmed its origin.
33. Hong Jin Hee, interviewed by the author, Seoul, February 1999.
34. Oh Hye Jung, interviewed by the author, Seoul, February 1999.
35. Ibid.
36. Choi Jin Wook, interviewed by the author, Seoul, February 1999.
37. Han Chang Kwon, interviewed by the author, Seoul, February 1999.
38. Chun In Bum, interviewed by the author, Hill 911, May 1997.
39. Ibid.
40. Ibid.
41. Ibid.

42. South Korean sources, interviewed by the author, Seoul, December 1996. They did not want their names used.

43. SaKong Il, interviewed by the author, Seoul, December 1996.

44. Kim Young Min, interviewed by the author, Seoul, December 1996.

45. Korean military sources, interviewed by the author, Seoul, December 1996. They did not want their names used.

46. Ro Jae Bong, interviewed by the author, Seoul, December 1996.

47. Don Oberdorfer, *The Two Koreas: A Contemporary History* (Reading, MA: Addison-Wesley, 1997), pp. 392-93. Oberdorfer traces not only the submarine incident but the entire course of the Geneva talks.

48. Yonhap News Agency, March 1, 1998.

49. Selig Harrison, interviewed by the author, Seoul, March 1998.

50. U.S. diplomat, interviewed by the author on a background basis, Seoul, March 1998.

51. South Korean diplomat, interviewed by the author on a background basis, Seoul, March 1998.

52. Gerald Segal, interviewed by the author, Seoul, April 1998.

53. Yonhap News Agency, April 18, 1998.

54. Richard Grant, interviewed by the author, Seoul, April 1998.

55. Korea Central News Agency, monitored in Seoul, April 29, 1998.

56. Yonhap News Agency, April 29, 1998.

57. Lho Kyung Soo, interviewed by the author, Seoul, April 1998.

58. Yun Sang Sop, interviewed by the author, Seoul, April 1998.

59. Pak Bo Hi, interviewed by the author, Seoul, April 1998.

60. Ibid.

61. Ibid.

62. Rev. Moon Sun Myung, speech, Seoul, February 4, 1999.

63. The government's Financial supervisory Service provided the statistics, February 1999.

64. Seo Pyong Kyu, interviewed by the author, Seoul, February 1999.

65. Yonhap News Agency, May 7, 1998.

66. Ibid.

67. Chung Se Yung, interviewed by the author, Panmunjom, June 16, 1998.

68. Chung Mong Hun, other Hyundai officials, interviewed by the author, Panmunjom, June 16, 1998.

69. Hyundai spokesman, interviewed by the author, Seoul, June 1998.

70. South Korean defense ministry officials, quoted in the Korean media, including Yonhap News Agency and Korea Broadcasting System, June 22-24, 1998.

71. Munwha Broadcasting System, June 23, 1998.

72. Korea Broadcasting System, June 23, 1998.

73. Korea Central News Agency, monitored in Seoul, June 23, 1998.

74. Korea Broadcasting System, June 25, 1998.

75. Ibid.
76. Korea Broadcasting System, June 26, 1998.
77. Defense ministry spokesman, June 26, 1998.
78. Korea Broadcasting System, June 25, 1998.
79. Oh San Yul, interviewed by the author, Seoul, June 1998.
80. Chung Mong Hun, press conference, Seoul, July 2, 1998.
81. Yonhap News Agency, July 2, 1998.
82. Chung Ju Yung, remarks, released by Hyundai group public relations office, July 2, 1998.
83. Munwha Broadcasting System, July 12, 1998.
84. The defense ministry provided the details, July 12, 1998.
85. Combined Forces Command, statement, Seoul, July 14, 1998.
86. David Steinberg, interviewed by the author, Seoul, July 1998.
87. Han Sung Joo, interviewed by the author, Seoul, July 1998.
88. Foreign ministry spokesman, interviewed by the author, Seoul, July 7, 1998.
89. Korea Central News Agency, Pyongyang, July 6, 1998. The NSPA was renamed the National Intelligence Service, partly as a result of the notoriety of this incident.
90. Yonhap News Agency, July 7, 1998.
91. Han Sung Joo, interviewed by the author, Seoul, July 7, 1998.
92. Yonhap News Agency, July 8, 1998.
93. Gennady Isaev, interviewed by the author, Seoul, July 8, 1998.
94. Tass News Agency, August 31, 1998.
95. Reuters, September 2, 1998.
96. Ministry of unification official, interviewed by the author, Seoul, September 1998. He asked that his name not be used.
97. Press office, Blue House, Seoul, August 31, 1998.
98. Kyodo News Agency, Tokyo, August 31, 1998.
99. Chung Young Tae, interviewed by the author, Seoul, August 31, 1998.
100. Lee Ho, interviewed by the author, Seoul, August 31, 1998.
101. Huh Moon Young, interviewed by the author, Seoul, August 31, 1998.
102. Western diplomat, interviewed by the author, August 31, 1998, Seoul, on a background basis.
103. Korea Central News Agency, September 2, 1998.
104. Yonhap News Agency, September 2, 1998.
105. Radio Pyongyang, September 4, 1998, monitored in Seoul.
106. Ibid.
107. Kenneth Quinones, interviewed by the author, Seoul, September 1998.
108. Kenneth Quinones, interviewed by the author, Seoul, September 1998.
109. Japanese diplomats offered this explanation in background briefings, September 1998 and March 1999.
110. Toshimitsu Shigemura, interviewed by the author, Seoul, September 1998.

111. *Rodong Sinmun*, reported by the Korea Central News Agency, Pyongyang, September 2, 1998.

## 11: Of Missiles and Nukes

1. Kang In Duk, speech, Seoul, September 8, 1998.
2. Cho Dong Ho, interviewed by the author, Seoul, September 1998.
3. Foreign businessman, interviewed by the author, Seoul, September 1998. He did notwant his name used.
4. Cho Eun Ho, interviewed by the author, Seoul, September 1998.
5. Roxley Pacific executive, interviewed by the author, Seoul, September 1998. He did notwant his name used.
6. Iris Ying, interviewed by the author, Seoul, 1998.
7. Robert Gallucci, interviewed by the author, Seoul, 1998.
8. Donald Gregg, interviewed by the author, Seoul, 1998.
9. Park Kun Woo, interviewed by the author, Seoul, October 1998.
10. Kim Gook Jin, interviewed by the author, Seoul, October 1998.
11. John Barry Kotch, interviewed by the author, Seoul, October 1998.
12. Lee Soh Hang, interviewed by the author, Seoul, October 1998.
13. Stephen Bosworth, speech, Seoul, October 1998.
14. Munwha Broadcasting System, November 15, 1998.
15. Korea Broadcasting System, November 15, 1998.
16. Charles Kartman, press conference, Seoul, November 19, 1998.
17. Ibid. Yonhap News Agency reported the North had requested $300 million for a single inspection, and the State Department later confirmed the figure.
18. Ibid.
19. Kim Dae Jung, press conference, Seoul, November 19, 1998.
20. Kim Gook Jin, interviewed by the author, Seoul, November 1999.
21. Hong Yong Pyo, interviewed by the author, Seoul, November 1998.
22. Charles Kartman, "Clarification," U.S. embassy, Seoul, November 21, 1998.
23. President Clinton, remarks, Blue House, November 21, 1998.
24. President Clinton, remarks, Osan Air Base, Korea, November 22, 1998.
25. United Nations Command, statement, December 13, 1998.
26. Ibid.
27. Kenneth Quinones, interviewed by the author, Seoul, December 1998.
28. Ibid.
29. Stephen Tharp, interviewed by the author, Panmunjom, December 1998.
30. Lennart Wendel, interviewed by the author, Panmunjom, December 1998.
31. Stephen Tharp, interviewed by the author, Panmunjom, December 1998.
32. Michael V. Hayden, interviewed by the author, Panmunjom, December 1998. Hayden was subsequently promoted to lieutenant general and named director of the National Security Agency.

33. Ibid.
34. Ibid.
35. Ibid.
36. John Barry Kotch, interviewed by the author, Seoul, December 1998.
37. William Cohen, remarks, press conference, Seoul, January 15, 1999.
38. William Cohen and Chun Yong Taek, joint communiqué, Seoul, January 15, 1999.
39. New China News Agency, Beijing, January 15, 1999.
40. Park Young Kyu, interviewed by the author, Seoul, January 15, 1999.
41. Han Sung Joo, interviewed by the author, Seoul, January 15, 1999.
42. Ashton B. Carter and William James Perry, *Preventive Defense: A New Security Strategy for America* (Washington, D.C.: Brookings Institution Press, 1999), p. 221.
43. *Korea Herald,* March 8, 1999. The embassy responded to the author's request for comment, March 8, 1999.
44. Chun Yong Taek, remarks, Seoul Foreign Correspondents' Club, March 1999.
45. William Perry, remarks, Kimpo International Airport, March 8, 1999.
46. Song Min Soon, interviewed by the author, Seoul, March, 1999.
47. Yonhap News Agency, March 17, 1999.
48. Ibid.
49. Briefings by Korean officials to Korean media, March 17, 1999.
50. Huh Moon Young, interviewed by the author, Seoul, March 1999.
51. Kyodo News Agency, Tokyo, March 19, 1999.
52. Ibid.
53. Keizo Obuchi, speech, official English translation, Korea University, Seoul, March 19, 1999.
54. Robert Einhorn, press conference, Seoul, March 31, 1999.
55. William Perry, remarks, Korean Political Science Association, Seoul, March 25, 1999.
56. L. Gordon Flake, remarks, Seoul, March 25, 1999.
57. Chun Hyon Joon, Yun Duk Min, Choi Won Ki, interviewed by the author, Seoul, May 1999.
58. Kim Dae Jung, remarks, Blue House, May 24, 1999.
59. Park Young Ho, interviewed by the author, Seoul, May 1999.
60. Lee Sang Chul, interviewed by the author, Seoul, May 1999.
61. Yevgeny Afanasiev, remarks, Institute of Foreign Affairs, Seoul, May 14, 1999.
62. Cha Yung Koo, interviewed by the author, Seoul, May 1999.
63. James Rubin, State Department, Washington, May 27, 1999.
64. William Perry, statement to the press, Seoul, May 29, 1999.
65. Cha Yung Ku, interviewed by the author, Seoul, June 1999.
66. Ibid.
67. Park Young Ho, interviewed by the author, Seoul, June 1999.

68. John Barry Kotch, interviewed by the author, Seoul, June 1999.

69. Yonhap News Agency, Seoul, June 14, 1999.

70. Ibid.

71. All three Korean television networks covered the episode throughout the day with video from their own reporters and the defense ministry, and live reporting of press conferences in Seoul.

72. Korea Broadcasting System, June 15, 1999.

73. United Nations Command, Seoul, June 15, 1999.

74. Blue House, press office, June 15, 1999.

75. United Nations Command, Seoul, June 15, 1999.

76. Aidan Foster-Carter, interviewed by the author, Seoul, June 1999.

77. Yoon Duk Min, interviewed by the author, Seoul, June 1999.

78. Kim Dae Jung, speech, International Olympics Committee, Seoul, June 16, 1999.

79. *Rodong Sinmun,* reported by the Korea Central News Agency, Pyongyang, June 16, 1999.

80. Choi Jin Wook, interviewed by the author, Seoul, June 1999.

81. Kim Yong Han, Lee In Hae, Lee Jong Kwon, interviewed by the author, Seoul, June 1999.

82. United Nations Command, Seoul, June 1999.

83. Park Young Ho, interviewed by the author, Seoul, June 1999.

84. NHK, June 17, 1999.

85. Korea Central News Agency, Pyongyang, June 20, 1999.

86. Yonhap News Agency, June 29, 1999.

87. Hyundai group, public affairs office, June 21, 1999.

88. Korea Broadcasting System, June 21-22, 1999.

89. Blue House, press office, June 2122, 1999.

90. Korea Central News Agency, Pyongyang, June 25, 1999.

91. Hyundai group, public affairs office, June 25-26, 1999.

92. Ibid.

93. Lee Yong Jun, interviewed by the author, Seoul, June 1999.

94. Blue House and defense ministry officials, interviewed by the author, Seoul, June 1999.

95. You Jin Kyu, interviewed by the author, Seoul, June 1999.

96. Sohn Young Hwan, Kim Chang Suu, interviewed by the author, Seoul, June 1999.

97. Park Young Ho, interviewed by the author, Seoul, July 1999.

98. Korean defense ministry officials, interviewed by the author, Seoul, July 1999.

99. Cohen spoke at a press conference at the American embassy, Tokyo, July 28, 1999.

100. *Rodong Sinmun,* reported by Korea Central news Agency, Pyongyang, August 5, 1999.

101. Tomohisa Sakanaka, interviewed by the author, Tokyo, August 1999.
102. Sadaaki Numata, interviewed by the author, Tokyo, August 1999.
103. *Yomiuri Shimbun,* August 5, 1999.
104. Korea Central News Agency, Pyongyang, August 10, 1999.
105. United Nations Command, Seoul, August 10, 1999.
106. Ibid.
107. Yonhap News Agency, August 29, 1999.
108. Xianhua News Agency, August 29, 1999.
109. Yonhap News Agency, September 2, 1999.
110. Reuters, September 12, 1999.
111. Ibid., September 17, 1999.
112. The White House and State Department cited the terms and extent of the lifting of sanctions, September 17, 1999.
113. Korea Central News Agency, Pyongyang, September 24, 1999.
114. Hyundai group, public affairs office, October 1, 1999.
115. William Perry, "Review of United States Policy Toward North Korea: Findings and Recommendations," State Department, Washington, D.C., October 12, 1999.
116. Ibid.
117. *Investor's Business Daily,* Los Angeles, November 4, 1999.
118. Charles Kartman, interviewed by the author, Seoul, December 1999.

## 12: Crisis of Identity

 *  Chang Sung Hee interpreted a number of interviews in this chapter.
 1. Yonhap News Agency, November 11, 1998.
 2. Ibid.
 3. Lee Han Byul, interviewed by the author, Seoul, September 1998.
 4. Park Sung Hwan, interviewed by the author, Seoul, September 1998.
 5. Cho Jae Hyun, interviewed by the author, Seoul, September 1998.
 6. Lew Young Ick, interviewed by the author, Seoul, August 1998.
 7. Kim Ki Hwan, interviewed by the author, Seoul, August 1998.
 8. Han Hae Soo, interviewed by the author, Seoul, October 1998.
 9. Shin Hyung Sook, interviewed by the author, Seoul, October 1998.
10. Park Chul Ho, interviewed by the author, Seoul, October 1998.
11. Park Moon Suk, interviewed by the author, Seoul, October 1998.
12. Hwang Hye Joung, interviewed by the author, Seoul, October 1998.
13. Kim Hyun Jong, interviewed by the author, Seoul, October 1998.
14. Yoshioki Nakamura, interviewed by the author, Seoul, March 1999.
15. Shin Hyoung Oen, interviewed by the author, Seoul, March 1999.
16. Japanese diplomat, interviewed by the author, Seoul, March 1999. He did notwant his name used.
17. Yun Jo Jin, interviewed by the author, Seoul, March 1998.

18. Kolleen Park, interviewed by the author, Seoul, March 1998.
19. Venerable Mu Hyu, interviewed by the author, Seoul, November 1998.
20. Frank Tedesco, interviewed by the author, Seoul, December 1998.
21. Peter Hyun, interviewed by the author, Seoul, August 1998.
22. Ibid.
23. Ibid.
24. Ibid.
25. Mother and landlady, interviewed by the author, Seoul, July 1998. They did notwant their names used.
26. Cho Kyu Hwan, interviewed by the author, Seoul, July 1998.
27. Lee Chang Jun, interviewed by the author, Seoul, July 1998.
28. Health and welfare ministry provided the statistics, May 1999.
29. Manager, Eastern Child Welfare Society, interviewed by the author, Seoul, July 1998. He did notwant his name used.
30. Ahn Sung Ki, interviewed by the author, Seoul, December 1998.
31. The author witnessed the demonstration, in front of the Kwangwhamun Building, Seoul, December 10, 1998.
32. Kang Je Gyu, interviewed by the author, Seoul, February 1999.
33. Kim Yun Jin, interviewed by the author, Seoul, February 1999.
34. Kim Young Baek, interviewed by the author, Seoul, May 1999.
35. Ibid.
36. Ibid.
37. Shin Dong Pyo, interviewed by the author, Seoul, May 1999.
38. Bank of Korea, report, May 20, 1999.
39. Hyun Oh Seok, interviewed by the author, Seoul, May 1999.
40. Kim Tae Ho, interviewed by the author, Seoul, May 1999.
41. Kim Hye Ok, interviewed by the author, Seoul, May 1999.
42. National Statistics Office, Seoul, May 20, 1999.
43. Chung Soo Young, interviewed by the author, Seoul, May 1999.
44. Whang Keong Bae, interviewed by the author, Seoul, May 1999.
45. Michael Camdessus, press conference, Seoul, May 20, 1999.
46. Lee Kyu Sung, interviewed by the author, Seoul, May 1999.
47. Chung Duck Koo, remarks, finance ministry, Kwachon, May 24, 1999.
48. Park Nei Hei, interviewed by the author, Seoul, May 1999.
49. Lee Kyong Hee, "Commentary: Our presidential hall of shame," *Korea Herald,* May 24, 1999.
50. Hubert Weiss, interviewed by the author, Seoul, December 1999.
51. Choi Jang Jip, interviewed by the author, Seoul, December 1991.

# Select Bibliography

## Books

American Chamber of Commerce in Korea. *Guide to Doing Business in Korea.* Seoul: 1998.

Bergsten, C. Fred, and SaKong Il, eds, *The Korea–United States Economic Relationship.* Washington, D.C.: Institute for International Economics; Seoul: Institute for Global Economics, 1997.

Breen, Michael. *The Koreans: Who They Are, What They Want, Where Their Future Lies.* London: Orion Business Books, 1998.

Carter, Ashton B., and William James Perry. *Preventive Defense: A New Security Strategy for America.* Washington, D.C.: Brookings Institution Press, 1999.

Clifford, Mark L. *Troubled Tiger: Businessmen, Bureaucrats and Generals in South Korea.* Rev. ed. Armonk, NY: M. E. Sharpe, 1998.

Cumings, Bruce. *Korea's Place in the Sun: A Modern History.* New York: W. W. Norton, 1997.

Delhaise, Philippe F. *Asia in Crisis: The Implosion of the Banking and Finance Systems.* Singapore: John Wiley & Sons Pte Ltd., 1998.

Downs, Chuck. *Over the Line: North Korea's Negotiating Strategy.* Foreword by James R. Lilley. Washington, D.C.: AEI Press, 1999.

East Asia Analytical Unit. *Korea Rebuilds: From Crisis to Opportunity.* Barton, ACT, Australia Department of Foreign Affairs and Trade, 1999.

Eberstadt, Nicholas. *The End of North Korea.* Washington, DC: AEI Press, 1999.

Eckert, Carter. *The Koch'ang Kims and the Colonial Origins of Korean Capitalism, 1876-1945.* Seattle and London: University of Washington Press, 1991.

Gertz, Bill. *Betrayal: How the Clinton Administration Undermined American Security.* Washington, D.C.: Regnery Publishing, 1999.

Grinker, Roy Richard, *Korea and Its Futures: Unification and the Unfinished War.* New York: St. Martin's Press, 1998.

Henderson, Callum. *Asia Falling? Making Sense of the Asian Currency Crisis and Its Aftermath.* Singapore: McGraw-Hill, 1998.

Jackson, Karl D., ed. *Asian Contagion: The Causes and Consequences of a Financial Crisis.* Boulder, CO: Westview Press, 1999.

Jwa, Sung Hee. "Globalization and New Industrial Organization: Implications for Structural Adjustment Policies," in *Regionalism versus Multilateral Trade Arrangements,* eds. Takatoshi Ito and Anne O. Krueger. Chicago: University of Chicago Press, 1997, chapter 11.

Kang Myung Hun. *The Korean Business Conglomerate: Chaebol Then and Now.* Berkeley: Institute of East Asian Studies, University of California, 1996.

Kim Dae Jung. *Selected Speeches.* Vol. 1, *Government of the People.* Seoul: Office of the President, Republic of Korea, 1999.

Kim Woo Choong. *Every Street is Paved With Gold: The Road to Real Success,* introduction by Louis Kraar. New York: William Morrow and Company, 1992.

Kirk, Donald. *Korean Dynasty: Hyundai and Chung Ju Yung.* Hong Kong: Asia 2000; Armonk, NY: M. E. Sharpe, 1994.

*Korea Company Handbook: Investment Guide.* Biannual, Spring and Autumn editions. Seoul: Asia-Pacific Infoserv.

Lee Kyong Hee. *Korean Culture: Legacies and Lore.* Seoul: Korea Herald, 1993.

————. *World Heritage in Korea.* Seoul: Organizing Committee of the Year of Cultural Heritage 1997, Samsung Foundation of Culture, 1997.

Mazaar, Michael J. *North Korea and the Bomb: A Case Study in Nonproliferation.* New York: St. Martin's Press, 1998.

McKinsey Seoul Office, McKinsey Global Institute. *Productivity-Led Growth in Korea.* Seoul and Washington, D.C.: McKinsey & Co., March 1998.

Merrill, John. *Korea : The Peninsular Origins of the War.* Dover: University of Delaware Press, 1989.

Michell, Anthony, and James P. Rooney. *One Million Jobs Project: A Starting Point.* Seoul: Euro-Asian Business Consultancy and Ssangyong Templeton, December 1998.

Oberdorfer, Don. *The Two Koreas: A Contemporary History.* Reading, MA: Addison-Wesley, 1997.

SaKong Il and Kwang Suk Kim, eds. *Policy Priorities for the Unified Korean Economy.* Proceedings of international symposium on Korean unification. Seoul: Institute for Global Economics, 1998.

Sigal, Leon V. *Disarming Strangers: Nuclear Diplomacy With North Korea.* Princeton, NJ: Princeton University Press, 1998.

Sohn Chan Hyun, and Yang Junsok, eds., *Korea's Economic Reform Measures under the IMF Program: Government Measures in the Critical First Six Months of the Korean Economic Crisis.* Seoul: Korea Institute for International Economic Policy, June 1998.

Soros, George. *The Crisis of Global Capitalism [Open Society Endangered].* New York: Public Affairs, 1998.

Steers, Richard M., *Made in Korea.* New York: Routledge, 1999.

Stueck, William. *The Korean War: An International History.* Princeton, NJ: Princeton University Press, 1995.

Yang Sung Chul. *The North and South Korean Political Systems: A Comparative Analysis.* Rev. ed. Seoul; Elizabeth, N.J.: Hollym International, 1999.


### Booklets, brochures, pamphlets, periodicals, and speeches

Akaba Yuji, Florian Budde, and Choi Jungkiu. "Restructuring South Korea's Chaebol." *McKinsey Quarterly* no. 4 (1998): 68-79.

American Chamber of Commerce in Korea. "Improving Korea's Business Climate: Recommendations from American Business." Seoul, 1999.

———. "Korea: U.S. Trade and Investment Issues 1998." Seoul, 1998.

Bosworth, Stephen. "The International Economic Situation and Future Prospects for Korea." Remarks, Federation of Korean Industries, Seoul, March 10, 1999.

———. "The Security Role of the United States in a Post-Korean Unification East Asia." Speech, Society for Unification Studies, Seoul, December 30, 1998.

Brown, Michael. "A Message from the President." *Journal,* American Chamber of Commerce in Korea, vol. 42, no. 9 (May-June 1997): 13.

Chon Chol Hwan. "Recent Economic Developments and the Monetary Policy Dispute." Speech, Seoul Foreign Correspondents' Club, November 4, 1998.

Camdessus, Michael, "Sustaining Asia's Recovery from Crisis." Speech, 34th South East Asian Central Banks Governors' Conference, Seoul, May 20, 1999.

Commercial Bank of Korea. "Results of Corporate Viability Assessment." Seoul, June 18, 1998.

Dodsworth, John R. "Korea: Chronology of the Crisis." Seoul: IMF Seoul Office, December 1998.

———. "The Korean Crisis: A Year of Hardship and Reform." Speech, Seoul Foreign Correspondents' Club, December 7, 1998.

———. "Strong Recovery or Disaster Scenario?" Statement, Seoul, May 1999.

European Union Chamber of Commerce in Korea. "Trade Issues 1998." Seoul, 1998.

Financial Supervisory Commission. "Agreement for the Restructuring of the Top 5 Chaebol." December 7, 1998.

———. "Corporate Restructuring—Performance and Future Plan." December 4, 1998.

———. "Korea: Memorandum on Economic Policies." March 10, 1999.

———. "Progress in Financial and Corporate Restructuring and Future Tasks." September 28, 1998.

Hong Soon Young. "Comprehensive Approach Toward North Korea." Speech, Seoul Foreign Correspondents' Club, March 17, 1999.

Jang, Ha Sung. "Corporate Governance and Economic Development: The Korean Experience." Included in "Corporate Governance and Economic Development," International Conference on Democracy, Market Economy and Development, Republic of Korea, World Bank, February 26-27, 1999.

*Joongang Ilbo.* "The Press Stands Up." Seoul, October 1999.

Jwa Sung Hee. "Property Rights and Economic Behaviors: Lessons for Korea's Economic Reform." Korea Development Institute, 1998.

Karacadag, Cem, and Barabara C. Samuels II. "How Markets Failed Asia." *International Economy* (November/December 1998): 34-37.

Khang, Hyun Sung. "North Koreans flee cannibals: Children are stalked and killed for food as starvation grips country." *Sunday Times,* London, January 10, 1999, p. 17.

"Kim Dae Jung's Policies on North Korea: Achievements and Future Goals." Ministry of Unification, March 25, 1999.

Kim Soon Kwon. "Development of Super-Maize for North Korea: Sufficiency and Sustainability." Seoul: International Corn Foundation, 1999.

Kim Tae Dong. "A Year under the IMF Program and Future Policy Initiatives." Speech, Seoul Foreign Correspondents' Club, December 2, 1998.

Kirk, Donald. "Viewpoint: Lambs to the Slaughter: Washington's pact with Pyongyang won't help the starving children." *Time,* Asia edition, vol. 153, no. 12 (March 29, 1999): 18.

———. "Pacifying Pyongyang." *The New Leader,* vol. 82, no. 11 (September 20-October 4, 1999), pp. 6-8.

Koh Wonjong. "Alarm bells ringing for the Daewoo group." *Nomura Korea,* October 29, 1998, pp. 1-3.

Korea Fair Trade Commission. "Major Efforts of the Korea Fair Trade Commission: A Year after the Economic Crisis." November 25, 1998.

Korea Institute for International Economic Policy. "Korea in Transition: Reforms Today, Rewards Tomorrow." Seoul, 1998.

Korea International Labour Foundation. "Unemployment Measures from Desperation to Reconstruction: Job Protection and Creation for the Unemployed." Seoul, 1999.

———. "Current Labor Situation in Korea." Seoul, 1999.

Korea Investment Service Center. "Investment Success in Korea." Seoul: Korea Trade-Investment Promotion Agency, 1999.

Korea Trade-Investment Promotion Agency (KOTRA). "New Directions: Investing and Living in Korea." Seoul, 1998.

Kotch, John Barry. "Korea's Security Dilemmas and the Diplomacy of an Emerging World Power." *Korea and World Affairs* vol. 21, no. 4 (Winter, 1997): 628-651.

———. "Patron to Partner: Reflections on the US-South Korean Relationship." *Harvard International Review* (Spring 1999): 18-21.

———. "Whither the Chaebol." *Journal,* American Chamber of Commerce in Korea (May-June 1999): 47-49.

Kremenak, Ben. "Korea's Road to Unification: Potholes, Detours, and Dead Ends." Center for International and Securities Studies at Maryland School of Public Affairs, University of Maryland at College Park, May 1997.

Lee Hun Jai. "Financial Reform and Its Impact on the Domestic Market." Speech, Seoul Foreign Correspondents' Club, April 1, 1999.

Lee Ki Ho. "Labor Policies in Korea—Recent Challenges and New Opportunities." Speech, Seoul Foreign Correspondents' Club, August 28, 1998.

Lee Kyu Sung. "Korea's Progress: Implications for Global Financial Stability." Speech, IMF/World Bank Annual Meeting, Washington, D.C., October 7, 1998.

———. "Korea's New Economic Landscape: Investment Opportunities for the 21st Century." CLSA Investors' Forum, Hong Kong, May 17, 1999.

Lim, James. "Something in the Heir? Korea's chaebol pay for the sins of the sons." *Asia, Inc.* Hong Kong (December 1997/January 1998): 33-37.

Marvin, Stephen E. "Death Throes." Ssangyong Investment & Securities, Korea Equity Research, Seoul, May 25, 1998.

———." South Korean economy: About as good as Chinese dikes." Jardine Fleming Research, Seoul, September 1998.

———. "South Korean strategy: Climb aboard but don't sit down." Jardine Fleming Research, Seoul, November 2, 1998.

———. "South Korean strategy: a tale of two rallies." Jardine Fleming Research, Seoul, February 1999.

Ministry of Finance and Economy, Republic of Korea. "The Road to Recovery in 1999: Korea's Ongoing Economic Reform." Seoul, February 1999.

———. "The Year in Review: Korea's Reform Progress." Seoul, November 1998.

———. "Financial Market Stabilization Package Related to Daewoo Group Workout Plan," Seoul, November, 1999.

———. "Korea: An Economy Transformed." Seoul, December 1999.

———. "Progress in Korea's Corporate Reform: Q&As." Seoul, September 1999.

Ministry of Labor, Republic of Korea. "Comprehensive Unemployment Policy." Seoul, February 1999.

Ministry of Unification, Republic of Korea. "Kim Dae-Jung's Policies on North Korea: Achievements and Future Goals." Seoul, March 25, 1999.

Moody's Investors Service, Global Credit Research. "White Paper: Moody's Rating Record in the East Asian Financial Crisis." New York, May 1998.

Obuchi, Keizo. "Japan - Republic of Korea Relations in the Coming Century—Creation of a New History." Speech, Korea University, Seoul, March 20.

Office of the President. "Overcoming a National Crisis, the Republic of Korea Rises Up Again: A Year of Trials and Challenges for President Kim Dae Jung." Seoul, February 1999.

Oh Kap Soo. "Experience and Future Direction of Structural Reform in Korea." *Journal,* American Chamber of Commerce in Korea (May-June 1999).

Organisation for Economic Co-operation and Development. "OECD Economic Surveys, 1998-1999: Korea." Paris, 1999.

Palais, James B. "Views on Korean Social History." Institute for Modern Korean Studies, Yonsei University, IMKS Special Lecture Series No. 2, 1998.

Perry, William. "Review of United States Policy Toward North Korea: Findings and Recommendations," State Department, Washington, D.C., October 1999.

Planning and Budget Commission. "Information Package on Privatization Policy of Korea." July 3, 1998.

Root, Hilton L., in collaboration with Mark Andrew Abdollahian, Greg Beier, and Jacek Kugler. "The New Korea: Crisis Brings Opportunity," Milken Institute, Santa Monica, February 1999.

Sen, Amartya. "Democracy and Social Justice." Seoul Conference on Democracy, Market Economy and Development, February 27-28, 1999.

Shim Jae Hoon. "North Korea: A Crack in the Wall." *Far Eastern Economic Review* vol. 162, no. 17 (April 29, 1999): 10-14.

Sohn, Byung Doo. "Progress in Corporate Restructuring and Its Prospects." Speech, Federation of Korean Industries, December 4, 1998.

———. "Reinventing the Korean Economy: Existing Challenges and Future Prospects." Speech, Seoul Foreign Correspondents' Club, June 3, 1998.

"South Korea: Diversity of voice and quality lacking." *Asian Intelligence* no. 10 (June 10, 1998): 11-12.

Steinberg, David I. "The Effects of the Economic Crisis on the Korean Middle Class." Speech, American Chamber of Commerce in Korea, Seoul, July 10, 1998.

———. "Korea: Triumph amid Turmoil." *Journal of Democracy* vol. 9, no. 2 (April 1998).

Stiglitz, Joseph. "Participation and Development: Perspectives from the Comprehensive Development Paradigm." Speech, World Bank, Seoul, February 27, 1999.

———. "The Korean Miracle: Growth, Crisis and Recovery." Speech, International Conference on Economic Crisis and Restructuring in Korea, Seoul, December 3, 1999.

Summers, Lawrence H. "Policy Challenges for Asia in 1999." Speech, American Chamber of Commerce in Korea, Seoul, February 25, 1999.

"Syngman Rhee's Independence Activities and Founding of the Republic of Korea." Conference papers, Institute for Modern Korean Studies, Graduate School of International Studies, Yonsei University, August 13-14, 1998.

"The Tripartite Commission, surmounting confrontation and conflict, is opening up a new labor-management culture." Tripartite Commission, Seoul, November 1998.

"The United States and the Two Koreas at the Crossroads: Searching for a New Passage." Korean Political Science Association, *Chosun Ilbo,* Korean-America Friendship Society, Asia Foundation, Seoul, March 26-27, 1998.

World Bank, Korea Country Management Unit. "The Republic of Korea and the World Bank: Partners in Economic Recovery." Seoul, February 1999.

# Index